Proteolytic Signaling in Health and Disease

Proteolytic Signaling in Health and Disease

Edited by

André Zelanis

Assistant Professor of Proteomics and Systems Biology, and Head of the Functional Proteomics Laboratory, Department of Science and Technology, Federal University of Sao Paulo, Sao Jose dos Campos, Brazil

ELSEVIER

ACADEMIC PRESS

An imprint of Elsevier

Academic Press is an imprint of Elsevier
125 London Wall, London EC2Y 5AS, United Kingdom
525 B Street, Suite 1650, San Diego, CA 92101, United States
50 Hampshire Street, 5th Floor, Cambridge, MA 02139, United States
The Boulevard, Langford Lane, Kidlington, Oxford OX5 1GB, United Kingdom

Library of Congress Cataloging-in-Publication Data
A catalog record for this book is available from the Library of Congress

British Library Cataloguing-in-Publication Data
A catalogue record for this book is available from the British Library

ISBN 978-0-323-85696-6

Publisher: Stacy Masucci
Acquisitions Editor: Peter B. Linsley
Editorial Project Manager: Sara Pianavilla
Production Project Manager: Maria Bernard
Cover Designer: Matthew Limbert

Typeset by STRAIVE, India

Contents

Contributors

Luiz G.N. de Almeida McCaig Institute for Bone and Joint Health; Department of Biochemistry and Molecular Biology, University of Calgary, Calgary, AB, Canada

Ulrich auf dem Keller Technical University of Denmark, Department of Biotechnology and Biomedicine, Kongens Lyngby, Denmark

Thamali Ayagama Department of Pharmacology, University of Oxford, Oxford, United Kingdom

Uilla Barcick Functional Proteomics Laboratory, Department of Science and Technology, Federal University of São Paulo, São Paulo, Brazil

Samuel J. Bose Department of Pharmacology, University of Oxford, Oxford, United Kingdom

Louise Bundgaard Technical University of Denmark, Department of Biotechnology and Biomedicine, Kongens Lyngby, Denmark

Rebecca A.B. Burton Department of Pharmacology, University of Oxford, Oxford, United Kingdom

Daniela Cajado-Carvalho Special Laboratory for Applied Toxinology, Butantan Institute/Center of Toxins, Immune-Response and Cell Signaling (CeTICS), São Paulo, Brazil

Maurício Frota Camacho Functional Proteomics Laboratory, Department of Science and Technology, Federal University of São Paulo, São Paulo, Brazil

Niels Olsen Saraiva Câmara Department of Immunology, Institute of Biomedical Sciences; Nephrology Division, Department of Medicine, Federal University of São Paulo, São Paulo, Brazil

Camila Rolemberg Santana Travaglini Berti de Correia Department of Genetics and Evolution, Federal University of Sao Carlos, Brazil

Patrícia Maria Siqueira dos Passos Department of Genetics and Evolution, Federal University of Sao Carlos, Brazil

Antoine Dufour McCaig Institute for Bone and Joint Health; Department of Biochemistry and Molecular Biology; Department of Physiology and Pharmacology, University of Calgary, Calgary, AB, Canada

Wolfgang Esser-Skala Computational Systems Biology Group, Department of Biosciences, University of Salzburg, Salzburg, Austria

Jessica de Alcantara Ferreira Special Laboratory for Applied Toxinology, Butantan Institute/Center of Toxins, Immune-Response and Cell Signaling (CeTICS), São Paulo, Brazil

Nikolaus Fortelny Computational Systems Biology Group, Department of Biosciences, University of Salzburg, Salzburg, Austria

Vanessa Morais Freitas Department of Cell and Developmental Biology, Institute of Biomedical Sciences, University of Sao Paulo, São Paulo, SP, Brazil

Daniela C. Granato Laboratório Nacional de Biociências, LNBio, Centro Nacional de Pesquisa em Energia e Materiais, CNPEM, Campinas, SP, Brazil

Hinrich P. Hansen Department of Internal Medicine I, University Hospital Cologne, CECAD Research Center, Cologne, Germany

Tâmisa Seeko Bandeira Honda Department of Immunology, Institute of Biomedical Sciences, University of São Paulo, São Paulo, Brazil

Pitter F. Huesgen Central Institute for Engineering, Electronics and Analytics, Jülich; Cluster of Excellence Cellular Stress Responses in Aging-Associated Diseases, CECAD, Medical Faculty and University Hospital Cologne; Institute of Biochemistry, Department of Chemistry, University of Cologne, Cologne, Germany

Ruy Gastaldoni Jaeger Department of Cell and Developmental Biology, Institute of Biomedical Sciences, University of Sao Paulo, São Paulo, SP, Brazil

Konstantinos Kalogeropoulos Technical University of Denmark, Department of Biotechnology and Biomedicine, Kongens Lyngby, Denmark

Maithreyan Kuppusamy Central Institute for Engineering, Electronics and Analytics, Jülich, Germany

Andrei Leitão Medicinal & Biological Chemistry Group, Institute of Chemistry of São Carlos, University of São Paulo, Brazil

Adriana F. Paes Leme Laboratório Nacional de Biociências, LNBio, Centro Nacional de Pesquisa em Energia e Materiais, CNPEM, Campinas, SP, Brazil

João Agostinho Machado-Neto Department of Pharmacology, Institute of Biomedical Sciences, University of São Paulo, São Paulo, Brazil

Isabela Fernanda Morales Martins Department of Genetics and Evolution, Federal University of Sao Carlos, Brazil

Milene Cristina Menezes Special Laboratory for Applied Toxinology, Butantan Institute/Center of Toxins, Immune-Response and Cell Signaling (CeTICS), São Paulo, Brazil

Leandro X. Neves Laboratório Nacional de Biociências, LNBio, Centro Nacional de Pesquisa em Energia e Materiais, CNPEM, Campinas, SP, Brazil

Caio Almeida Batista de Oliveira Department of Genetics and Evolution, Federal University of Sao Carlos, Brazil

Barbara Nunes Padovani Department of Immunology, Institute of Biomedical Sciences, University of São Paulo, São Paulo, Brazil

Murilo Salardani Functional Proteomics Laboratory, Department of Science and Technology, Federal University of São Paulo, São Paulo, Brazil

Adriane Sousa de Siqueira Department of Cell and Developmental Biology, Institute of Biomedical Sciences, University of Sao Paulo, São Paulo, SP; School of Health Sciences, Positivo University, Curitiba, PR, Brazil

Valentine Spagnol Department of Genetics and Evolution, Federal University of Sao Carlos, Brazil

Felipe Roberti Teixeira Department of Genetics and Evolution, Federal University of Sao Carlos, Brazil

Dilza Trevisan-Silva Centre of Excellence in New Target Discovery (CENTD), Butantan Institute, São Paulo, Brazil

Luciana D. Trino Laboratório Nacional de Biociências, LNBio, Centro Nacional de Pesquisa em Energia e Materiais, CNPEM, Campinas, SP, Brazil

André Zelanis Functional Proteomics Laboratory, Department of Science and Technology, Federal University of São Paulo, São Paulo, Brazil

Preface

Proteases have leading roles in biology, and they caught my attention since the very beginning of my academic career, when I started studying the composition of snake venoms and the role of proteases in the pathophysiology of the envenomation. It turns out that, to my surprise, the main biological events observed after a hemorrhagic process (which are mainly caused by the metalloproteases present in snake venoms) are not only triggered by proteases but also sustained and amplified by such enzyme class. In fact, snake venom proteases were shaped by evolution, allowing the snakes to deal with their distinct biological constraints. Snake venom proteases were the starting point of my journey in the protease field. Since then, I have been studying the activity of proteases in different biological contexts and keep fascinated about the implications of their activities in biology.

Classification, evolution, and mechanistic studies on proteases are important sources of information for the understanding of their functions. However, a comprehensive understanding of the biological role of a given protease is only revealed by mapping the alterations triggered by them in signaling pathways. Yet this is quite challenging as there is a dynamic interplay of interactions in biological circuits; therefore, a substrate in one reaction may become an activated protease in the other (the activation of protease zymogens, in cascade-like pathways, clearly illustrates such a feature). Consequently, with everything in biology, understanding a given phenomenon often implies the understanding of various seemingly unrelated events.

From the very first glimpse of life, during egg fertilization, to cell death (in apoptosis), proteases are ubiquitous players in several metazoan signaling pathways. They all have in common the chemical reaction catalyzed (the hydrolysis of peptide bonds); however, the profusion of their biological effects is not easily anticipated and, more importantly, diverges strikingly even among protease isoforms and physiological states (i.e., health and disease states). Several factors contribute to this: structural diversity, which includes the existence of multidomain architecture in some enzymes (which is strongly correlated with their substrate diversity), the interaction with inhibitors, and more importantly, the fate of signaling pathways after the beginning of a proteolytic event.

Preface

Biological signaling processes have long been appreciated by investigators from all fields of biology, from homeostasis to pathological conditions. In this context, the study of proteolytic signaling in biological systems has been tremendously improved as high-throughput analytical approaches are emerging. Overall, *Proteolytic signaling in health and disease* intends to provide information on fundamental physiological processes regulated by proteases, from protein turnover and autophagy to antigen processing and presentation through MHC molecules. Pathological conditions will also be covered, including inflammation, wound healing, and cancer. Up-to-date analytical methods to study proteolytic signaling events will also be presented and discussed in this book. Collectively, the information contained in this book will make easier understanding of the set of interconnected events in which proteases (and their inhibitors) have leading roles, the so-called protease web, an important concept that wraps up the information contained in this book.

The interconnection nature of biological signaling events in which proteases have a role demonstrates that, indeed, the whole is something besides the parts. In other words, in essence, proteolytic signaling is the emergence of properties, from parts of a whole (interconnected) biological system and, in this context, cleaved substrates are as important as the proteases themselves for the proper biological signaling to occur.

Finally, I want to warmly acknowledge all the authors of this book for joining me in this endeavor and to have dedicated an important amount of their time to write their chapters, despite the worldwide pandemic SARS-COV-2 situation. Many thanks.

I hope you enjoy the ride on proteolytic signaling as much as we do!

CHAPTER 1

Proteolytic signaling: An introduction

Uilla Barcick, Maurício Frota Camacho, Murilo Salardani, and André Zelanis
Functional Proteomics Laboratory, Department of Science and Technology, Federal University of São Paulo, São Paulo, Brazil

Introduction

This chapter is intended to be a concise introduction to the subject of proteolytic signaling; the concept will be discussed considering some selected examples. There is no attempt to perform an exhaustive and descriptive explanation of signaling pathways in which proteases are central players; rather the focus is on the biological implications of such a particular signaling process. Specific biological processes that are somehow triggered or regulated by proteolysis will be presented and discussed in a comprehensive manner in the following chapters.

Proteases: A brief (and incomplete) introduction

Proteases are hydrolytic enzymes that are encoded by genomes from essentially every organism, from virus and bacteria to plants and vertebrates. Indeed, proteases account for almost 2% of every known genome [1]. Such enzymes are key effectors of several biological processes, ranging from the first glimpse of life in egg fertilization—a process in which acrosomal proteases in spermatozoon displays a role in breaking the egg's glycoproteins matrix, in the zona pellucida—to cell death, in apoptosis (a regulated pathway mediated by a class of proteases termed caspases). Even though many biological processes rely on protease function, two important aspects related to protease activity need to be clarified beforehand: **protein degradation** and **proteolytic processing** are both performed by proteases, though they are not synonyms.

Protein content in any living organism is a result of the fine-tuned balance between two opposite rates: the rate of protein synthesis and the rate of protein degradation. In this respect, **protein degradation** is a "broad" term which, in eukaryotes, is mainly accomplished by the ubiquitin-proteasome pathway (discussed in detail in Chapter 2) and by lysosomal proteases, which are available in specialized cellular compartments, termed lysosomes. Therefore, under normal physiological conditions, these processes must be strictly coordinated to the cellular

Proteolytic Signaling in Health and Disease. https://doi.org/10.1016/B978-0-323-85696-6.00013-5

needs. On the contrary, **proteolytic processing** (or limited proteolysis) is a particular type of proteolysis where the cleavage of substrates is restricted to specific portions, frequently altering the activity of target proteins. Such a process is regarded as a posttranslational protein modification as it occurs in mature proteins (i.e., after their translation within living cells). Two main processes are involved in the modulation of protease activity: (i) the activation of protease precursors (zymogens) by limited proteolysis and (ii) their inactivation by the interaction with inhibitors. Hence, the interplay among proteases and their inhibitors, the protease web [2] (discussed in Chapter 11), has pivotal implications for both health and disease states in any living organism.

The chemical reactions by which a protein is broken down into peptide(s) vary among the distinct classes of proteases (i.e., serine, metallo, cysteine, aspartic, threonine, and glutamic proteases), and mechanistic details of such processes are beyond the scope of this book; however, a major determinant of any protease activity is its three-dimensional structure including, in multidomain proteases, the presence of ancillary domains (in addition to the protease domain), which are important for substrate targeting, kinetic properties, and cellular localization (Fig. 1A). In this context, the key point to understand protease roles in biological systems is to underscore its repertoire of substrates, the protease "degradome" (discussed in detail in Chapters 10 and 11), which may vary tremendously even among close-related protease families. Since the primary specificity of a protease is mainly determined by structural constraints related to the amino acids at the peptide bond that undergoes cleavage (the scissile bond), protease activity upon its substrates may be viewed as the protease's "footprint" (Fig. 1B) and this feature is the fundamental basis for high-throughput analytical approaches to study protease cleavage sites in complex biological contexts (discussed in detail in Chapter 10).

Limited proteolysis and proteolytic signaling are seamlessly connected

Since some key points were introduced, we can now focus on the definition of **proteolytic signaling**, which can be described as a biological signaling event that is not only triggered and regulated by proteolysis; it is a signaling pathway, in which the cleavage event may significantly affect the fate of target proteins and, more importantly, the biological outcome. Unlike signaling pathways in which some reactions may be reverted by enzymes acting in opposing reactions, such as pathways regulated by kinases and phosphatases, the uniqueness of proteolytic signaling is mainly related to its irreversibility: proteases hydrolyze substrates, thereby resulting in the physical separation of portions in target proteins. Furthermore, cleaved substrates may affect downstream signaling pathways, acting in distinct ways according to the biological context (activation/inactivation of protein precursors or modulation of protein functions, for example). It is important to highlight, however, that proteolytic signaling pathways do not comprise proteolysis only; rather, proteolysis is an

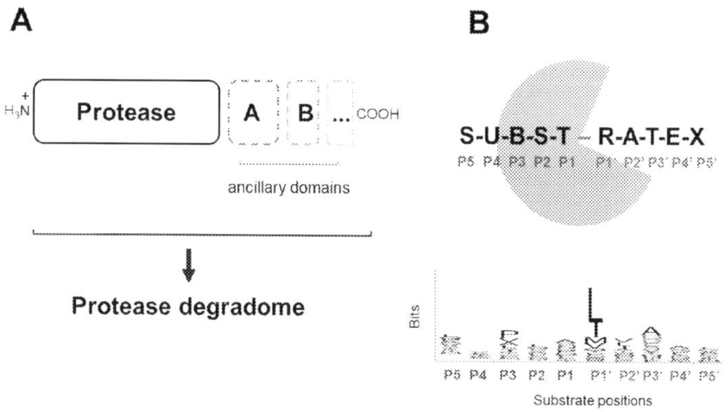

Fig. 1
Protease structure has main implications on its functions. (A) Structure of a hypothetical mature multidomain protease, evidencing its ancillary domains (A, B, and so on) which, in general, are involved in substrate targeting, kinetic properties, and cellular localization. Such additional domains are responsible for the substrate repertoire (degradome) of the protease. (B) Upper panel: the primary specificity of a protease (*gray*) is ruled by structural constraints at the scissile bond (displayed in *red* (dark gray in the print version), in the upper panel). Lower panel: the evaluation of the frequency of amino acids at P5 to P5′ in substrates (illustrated by the *size of the letters* in the *y*-axis) denotes the subsite specificity of a protease and might be regarded as the "protease's footprint" and (B) presents the nomenclature by Schechter and Berger in which residues C-terminal to the cleavage site (scissile bond) are referred to as prime (P′), whereas N-terminal residues are referred to as non-prime (P) [3]. (For interpretation of the references to color in this figure legend, the reader is referred to the web version of this article.)

irreversible part of the process, although downstream reactions may eventually be performed by other proteins. For example, it has been shown that in some pathological conditions such as cancer, the cross talk between kinases and proteases is central for cancer progression [4]. The phosphorylation of proteases affects several important aspects of protease function, including their activation/inactivation, cellular location, changes in interaction partners, and their half-life. On the contrary, proteolytic processing of kinases also displays central regulatory roles, such as the removal of inhibitory domains, ectodomain shedding, and the generation of novel biologically active fragments [4].

There are several biological examples illustrating how limited proteolysis and proteolytic signaling are seamlessly connected; nevertheless, when considering these two interconnected events, there are three fundamental aspects that must be considered: [1] ***what*** substrate(s) is(are) cleaved [2], ***where*** it is cleaved and, finally, [4] the ***biological context*** underlying the signaling event. The following selected examples will be briefly discussed, along with their implications in biological systems.

N-terminal protein processing and proteolytic signaling

Proteases are ubiquitously required by several different signaling pathways in living organisms. For example, in both prokaryotes and eukaryotes, a number of intracellular proteins require the proteolytic processing of the initiator methionine for their maturation [5, 6]. Such a process is performed by methionine aminopeptidases (MetAPs) and often occurs while protein synthesis is still taking place, by the ribosomes in the cytosol; therefore, it is called a "cotranslational" modification process. In fact, the N-terminal region of MetAPs presents zinc finger-like domains that are important for proper ribosome association [7]. Although the proteolytic removal of initiator methionine in a large subset of nascent proteins is a process conserved from prokaryotes to eukaryotes, the biological roles of such aminopeptidases are beyond the maturation of nascent proteins. The inhibition of the human methionine aminopeptidase 1 activity in tumoral cells led to an accumulation of cells in the G_2/M phase of cell cycle [8], suggesting a role for this aminopeptidase in the progression of the cell cycle. Interestingly, the proteolytic processing event catalyzed by MetAPs is significantly linked to other important signaling pathways related to protein turnover [9, 10]. In the middle of the 1980s, the group led by Prof. Alexander Varshavsky showed that the identity of the N-terminal residue of a protein is intrinsically related to its half-life [11], such a feature was named the "N-end rule." In this respect, there are some N-terminal residues that may be regarded as degradation signals (N-degrons) that are, in turn, recognized by N-recognins, which are central components of the N-end rule degradation pathway. Consequently, amino acid residues at the N-terminus of a protein are referred to as "stabilizing" and "destabilizing" when proteins bearing such residues have long or short half-lives, respectively. In this context, it was observed that when amino acid residues at the second position that is normally not processed by MetAPs become exposed, they are recognized by an N-end rule-specific ligase and targeted to degradation via proteasome. On the contrary, residues that are often exposed by MetAPs are refractory to N-end-rule-mediated degradation. Therefore, the link between the MetAP specificity and protein turnover through the N-end rule has important implications for protein half-life, since the generation of "stabilizing" new N-terminus may prevent premature degradation by the N-end rule pathway [9]. Interestingly, in addition to protein turnover, cleavage of target proteins triggered by the N-degron-mediated degradation may also lead to important cellular events, such as inflammatory responses. A remarkable example is the downstream inflammation events triggered by the proteolytic activity of the enteroviral 3C cysteine protease, one of the main proteases encoded by human rhinoviruses. The 3C protease has an important role in cleaving protein precursors encoded by viral genome into individual components [12]. Interestingly, an expressive inflammatory response is initiated after the cleavage of the NACHT, LRR, and PYD domains-containing protein 1 (NLRP1) by the 3C protease. The protein NLRP1 acts as a sensor component of the inflammasome, a complex of multiprotein oligomers of the innate immune system, responsible for the activation of inflammatory responses. By cleaving NLRP1 at the Glu130-Gly 131 peptide bond, 3C protease exposes a neo-N-terminus which, in turn, becomes a glycine N-degron. The subsequent degradation of the processed NLRP1 by the proteasome leads to the generation of a C-terminal fragment that can activate the inflammasome, eventually contributing to the inflammatory disease of the airway caused by human rhinoviruses [13]. These brief examples illustrate how two seemingly unrelated biological events (protein maturation and protein turnover) are connected via proteolytic processing as well as some implications of the proteolytic events in each context.

Cleavage of protein precursors and proteolytic signaling

There are many situations where cleaved substrates may perform distinct biological functions depending on the cleavage site or on the biological context (i.e., a disease or a pathologic state). Furthermore, changes in the substrate structure after proteolytic cleavage may give rise to a plethora of effects in signaling pathways. Limited proteolysis increases protein diversity in terms of both structural features (due to structural alterations after proteolytic processing) and functional aspects (the "new" protein species can now execute distinct functions to respond to specific biological contexts). Indeed, this may be viewed as a parsimonious scenario for living organisms: a given biological function may be executed by a processed substrate (which was already available, although not functional) instead of synthesizing a new protein species to perform such a task. Therefore, in some cases, the proteolytically processed products may be functionally viewed as a completely distinct protein in comparison with its nonprocessed counterpart. Indeed, the alternative (new) functions of processed substrates increase the biological reachability of input signals in biological circuits (Fig. 2).

There are numerous examples where the imbalance in biological circuits due to proteolytic signaling results in striking outcomes. Perhaps one of the better examples of proteolytic signaling is observed is the classical sequence of proteolytic events (and their corresponding biological consequences) during the coagulation of blood. Such a pathway involves a series of subsequent proteolytic processing steps, performed mainly by sequentially activated zymogen form of serine proteases that ultimately lead to the cleavage of alpha and beta chains of the protein fibrinogen by the serine protease thrombin. As the coagulation process goes on, the successive cleavages of serine protease zymogens eventually amplify the initial signal by several orders of magnitude. In normal physiological conditions, the cleavage of Aα and Bβ fibrinogen chains by thrombin results in the generation of fibrinopeptides A and B, respectively, which in turn activates the transglutaminase (activated factor XIII, FXIIIa), which cross-links fibrin, thereby rendering the fibrin clot stable and insoluble. Interestingly, some thrombin-like enzymes isolated from snake venoms can directly cleave fibrinogen, however, usually releasing only fibrinopeptide A or B [14]. It turns out that, in the coagulation disorders observed after envenomation by some snakes, the resulting fibrin clot is not cross-linked by FXIIIa [14, 15]. Consequently, thrombin-like enzymes from snake venoms act

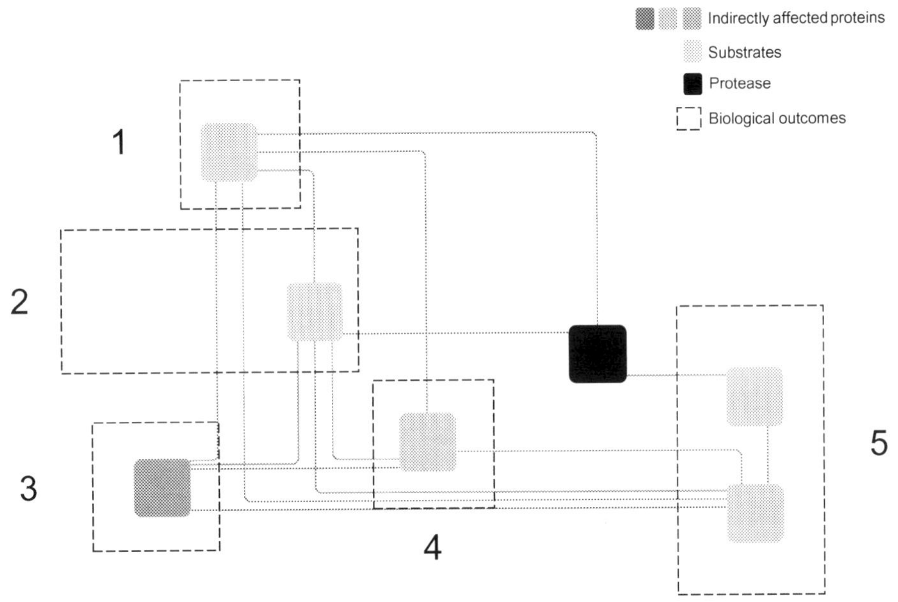

Fig. 2
Biological reachability is increased by means of proteolytic processing of substrates in biological circuits. Proteases can modulate biological circuits through their direct activity upon substrates (*gray squares*) as well as indirectly (*colored squares*), when cleaved proteins eventually affect downstream reactions, resulting in distinct outcomes (*shaded rectangles*).

as defibrinogenating agents; the activity of such serine proteases upon fibrinogen turns the blood nearly unclottable in vivo. This example illustrates the multifunctional roles of a single substrate after its cleavage by functionally related enzymes (i.e., thrombin and snake venom thrombin-like enzymes) in distinct pathophysiological conditions. Additionally, the venom of some snake species is rich in metalloproteases that are able to degrade extracellular matrix proteins, thereby promoting significant tissue damage in which hemorrhage is typically observed [16]. Synergistically, the proteolytic signaling triggered by snake venom proteases has important biological implications for prey subjugation, for example (further discussion on proteolytic signaling in snake envenomation will be presented in Chapter 12). In fact, these examples are just a snippet of a more complex biological scenario, which also demands the participation of several proteins (other than proteases).

Proteolytic signaling in health and disease

As a biological product derived from an intricate process of protein synthesis, protease expression is the result of the regulation of several pathways (transcription and translation, for example) acting coordinately and, more importantly, it is context-dependent. Furthermore, the main implications of protease activity in biological systems essentially rely on the fine-tuning mechanisms of protease inhibition, their biological availability (i.e., their expression status), and heritable diseases (mutations in protease genes, for example); proteolytic signaling, therefore, may result in distinct outcomes depending on the balance among all of these features which, in turn, mirrors the physiological state of the organism. As mentioned earlier, a number of ordinary physiological processes rely on proteolytic signaling. A classic example is the regulation of blood volume and systemic vascular resistance, which is performed by the renin-angiotensin system. By cleaving angiotensinogen at Leu43-Val44 peptide bond, the aspartic protease renin releases the decapeptide, angiotensin I. Further C-terminal removal of the two amino acid residues in angiotensin I by the zinc carboxy metalloprotease, angiotensin-converting enzyme (ACE) generates the octapeptide, angiotensin II, which promotes vasoconstriction [17]. In addition, ACE also degrades the potent vasodilator peptide, bradykinin, boosting the vasoconstrictor signaling initiated by renin [18]. Interestingly, an array of biologically active peptides generated from angiotensinogen have been described so far, all of them derived from the activity of proteases (i.e., renin, ACE, or ACE2) and with distinct functional implications in normal and disease conditions [19, 20].

Proteolytic signaling is also involved in the significant alterations in pathological conditions found in complex diseases such as cancer, for example. On this basis, proteases display a significant role in the process of metastasis, contributing to the effective spread of cancer cells to distant sites [21]. For example, increased activity of matrix metalloproteases (MMPs), namely MMP-1, MMP-2, and MMP-3, was reported in the set of secreted proteins derived from human gastric cancer-associated myofibroblasts [22], an important feature contributing to tissue remodeling of the cancer microenvironment. Qualitative and quantitative changes in the expression of proteases (mainly MMPs and ADAMs—A disintegrin and metalloproteases) are associated with the progression of melanoma, from nevi to radial and vertical growth [23]. Sandri and coworkers [24] showed that the resistance to the B-raf kinase inhibitor vemurafenib induced changes in the tumor microenvironment by the upregulation of MMP-2 and downregulation of its inhibitor, TIMP-2, in melanoma cells, with an increase in cell invasiveness. Laurent-Matha and coworkers [25] showed that proteolysis of the inhibitor of cysteine proteases, cystatin-C, by cathepsin D in the breast cancer environment enhanced extracellular proteolytic activity of cathepsins, contributing to cancer progression. Tumor growth and development depend on the cross talking between tumor and associated (stromal) cells. In this context, the interplay of surrounding nonmalignant stroma and/or infiltrating cells with tumoral cells has a pivotal role in the signaling events that take place within the tumor microenvironment. The activation of MMPs within the tumoral microenvironment, for example, is the result of, among other signaling events, proteolytic cascades generating autocrine feedback which includes the degradation of inhibitors, the activation of pro-MMPs, self-activation, and so on [26].

A close inspection on protease degradomes under distinct biological conditions (e.g., the profiling of proteolytic events in proteins secreted by a "normal" cell and its transformed phenotype) might shed light on the potential biological role of proteolytic signaling during oncogenesis [27]. Hence, the mapping and annotation of cleavage sites may reveal novel processed forms of substrates that, in turn, may be associated with the phenotypic plasticity found in the disease under investigation. Moreover, although the assignment of the protease(s) responsible for the proteolytic processing events in complex biological samples is challenging, the identification of proteolytic signaling exclusive to disease states are themselves "signatures" of a pathological state.

Overall, proteolytic signaling is more than meets the eye. More importantly, it is not only about protease activity; rather, it must be understood in light of the fate of processed substrates and their corresponding roles in biological circuits—the slight imbalance in protease web may answer for substantial deviations from health to disease states.

Outlook and perspectives

As an essentially irreversible biological event, proteolytic signaling represents an additional layer of complexity in the study of signaling pathways, which emerges not only from the dynamic interactions among proteins, or by the landscape of genetic mutations and/or altered protein expression, but also by means of the degradome status of a given set of proteins and under defined biological contexts. Since neo-N-termini are generated after proteolytic processing, such set of peptides/processed proteins may be targeted by analytical approaches, aiming at the prospection of markers in distinct physiological conditions (i.e., health and disease) and in several biological matrices such as tissue samples, plasma, saliva, and urine, for instance. The technology and analytical platforms to study proteolytic signaling events are evolving fast, allowing for the comprehensive profiling of protease degradomes in a timescale faster than ever seen before—this is critical for translational medicine, where personalized therapies are proving to be more effective than the traditional axiom: "treating the disease, not the patient."

Acknowledgment

The work in the author's laboratory is supported by grants from Fundação de Amparo à Pesquisa do Estado de São Paulo (FAPESP, grants # 2014/06579-3, 2019/10817-0, and 2019/07282-8).

References

[1] Rawlings ND, Waller M, Barrett AJ, Bateman A. MEROPS: The database of proteolytic enzymes, their substrates and inhibitors. Nucleic Acids Res 2014;42(D1).

[2] Overall CM, Kleifeld O. Validating matrix metalloproteinases as drug targets and anti-targets for cancer therapy. Nat Rev Cancer 2006;6(3):227–39.

[3] Harper E, Berger A. On the size of the active site in proteases: pronase. Biochem Biophys Res Commun 1972;46(5):1956–60.
[4] López-Otín C, Hunter T. The regulatory crosstalk between kinases and proteases in cancer. Nat Rev Cancer 2010;10:278–92.
[5] Ball LA, Kaesberg P. Cleavage of the N-terminal formylmethionine residue from a bacteriophage coat protein in vitro. J Mol Biol 1973;79(3):531–7.
[6] Ben-Bassat A, Bauer K, Chang SY, Myambo K, Boosman A. Processing of the initiation methionine from proteins: properties of the Escherichia coli methionine aminopeptidase and its gene structure. J Bacteriol 1987;169(2):751–7.
[7] Addlagatta A, Hu X, Liu JO, Matthews BW. Structural basis for the functional differences between type I and type II human methionine aminopeptidases. Biochemistry 2005;44(45):14741–9.
[8] Hu X, Addlagatta A, Lu J, Matthews BW, Liu JO. Elucidation of the function of type 1 human methionine aminopeptides during cell cycle progression. Proc Natl Acad Sci U S A 2006;103(48):18148–53.
[9] Bradshaw RA, Yi E. Methionine aminopeptidases and angiogenesis. Essays Biochem 2002;65–78.
[10] Varshavsky A. The N-end rule pathway and regulation by proteolysis. Protein Sci 2011;20:1298–345.
[11] Bachmair A, Finley D, Varshavsky A. In vivo half-life of a protein is a function of its amino-terminal residue. Science (80–) 1986;234(4773):179–86.
[12] Palmenberg AC. Proteolytic processing of picornaviral polyprotein. Annu Rev Microbiol 1990;44:603–23.
[13] Robinson KS, DET T, Sen TK, Toh GA, Ong HH, Lim CK, et al. Enteroviral 3C protease activates the human NLRP1 inflammasome in airway epithelia. Science (80–) 2020;, eaay2002.
[14] Serrano SMT, Maroun RC. Snake venom serine proteinases: sequence homology vs. substrate specificity, a paradox to be solved. Toxicon 2005;45(8):1115–32.
[15] Alvarez-Flores MP, Faria F, de Andrade SA, Chudzinski-Tavassi AM. Snake venom components affecting the coagulation system. In: Snake Venoms; 2016. p. 1–20.
[16] Asega AF, Menezes MC, Trevisan-Silva D, Cajado-Carvalho D, Bertholim L, Oliveira AK, et al. Cleavage of proteoglycans, plasma proteins and the platelet-derived growth factor receptor in the hemorrhagic process induced by snake venom metalloproteinases. Sci Rep 2020;10(1).
[17] Lu H, Cassis LA, CWV K, Daugherty A. Structure and functions of angiotensinogen. Hypertens Res 2016;39:492–500.
[18] Imig JD. ACE inhibition and bradykinin-mediated renal vascular responses: EDHF involvement. Hypertension 2004;43:533–5.
[19] Jiang F, Yang J, Zhang Y, Dong M, Wang S, Zhang Q, et al. Angiotensin-converting enzyme 2 and angiotensin 1-7: novel therapeutic targets. Nat Rev Cardiol 2014.
[20] Danilczyk U, Eriksson U, Oudit GY, Penninger JM. Physiological roles of angiotensin-converting enzyme 2. Cell Mol Life Sci 2004;61:2714–9.
[21] Barberis I, Martini M, Iavarone F, Orsi A. Available influenza vaccines: immunization strategies, history and new tools for fighting the disease. J Prev Med Hyg 2016;57(1):E41–6.
[22] Holmberg C, Ghesquière B, Impens F, Gevaert K, Kumar JD, Cash N, et al. Mapping proteolytic processing in the secretome of gastric cancer-associated myofibroblasts reveals activation of MMP-1, MMP-2, and MMP-3. J Proteome Res 2013;12(7):3413–22.
[23] Moro N, Mauch C, Zigrino P. Metalloproteinases in melanoma. Eur J Cell Biol 2014;93:23–9.
[24] Sandri S, Faião-Flores F, Tiago M, Pennacchi PC, Massaro RR, Alves-Fernandes DK, et al. Vemurafenib resistance increases melanoma invasiveness and modulates the tumor microenvironment by MMP-2 upregulation. Pharmacol Res 2016;111:523–33.
[25] Laurent-Matha V, Huesgen PF, Masson O, Derocq D, Prébois C, Gary-Bobo M, et al. Proteolysis of cystatin C by cathepsin D in the breast cancer microenvironment. FASEB J 2012;26(12):5172–81.
[26] Kessenbrock K, Plaks V, Werb Z. Matrix metalloproteinases: regulators of the tumor microenvironment. Cell 2010;141:52–67.
[27] Liberato T, Fukushima I, Kitano ES, Serrano SMT, Chammas R, Zelanis A. Proteomic profiling of the proteolytic events in the secretome of the transformed phenotype of melanocyte-derived cells using Terminal Amine Isotopic Labeling of Substrates. J Proteomics 2019;192:291–8.

CHAPTER 2

Ubiquitin ligases: Proteolytic signaling, protein turnover, and disease

Patrícia Maria Siqueira dos Passos[a], Camila Rolemberg Santana Travaglini Berti de Correia[a], Caio Almeida Batista de Oliveira[a], Valentine Spagnol, Isabela Fernanda Morales Martins, and Felipe Roberti Teixeira

Department of Genetics and Evolution, Federal University of Sao Carlos, Brazil

Ubiquitin-proteasome system (UPS) and E3 ligases

The ubiquitin-proteasome system (UPS) is the major responsible for intracellular proteolysis having a central role in regulating many cellular processes such as cell cycle progression, apoptosis, immune response, differentiation, signal transduction, and protein quality control [1]. This system was described by Aaron Ciechanover, Avram Hershko, and Irwin Rose, who received the Nobel Prize in Chemistry in 2004.

The ubiquitin is a small protein composed of 76 amino acids with approximately 8.5 kDa, which is found ubiquitously in most eukaryotic cells [2]. It is attached to a protein substrate through an isopeptide bond formed between the carboxyl group (COO −) of the ubiquitin's glycine 76 and the amino group of the protein target. This posttranslational modification called ubiquitination involves a three-step enzyme cascade: initially, ubiquitin-activating enzyme (E1) catalyzes ubiquitin activation in an ATP-dependent reaction; secondly, activated ubiquitin is transferred to ubiquitin-conjugating enzyme (E2) via a thioester linkage between E2 cysteine in active site and C-terminal of ubiquitin; finally, ubiquitin ligases (E3) that specifically binds to the substrate recruits the E2 ~ Ub and catalyzes the transferring of ubiquitin to the substrate [1] (Fig. 1).

The human genome encodes two E1 enzymes, 37 E2 enzymes, and more than 600 E3 ligases, which are classified in two major families based on their specific domains: HECT (homologous to E6-associated protein C-terminus) domain or a RING (really interesting new gene) finger domain. HECT and RING E3 ligases have different mechanisms to ubiquitinate their substrates. HECTs catalyze substrate ubiquitination in a two-step reaction:

[a] These authors contributed equally for this work.

Proteolytic Signaling in Health and Disease. https://doi.org/10.1016/B978-0-323-85696-6.00010-X

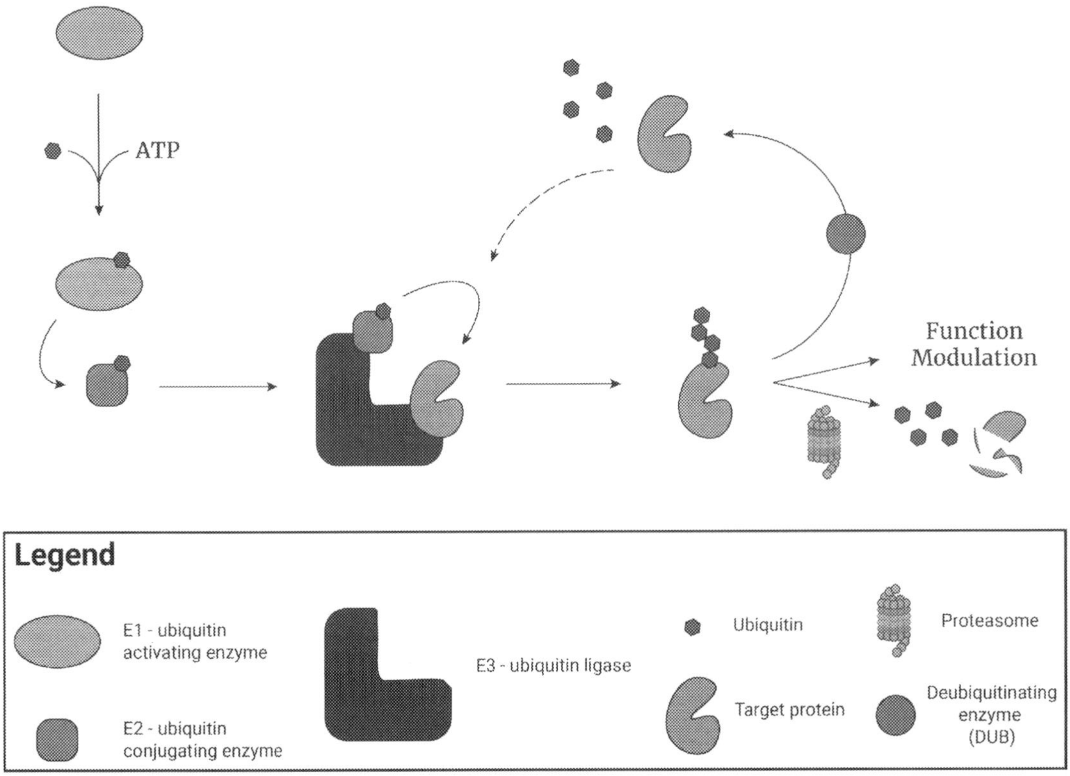

Fig. 1

The ubiquitin-proteasome system. Ubiquitin is activated by E1 in an ATP-dependent reaction and transferred to an E2. Finally, the E3 ubiquitin ligase transfers the ubiquitin from E2 to a substrate. Ubiquitination either can direct the polyubiquitinated substrate to the proteasome for degradation or can alter the function of the target protein. Furthermore, ubiquitinated substrate can be deubiquitinated by deubiquitinating enzymes.

in the first step, HECT conserved cysteine residue accepts activated ubiquitin from an E2 by a transthiolation reaction; in the second step, the ubiquitin is transferred from the E3 intermediate to a lysine on the target substrate, whereas RING E3s transfer ubiquitin directly from E2 to substrate [3] (Fig. 2).

Most of the human E3 ligases are RING type with the Cullin RING ligases (CRLs) superfamily being the most abundant. They are composed of Cullin protein (Cul-1, -2, -3, -4a, -4b, -5, -7, or -9), a small RING protein (Rbx1/Roc1/Hrt1), and either an adaptor protein(s) that binds to an interchangeable substrate recognition protein [4]. The most studied class of CRLs is the SCF (SKP, Cullin, F-box), which is composed of SKP1 (S-phase kinase-associated protein 1), Cullin 1, Rbx1 (RING box protein 1), and a member of F-box protein family (Fig. 3A). Rbx1 contains a RING domain that interacts with Cullin 1 and recruits E2 ~ Ub to be transferred to the substrate. Cullin 1 is a scaffold of the complex, interacting with Rbx1 at C-terminus and SKP1 at N-terminus. SKP1 works as an adapter linking the

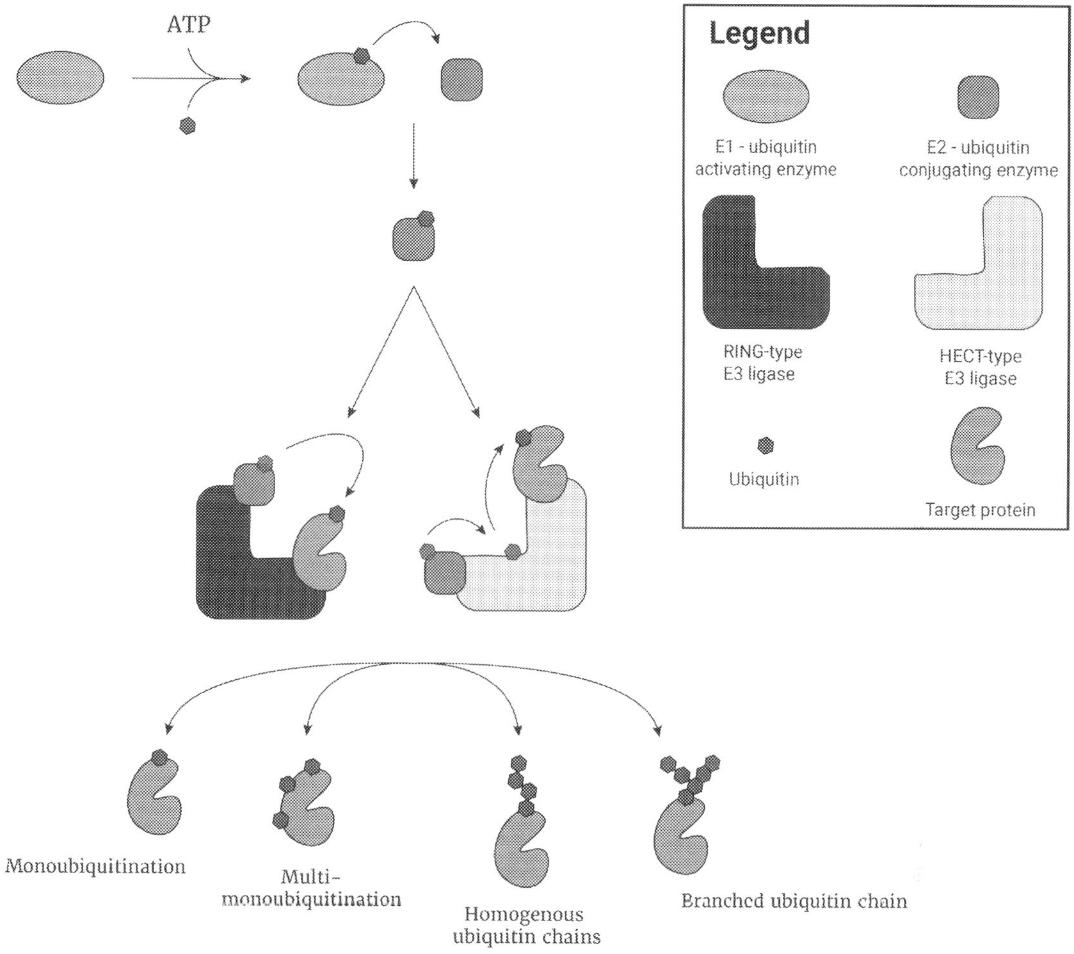

Fig. 2
Different mechanisms of ubiquitination by two classes of E3 ubiquitin ligases: HECT and RING E3 ligases. HECT E3 ligases catalyze substrate ubiquitination in a two-step reaction: first, ubiquitin is transferred from E2 to HECT E3, and then, ubiquitin is transferred to substrate, whereas RING E3s transfer ubiquitin directly from E2 ~ Ub to substrate.

F-box protein to Cullin 1, connecting it with the rest of the complex. Finally, F-box proteins are characterized by a 42–48 amino acid F-box motif that binds SKP1 and also a carboxy-terminal domain that interacts with substrates, which allow the specificity of ubiquitination [5].

There are 69 F-box proteins in humans that are classified into three families based on its substrate-interacting domain: FBXW, with WD40 repeated domain; FBXL, with Leucine-Rich Repeats (LRR); and FBXO, with either another or no other motif [6] (Fig. 3B). The SCF complexes regulate cell cycle, immune response, signaling pathways, and development stages through the ubiquitination of their targets, such as cyclins and β-catenin.

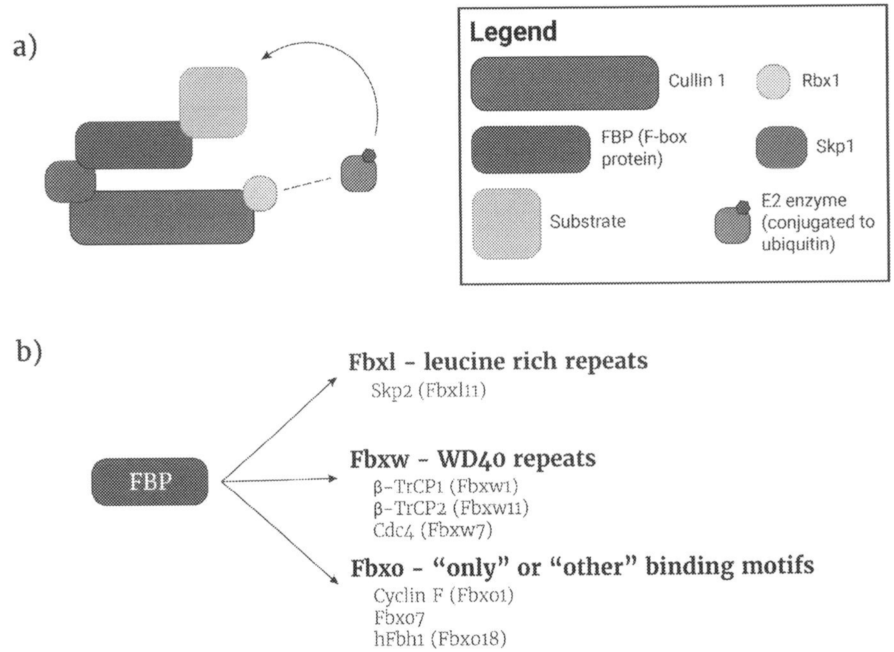

Fig. 3

(A) The SCF (SKP1, Cullin 1, F-box) E3 ubiquitin ligases. SCF complexes are composed of SKP1, Cullin 1, Rbx1 and an F-box protein. Rbx1 interacts with Cullin 1 and recruits E2 ~ Ub to be transferred to the substrate. Cullin 1 is a scaffold of the complex, interacting with Rbx1 and SKP1. The F-box proteins utilize the F-box domain to interact with SKP1. (B) The three classes of F-box proteins and the most studied proteins of each class.

Deregulation of SCF-dependent proteolysis can lead to neoplastic transformation, in which mutations of SCF subunit are found in several breast and ovarian cancer cell lines [7].

The target protein can be monoubiquitinated or multi-monoubiquitinated, where a ubiquitin is added in multiple lysine residues. Moreover, the first ubiquitin attached to substrate can prime the attachment of another ubiquitin into the first through their seven lysine residues (Lys6, Lys11, Lys27, Lys29, Lys33, Lys48, and Lys63), resulting in different types of ubiquitin chains. Each one of them has a unique tertiary structure resulting in different cascade events in the cells [8] (Fig. 4). Monoubiquitination is involved in cellular recycling, lysosome degradation of cell-surface receptors, DNA repair, histone regulation, and others. Ub chains linked via Lys48 or 63 are the most well characterized, while Lys48 polyubiquitinated destinates the substrate for 26S proteasome degradation, the Lys63 polyubiquitination has a nondegradative function in different signaling pathways such as endocytosis, DNA repair, and NF-κB [9].

The ubiquitination is a reversible process, where ubiquitin can be removed from the substrate by deubiquitinating enzymes (DUBs). There are approximately 79 DUBs in the human

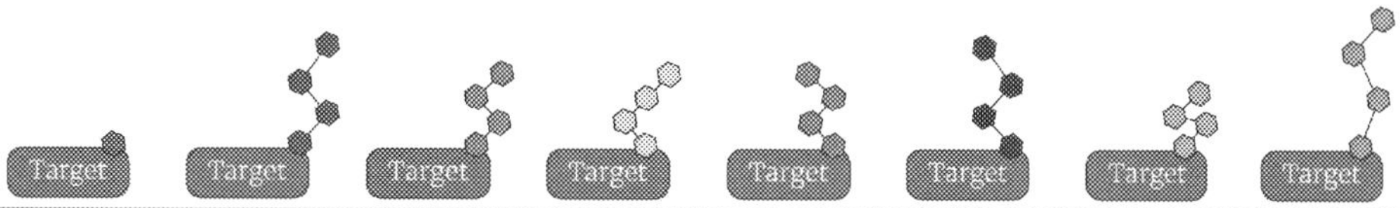

Linkage type	Mono-ubiquitin	K-6 chain	K-11 chain	K-27 chain	K-29 chain	K-33 chain	K-48 chain	K-63 chain
Function	·Protein turnover ·DNA repair ·Histone regulation ·Cell-surface receptors degradation	·Mitophagy ·Mitochondrial damage response	·Protein turnover ·Mitophagy ·Trafficking ·Degradation	·Autoimmunity ·Scaffold ·Degradation ·Mitochondrial damage response	·Degradation	·Trafficking ·Adaptive immunity	·Protein turnover ·Degradation by 26S proteasome	·Gene activation ·Innate immunity ·Scaffold ·DNA damage response

Fig. 4
Representation of monoubiquitinated substrate and different polyubiquitin chain types via Lys6, Lys11, Lys27, Lys29, Lys33, Lys48, Lys63, and their respective cellular functions.

genome responsible for removing ubiquitin molecules from the target protein. DUBs are also called as deubiquitylating or deubiquitinating enzymes and can be subdivided into five families: ubiquitin C-terminal hydrolases (UCHs), ubiquitin-specific proteases (USPs), ovarian tumor proteases (OTUS), Josephins, and JAB1/MPN/MOV34 metalloenzymes. DUBs' activities are specific for both substrates and particular ubiquitin chain types, and this specificity is not restricted to a particular DUB family [10].

In general, DUBs catalyze a proteolytic reaction between a Lys ε-amino group and a carboxy group corresponding to the C-terminal of ubiquitin in a single-step chain removal through ubiquitinated target sequences in substrates. These DUBs can both remove ubiquitin completely from substrates or leave them monoubiquitinated, in which most promiscuous DUBs remove ubiquitin completely, while linkage-specific DUBs might leave them monoubiquitinated. Once a substrate is monoubiquitinated, it can be extended again, including different types of ubiquitin chains; therefore, DUBs can also promote ubiquitin chain editing [10] (Fig. 5).

Proteasome

The proteasome is the major ATP-dependent protease in eukaryotic cells being the most downstream element of the UPS, responsible for the degradation of polyubiquitinated proteins. It plays a central role in many biological processes, such as cell cycle progression, signal transduction, and protein quality control. This 2.5 MDa protease is constituted by at least 32 different subunits that form a 20S proteolytic core (PC) and one or two 19S regulatory particles (RPs) that serve as a proteasome activator with a molecular mass of approximately 700 kDa (called PA700) (Fig. 6). The 19S RP binds to one or both ends of the latent 20S proteasome to form an enzymatically active proteasome. The proteasome works via a multistep mechanism where substrate recognition, deubiquitination, unfolding, and translocation occur into 19S RP, while peptide bond cleavage occurs in PC [11].

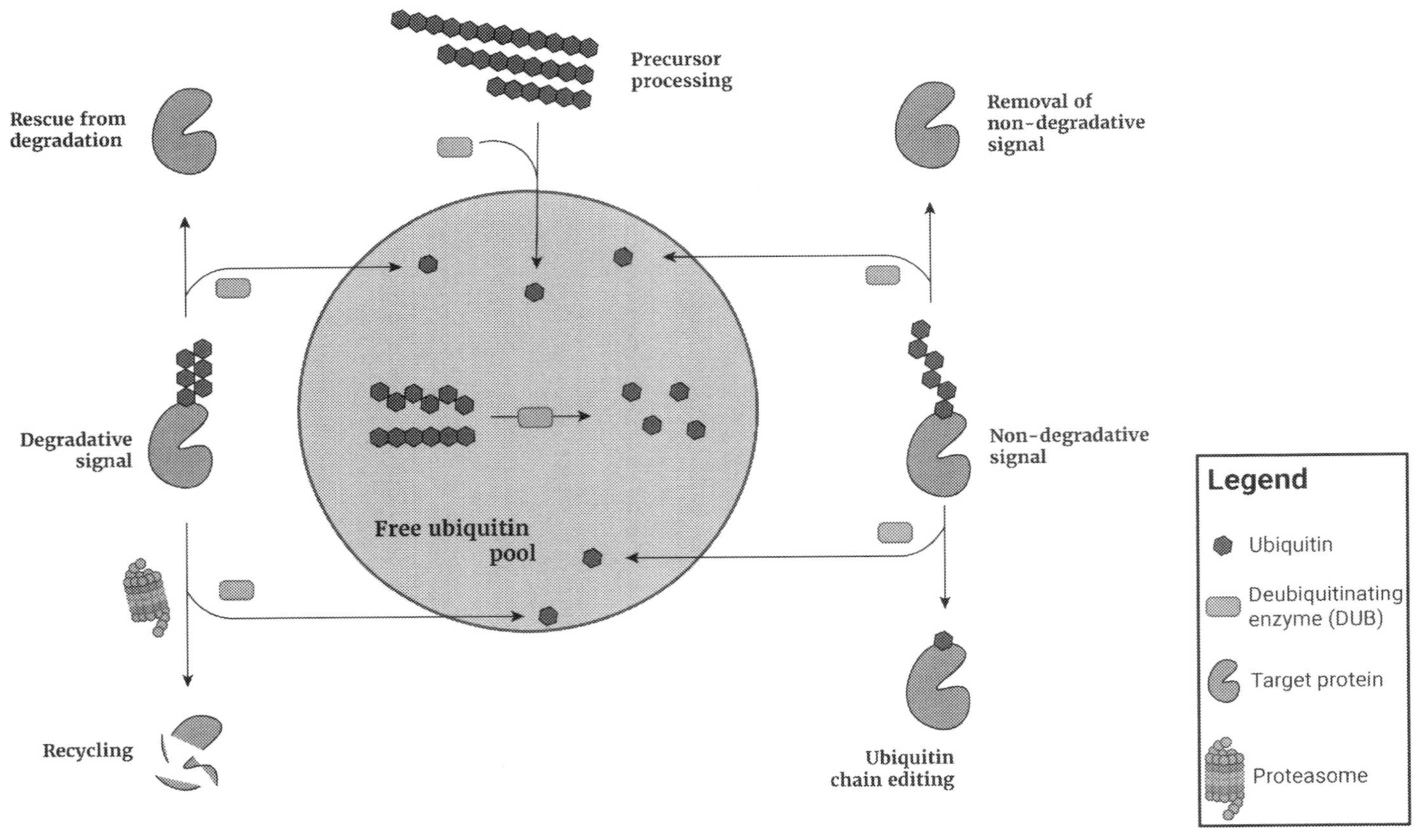

Fig. 5

Central roles of DUBs. Ubiquitin is transcribed and translated as a linear fusion protein consisting of multiple copies of ubiquitin and DUBs digest them in free ubiquitin. DUBs remove a nondegradative ubiquitin signal from the substrates regulating its function, and also it is responsible for removing the substrate ubiquitin degradative signal, which releases ubiquitin, recycling it to be used into the cells. Also, DUBs can promote ubiquitin chain edition exchanging types of ubiquitin linkage from substrate. Targeted protein can be rescued from degradation by DUBs by removing or changing its ubiquitin signal.

In Fig. 6, we can observe that the RP 19S particle consists of 20 subunits in mammals that are subdivided into the lid (Rpn 1–13, Rpn15) and base, which is composed of six ATPases subunits from the AAA + family (Rpt1–6). Polyubiquitinated proteins are recognized by Rpn10 and Rpn13 subunits and Rad23, Dsk2, and Ddi1 through their Ub-like (UBL) domains and Ub-associated (UBA) domains, respectively. The ubiquitin chains of polyubiquitinated substrates are first removed by the Rpn11 and unfolded by the hexameric ring (Rpt1–6) and subsequently translocated to the inner of CP 20S complex, which is composed by 28 subunits arranged in four hetero-heptameric rings (α1–7; β1–7; β1–7; α1–7) that are highly conserved in eukaryotes [11]. The two α rings complexes form a narrow channel that only allow the traffic of denatured proteins through it. The two inner β rings are the catalytic chamber of 20S CP, which contains three enzymatically active sites by β1, β2, and β5 classified in caspase-like, trypsin-like, and chymotrypsin-like, respectively. The final products of proteasome digestion are a heterogeneous mixture of peptides that in some cases can be biologically active and functional molecules (Fig. 6).

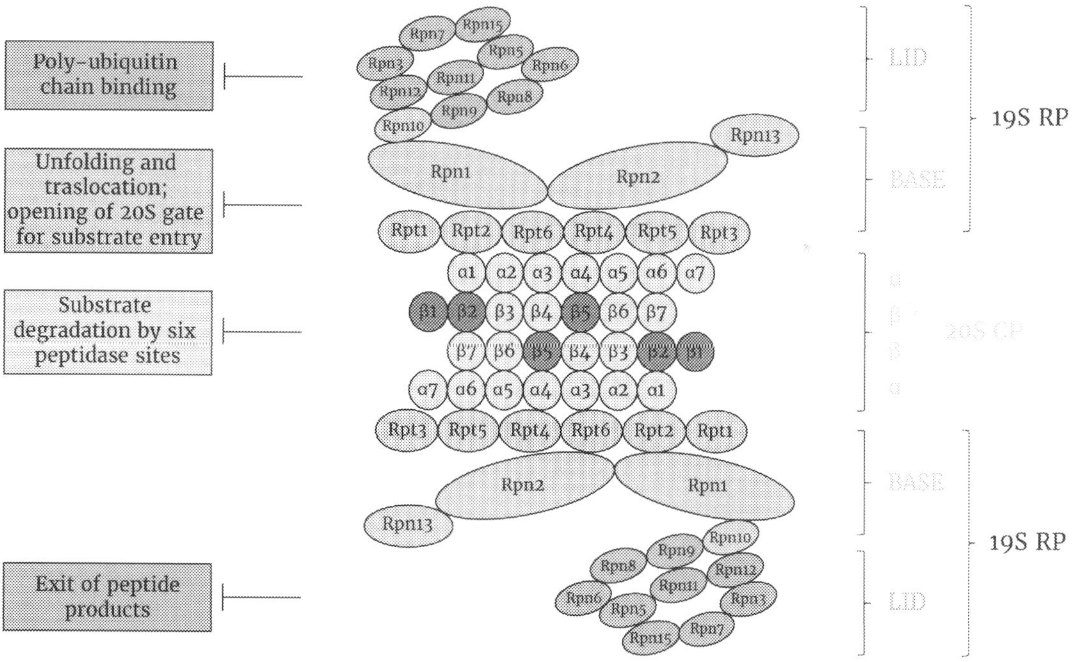

Fig. 6

Structure and function of 26S proteasome subunits. 19S regulatory particles (RP) act as a proteasome activator and are subdivided into the lid (Rpn 3, 5–9, 11, 12, and 15); base, which is composed of Rpn 1, 2, 10, and 13; and six ATPase subunits (Rpt1–6). The 20S proteolytic core (PC) is constituted by 28 subunits arranged into four hetero-heptameric rings (α1–7; β1–7; β1–7; α1–7). 20S CP contains three enzymatically active sites by β1, β2, and β5 classified into caspase-like, trypsin-like, and chymotrypsin-like, respectively (highlighted in *red* (dark gray in the print version)). (For interpretation of the references to color in this figure legend, the reader is referred to the web version of this article.)

Cell signaling regulation by ubiquitination

Ubiquitin is a key regulator in different cell signaling pathways such as mitophagy, NF-κB and interferon signaling pathway, cell cycle, and DNA repair. In this section, we will discuss the role of ubiquitination in each of these processes (Fig. 7).

NF-κB signaling pathway

The NF-κB protein family was discovered as transcription factors that bind to an enhancer of kappa light genes in B cells—therefore, the name κB, which is the name of the DNA region they bind. This family comprises five members, responsible for the expression of a series of genes related to immunity response, acting on cell survival and immunity. All of the five (p50, p52, p65/RelA, c-Rel, and RelB) possess a Rel homology domain (RHD), which enables the formation of heterodimers—the functional structure of NF-κB proteins. The RHD is also important in controlling the activity of these transcription factors, since NF-κB inhibitors (IκBs) also bind to this region when the signaling pathway is inactive, preventing dimerization and nuclear translocation from cytoplasm [12, 13].

Ubiquitin stands out as one of the main regulators of NF-κB signaling pathway being responsible for its activation or inhibition. When the canonical pathway is stimulated via tumor necrosis factor (TNF), interleukins (ILs), or pathogens products such as

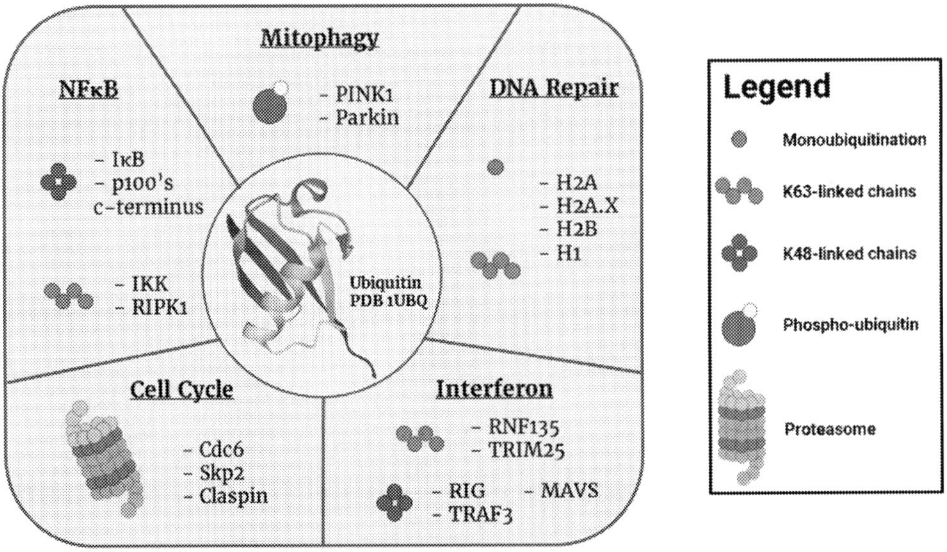

Fig. 7
Cellular processes regulated by ubiquitination. Examples of targets in each process are indicated with the ubiquitin types conjugated by the E3 ubiquitin ligases.

lipopolysaccharide (LPS), the signal transduction leads to phosphorylation of IκBs by the IκB kinase complex (IKK). The phosphorylated inhibitors are recognized and ubiquitinated by the E3 ligase SCF(β-TRCP)—also known as SCF(Fbxw1)—which are then destinated to by the proteasome (Fig. 8). Activation of the noncanonical pathway also relies on the ubiquitination of own NF-κB proteins. p100, for example, requires a proteasomal treatment that only degrades its C-terminus region, which acts as an intramolecular inhibitor—it is similar to IκBs; once this region is removed by the 26S proteasome, the p52 protein is active [12].

Ubiquitination is central in the control of canonical and noncanonical pathways, as several components of these pathways suffer modifications through ubiquitin. The IKK, for example, is only fully active when polyubiquitinated by K63—and this process is mediated by the E2 complex Ubc13/Uev1A and the TRAF6 protein, which contains a RING domain in its N-terminus region that is able to facilitate the formation of those chains on IKK. Activation of the pathway through IL-1β (consequently through IL-1 receptor, IL-1R) is completely ablated when the wild-type ubiquitin pool is replaced by a K63R-mutated ubiquitin within the cell; more than that, the K63 ubiquitin chains were first discovered as an activator of the NF-κB pathway [12, 14].

The activation of the pathway can also be reached through a very particular mechanism of ubiquitination that, so far, has only been described to be performed by one E3 ligase: linear ubiquitin chains, catalyzed by linear ubiquitin chain assembly complex (LUBAC). Differently from the ubiquitin chains formed by the addition of a ubiquitin onto another ubiquitin's lysine, isopeptide bonds of linear chains occur with the methionine 1 residue of ubiquitin. LUBAC is recruited after IL-1R or tumor necrosis factor receptor (TNFR) activation, and depending on the type of receptor stimulated, different response will occur. Activation of TNFR leads to recruitment and K63 polyubiquitination of RIPK1, which is the signal for LUBAC; in turn, LUBAC produces linear chains on RIPK1 and attracts NEMO, the regulatory subunit of the IKK complex, which is also a target for LUBAC's linear chains, responsible for activating IKK and causing a later degradation of IκBs by proteasome recognition of K48 chains added by SCF(β-TRCP). On the contrary, if the cascade starts via IL-1R, LUBAC will recognize ubiquitinated IRAK1, which was previously recruited by already modified Myd88 and IRAK4. In this situation, LUBAC is not only able to produce linear chains but also extends already existing K63 chains, forming hybrid chains. The simultaneous presence of TAB1-TAB2/3-TAK1 in this context completes the requirements for the activation of IKK complex, which ultimately will cause the expression of the genes controlled by this pathway [12, 15].

Deubiquitinating enzymes (DUBs) also play a major role in modulating NF-κB pathway activation avoiding a constant stimulation of NF-κB controlled genes, which could lead to chronic inflammation. Among them, A20 is one of the most important, as it is specific for K63 chains on RIPK1 and NEMO, attenuating inflammatory response. Interestingly, A20

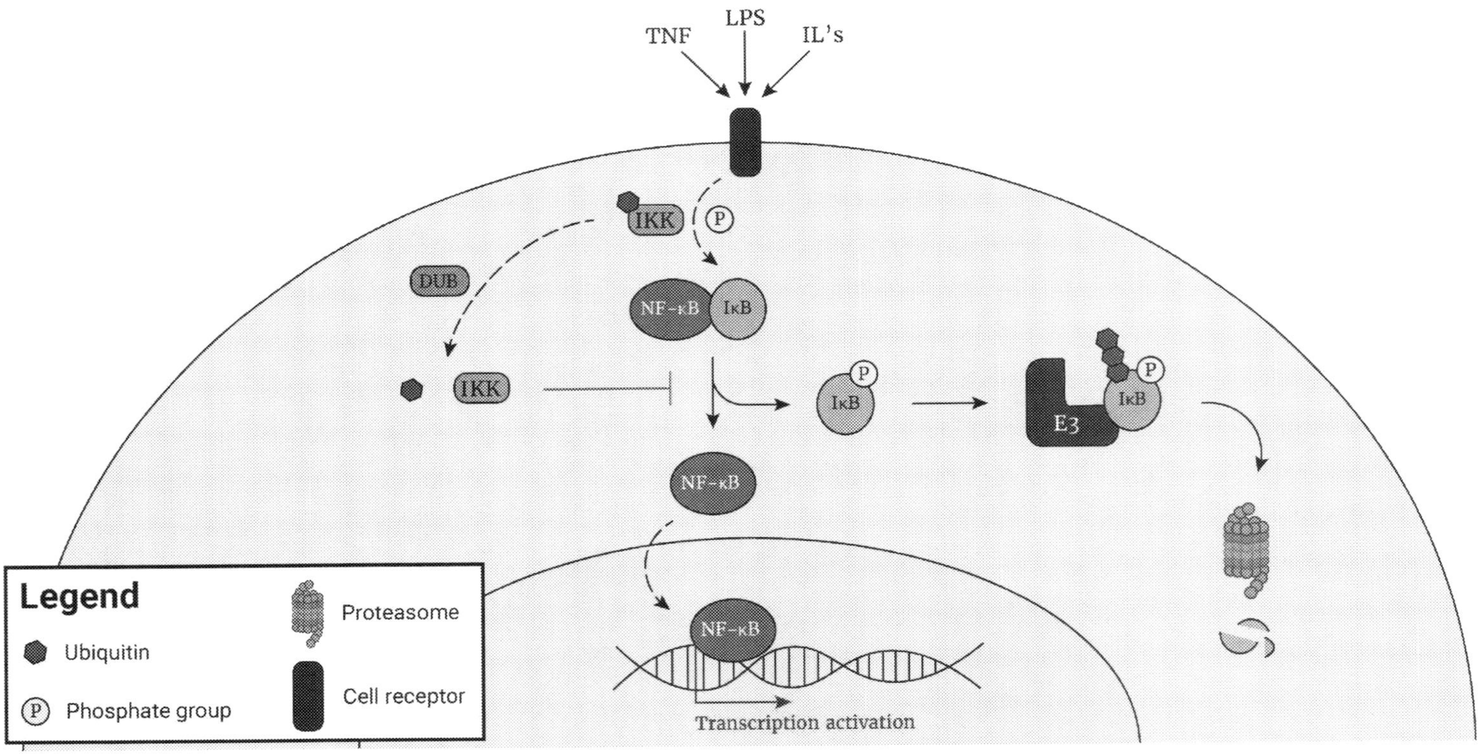

Fig. 8

General signal transduction cascade of NF-κB signaling pathway. After the activation of a cell receptor (by TNF, LPS, or ILs), IKK—which is active only after nondegrading ubiquitination—phosphorylates IκB, releasing NF-κB transcription factor from inhibition. While IκB is recognized by an E3 ligase, ubiquitinated, and degraded by the proteasome, the transcription factor is able to migrate to the nucleus and activate the expression of controlled genes.

has also an E3 ubiquitin ligase activity introducing K48 ubiquitin chains on those proteins, leading to their degradation and counteracting NF-κB activation. The absence of A20 causes a prolonged stimulation of NF-κB signaling through prolonged phosphorylation of IKK and increased degradation of IκBs. Other DUBs are also essential for NF-κB control, as, for example, OTULIN, which is able to directly counteract LUBAC action, hydrolyzing linear ubiquitin chains and diminishing pathway activation [12, 14, 15].

Mitophagy

Mitochondria are highly dynamic and multifunctional organelles that participate in a broad array of functions within eukaryotic cells. Besides generate the bulk of ATP in most cell types through oxidative phosphorylation, mitochondria also metabolize and synthesize complex macromolecules as lipids, amino acids, and nucleotides; regulate calcium homeostasis and cellular redox balance; control apoptosis; and play a key role in the innate response to viral infections, helping to induce antiviral and anti-inflammatory pathways [16]. Mitochondria are particularly susceptible organelles because their role in energy production exposes them to damage due to high levels of reactive oxygen species (ROS), a product of energy generation that can cause mutations in mitochondrial DNA (mtDNA). For this reason, mitochondria have multiple systems of quality control to ensure the maintenance of a healthy and functional mitochondrial network to meet the demands of the cell [17].

Mitochondria experiencing global/widespread damage undergo a selective form of autophagy, called mitophagy [18]. This process requires the ubiquitination of damaged mitochondria, which is mediated in large part by the E3 ubiquitin ligase Parkin. Most prominent of the genes involved in mitochondrial maintenance is *PINK1* that encodes the neuroprotective protein PTEN-induced serine/threonine kinase 1 (PINK1) and *PRKN* that encodes the E3 ubiquitin ligase Parkin. Mutations in *PRKN* and *PINK1* genes are linked to autosomal recessive early-onset Parkinson's disease [18, 19].

PINK1 is a mitochondrial-targeted protein that contains an N-terminal mitochondrial targeting sequence (MTS) and a transmembrane domain (TM). In healthy mitochondria, PINK1 is constitutively imported via translocase of the outer membrane (TOM) and translocase of the inner membrane (TIM) to the inner membrane (IMM). Translocation of the MTS through the TIM complex is energetically driven by the electrical membrane potential ($\Delta\Psi m$) across the IMM. PINK1 reaches the matrix, where it is sequentially cleaved by two proteases, mitochondrial processing peptidase (MPP) and presenilin-associated rhomboid-like (PARL). The truncated protein produced is retro translocated to the cytosol, where it is degraded via proteasomal pathway (Fig. 9, left panel). Under physiological conditions, PINK1 levels are quite low because it is rapidly degraded [18, 19].

Loss of mitochondrial membrane potential by damages inhibits PINK1 import through the TIM complex and its proteolytic cleavage inside the matrix, leading to insertion of its

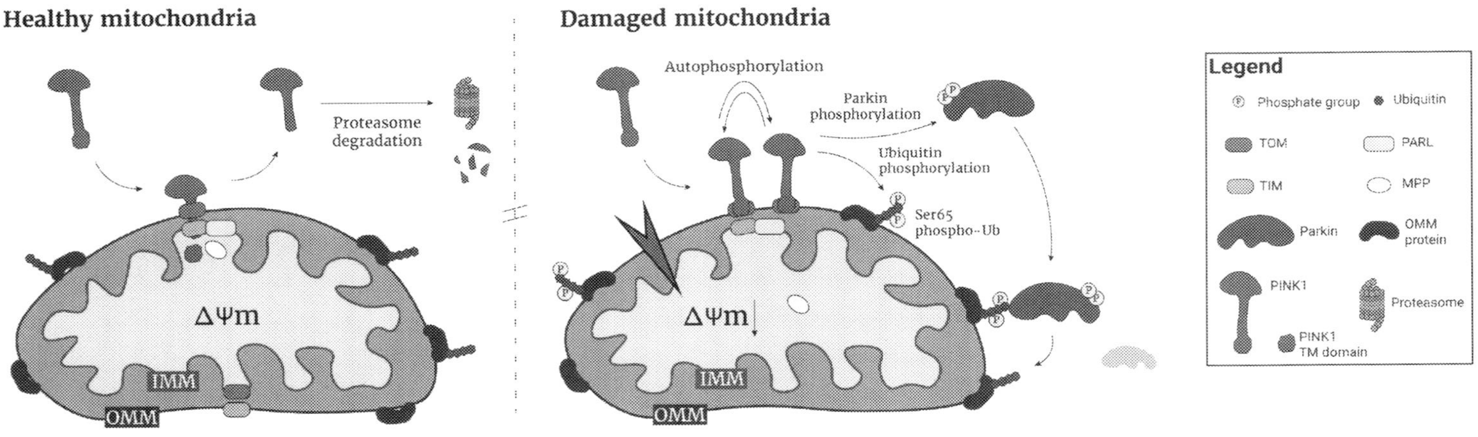

Fig. 9

(*Left panel*) In healthy mitochondria, PINK1 is constitutively imported into mitochondria, processed, and degraded by the proteasome. (*Right panel*) Upon mitochondrial stress or damage, PINK1 import to IMM is compromised and it is stabilized on the OMM, where it phosphorylates both Parkin and ubiquitin from OMM proteins. Parkin has a high affinity for phosphoUb and translocates to the mitochondrial surface, where ubiquitinates more OMM proteins. The newly incorporated ubiquitin is still phosphorylated by PINK1, which recruits more Parkin from the cytosol to the damaged mitochondria, in a positive feedback loop.

transmembrane domain (TM) into the outer mitochondrial membrane (OMM). At the OMM, PINK1 phosphorylates both Parkin and ubiquitin from OMM proteins at Ser65, producing the Ser65-phosphoUb. The E3 ubiquitin ligase Parkin has a high affinity for phosphorylated ubiquitin and it is translocated from the cytosol to the surface of damaged mitochondria under the increase of phosphoUb concentration (Fig. 9, right panel). The identification of phosphorylated ubiquitin has triggered enormous excitement and research into this and other chemical modifications functionally important of ubiquitin [18, 19].

Structural and biochemical studies revealed that the E3 Parkin is autoinhibited in several ways, but mainly due to the self-inhibition of its N-terminal Ubl domain that blocks the E2 binding site [20, 21]. A current model for PINK1-mediated activation of Parkin suggests that PINK1 phosphorylates ubiquitin attached to OMM proteins and the autoinhibited cytosolic Parkin is recruited by nanomolar affinity to sites of PINK1 activity. The Ser65-phosphoUb acts as an allosteric regulator, leading to significant conformational changes at Parkin that release the Ubl domain from the Parkin core, unblocking the E2 binding site, but also enable PINK1 to phosphorylate the Ubl domain. Parkin phosphorylation stabilizes it in its active form. Active Parkin ubiquitinates OMM proteins, and the newly incorporated ubiquitin is still phosphorylated by PINK1, which recruits more Parkin from the cytosol to the damaged mitochondria, in a positive feedback loop. Mitophagy receptors NDP52 and OPTN bind to ubiquitinated mitochondrial proteins and recruit the autophagy machinery to mitochondria. The phagophore engulfs mitochondria and fuses with the lysosome to degrade and recycle its contents [18–21].

Interferon

One of the main mechanisms of innate immunity against viruses is the induction of type I Interferon (IFNs) by infected cells. These are parts of a large family of cytokines, which are named this way for their ability to interfere in the viral infection, whose main function is to inhibit pathogen's replication [22]. When pathogen-associated molecular patterns (PAMPs) are recognized by pattern recognition receptors (PRRs), such as Toll-like receptors (TLRs), RIG-1-like receptors (RLRs), and cytosolic DNA receptors, several signaling pathways are activated stimulating the transcription of IFN gene through activation of the transcription factor (IRF) 3. Through ubiquitination, RING E3s are responsible for accurately regulating this process to amplify or decrease antiviral response [23, 24].

Viral nucleic acids are the most potent stimuli for inducing the synthesis of IFNs. RLRs, such as RIG-1 and MDA5, which contain caspase activation and recruitment domain (CARD), detect viral double-stranded RNA in the cytosol by RNA helicase domains and are associated with mitochondrial antiviral-signaling (MAVS) proteins and TRAF proteins, for activation of the TANK-binding kinase 1 (TBK1) that promotes phosphorylation and activation of IRF3, leading to the induction of interferon type I [25–28] (Fig. 10).

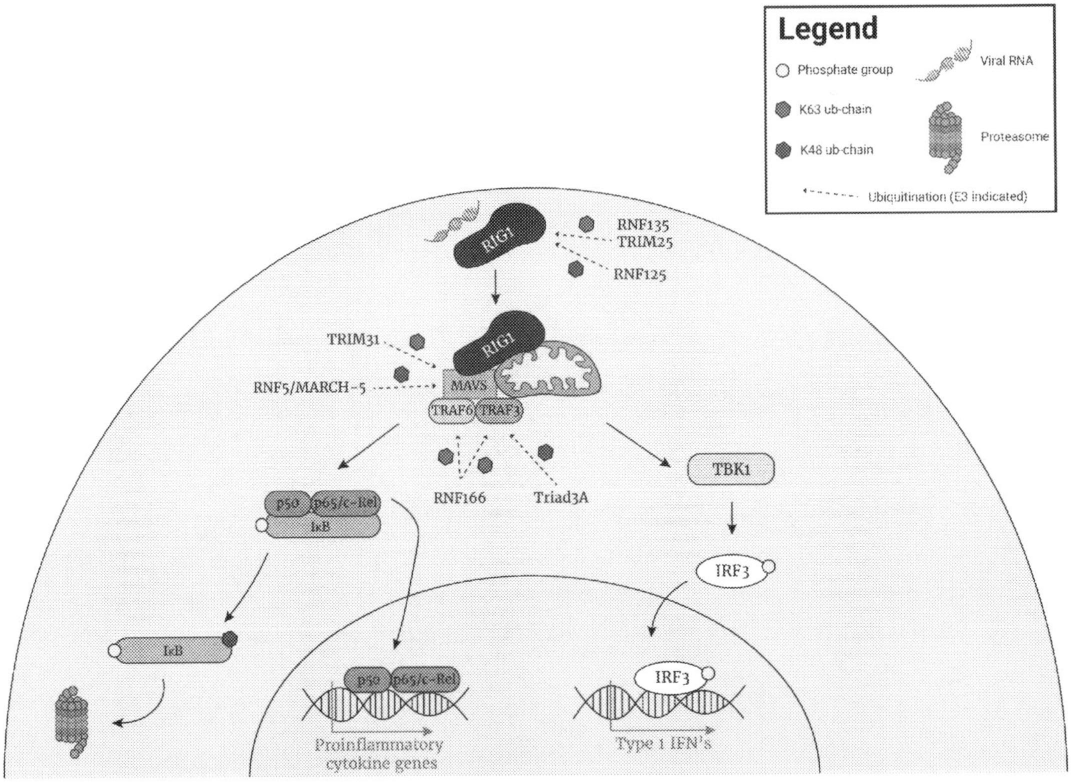

Fig. 10

Regulation of RING E3 ligases in RLR signaling. RIG-1 recognizes viral RNA, and it is K63-linked polyubiquitinated by RNF135 and TRIM25 inducing its interaction with MAVS adaptor protein. This recruits TRAF3 and TRAF6, being TRAF6 responsible for recruiting NEMO and, consequently, TBK1, which will activate IRF3. TRIM31 acts as a positive regulator of the pathway by promoting the aggregation of MAVS, while RNF166 by performing TRAF3 and TRAF6 ubiquitination. On the contrary, RNF125, RNF5/MARCH-5, and Triad3A are negative regulators by inducing RIG, MAVS, and TRAF3, respectively, to proteasome degradation.

However, for the interaction between RLRs and MAVS, it is necessary to its ubiquitination by TRIM25 and RNF135, both RING E3s. RNF135 transfers K63-linked polyubiquitin chains to the C-terminal domain of RIG-1, leading to the exposure of the CARD's domains, which will later be ubiquitinated by TRIM25, also via K63-linked polyubiquitin, thereby enabling CARD-CARD interactions between RIG-1 and MAVS [25, 29]. This activation induces the prion-like polymerization of MAVS, recruiting and activating TRAF3 and TRAF6. TRAF6, in turn, synthesizes polyubiquitin chains that will be recognized by NEMO, activating NF-κB, and the expression of proinflammatory cytokines. Moreover, NEMO is responsible for recruiting TBK1, inducing the expression of interferon type I, by activating the IRF3 [26, 29].

In addition to the functions mentioned earlier, other RING E3s perform important functions in this signaling pathway in order to stimulate or inhibit the production of interferon. For example, TRIM31 acts as a positive regulator of the antiviral response, by promoting the K63-linked ubiquitination of MAVS, generating greater aggregation of these adaptor proteins and, consequently stimulating the production of IFN [30]. For the same purpose, RNF166 increases the ubiquitination of TRAF3 and TRAF6 [31]. On the contrary, some RING E3s such as RNF125, RNF5/MARCH-5, and Triad3A act by negatively regulating this signaling pathway through the K48-linked ubiquitination of RIG, MAVS, and TRAF3, respectively, leading them to proteasomal degradation [32, 33].

Cell cycle

The cell cycle consists of an extremely orchestrated sequence of events that control the cellular life progression in a timely-ordered manner. The main effectors of these events are cyclins and cyclin-dependent kinases (CDKs), which are counteracted by the CDK inhibitors (CKIs). The combined activity of those different effectors modulates transition between different cellular phases, from G1 to S or from G2 to mitosis, for example. However, cyclins, CDKs, and CKIs are also targets of several controlling mechanisms, in which ubiquitination is highlighted with an essential role in the development of the cell cycle [34, 35].

Among more than 600 different E3 ligases expressed in humans, the RING family stands out in the context of cell cycle control, drawing attention to two classes of CRLs (Cullin RING ligases): SCFs and APC/Cs (anaphase-promoting complex/cyclosome) ligases. Both are multiprotein complexes where SCF contains an F-box protein as a substrate-interacting protein, while APC/Cs bind the substrate via one of two subunits in somatic cells, Cdc20 or Cdh1 (Fig. 11). However, a remarking difference between these two classes of CRLs is the cell phase that they are active: SCFs act throughout the entire cycle, while APC/Cs are mainly active between mitosis and the next G1 [34, 35] (Fig. 12).

During the end of a mitotic process, the APC/C(Cdh1) is the main E3 ligase establishing a new G1-phase, being responsible for ubiquitination and proteasomal degradation of important factors for replication and mitosis, such as Cdc6, SKP2, Cdc25A, and cyclins A and B. Cdc6 is a protein that mediates the formation of prereplication complexes (pre-RCs), binding to the origin recognition complex (ORC); Skp2, or Fbxl1, is kept at low levels in order to prevent the formation of SCF(Skp2), which targets CKIs and causes the cell cycle progression; Cdc25A removes phosphate groups attached to Cdk1/2, consequently activating them and the cell cycle; cyclins A and B are extremely important in the mitotic entry, the reason why they are targeted by APC/C(Cdh1). When the cycle progresses and G1-phase ends, the S-phase can start, APC/C(Cdh1) promotes ubiquitination of a specific APC/C E2, UbcH10, and autoubiquitination, decreasing its activity. In parallel, CDKs phosphorylate Cdh1, disrupting

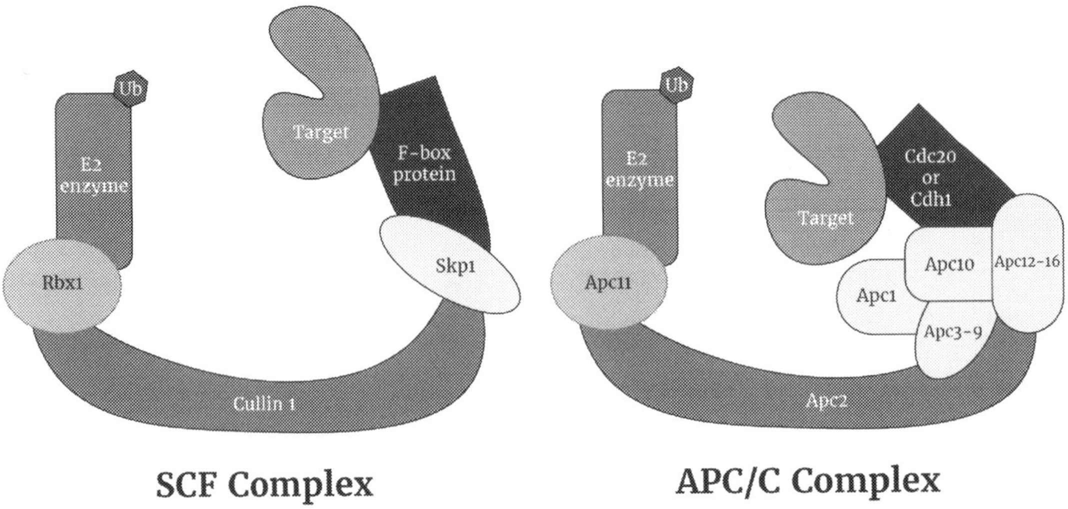

Fig. 11
Structure comparison of two classes of RING E3 ligases, SCF and APC/C. Scaffold proteins are indicated in *brown* (dark gray in the print version) (Cullin 1 and Apc2), RING-containing proteins in *orange* (gray in the print version) (Rbx1 and Apc11), adaptors in *light yellow* (light gray in the print version) (SKP1 and Apc1/3/9/10/12/16), and substrate-interacting proteins in *dark blue* (dark gray in the print version) (F-box and Cdc20/Cdh1). (For interpretation of the references to color in this figure legend, the reader is referred to the web version of this article.)

its interaction with the APC/C complex—an interaction that will only be restored at the end of the next mitosis [35, 36]. APC/C(Cdh1) is also responsible for attaching unusual ubiquitin linkage chains to its targets, more specifically K11 chains—which also cause targets to be degraded via proteasome. Synthesis of K11 chains by APC/C(Cdh1) is done with the assistance of two dedicated E2s, UBE2C and UBE2S, the first responsible for working on chain initiation and the latter for chain elongation [37] (Fig. 12).

The SCF(SKP2) acts in early interphase, being activated in late G1—after a decrease in Cdh1 activity—and targeting cell cycle inhibitors promoting cell cycle progression [34, 35]. One of its targets is p27, an inhibitor of cyclin E-Cdk2 complex. SCF(SKP2) also targets other CKIs, causing a strengthening in cyclin-CDKs activity and stimulating cycle progression [36]. Thus, SKP2 is seen as an oncogenic protein and APC/C(Cdh1) a tumor suppressor by promoting SKP2 degradation [35]. After the entry and progression of S-phase, cyclins D1 and E are ubiquitinated by two SCF complexes, SCF(Fbxo4)-α/B-crystalline and SCF(Fbxw7), respectively. During this phase and later G2, the activity of SCFs is essential in DNA damage checkpoints: when such an event occurs, Cdc25A is phosphorylated and later ubiquitinated by SCF(β-TRCP/Fbxw1), arresting cell cycle progression to allow the DNA repair. After checkpoint recovery, Claspin—a protein that promotes Cdc25A phosphorylation—is phosphorylated and this modification recruits it to interact with SCF(β-TRCP), which is

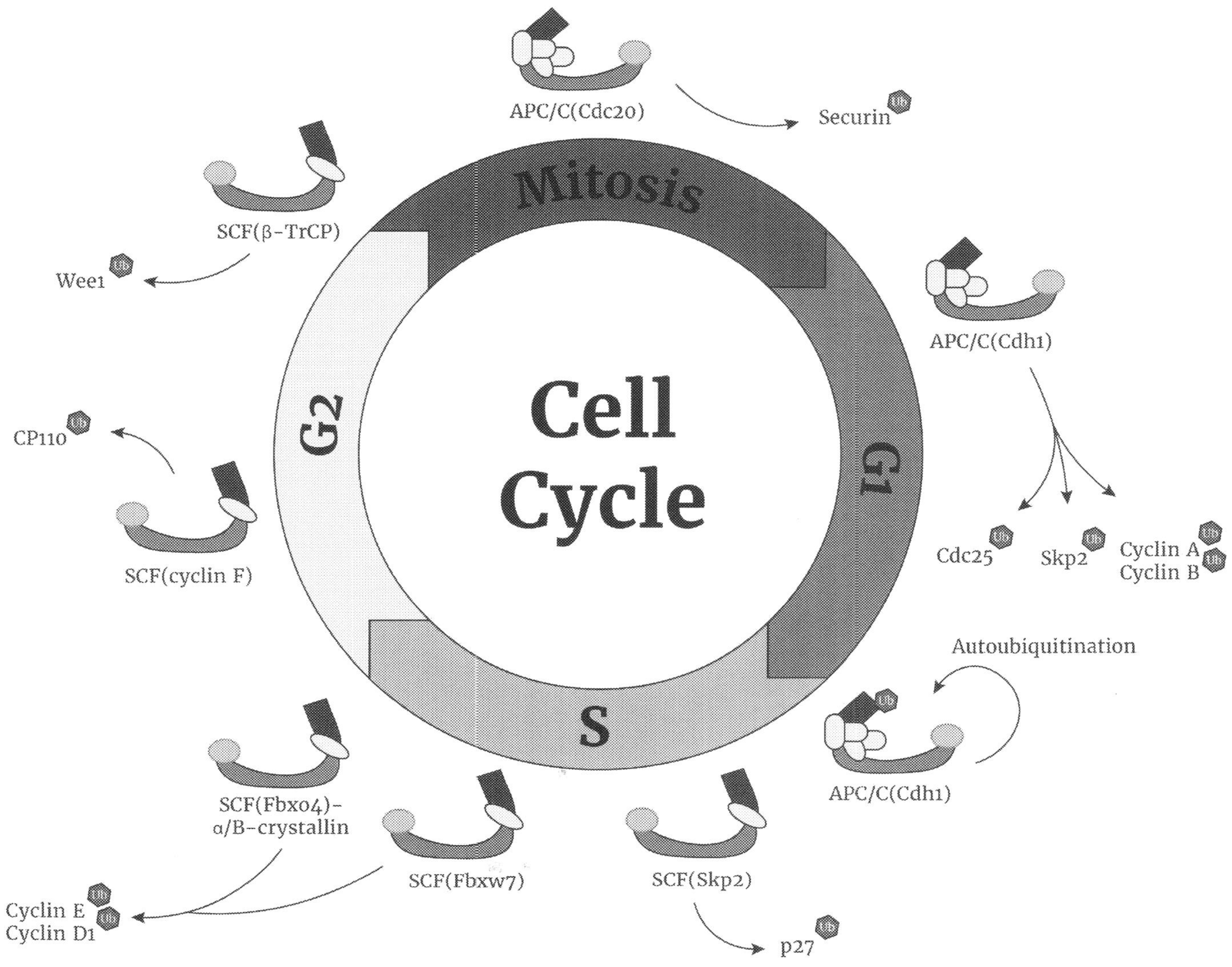

Fig. 12

Examples of SCF and APC/C complexes active throughout the cell cycle. Position of each E3 relates to the stage, in which they are most active during the cycle. *Arrows* indicate some of their targets.

polyubiquitinated and degraded by proteasome. eEF2K, an inhibitor of translation factor eEF2, is also targeted by this SCF, resuming protein synthesis and cycle progression [35] (Fig. 12).

After DNA replication, secure transitions to G2- and later to M-phases require the activity of different SCF complexes. SCF(cyclin-F) is active after DNA is duplicated, ubiquitinating RRM2, a critical enzyme in deoxyribonucleotide synthesis. In addition, SCF(cyclin-F) also targets CP110 for degradation, a protein responsible for centrosome replication, avoiding overduplication of this chromosome region [35, 37]. In late G2, SCF(β-TRCP) promotes cycle progression and mitotic entry as it targets for the degradation of a Cdk1 inhibitor, Wee1. This activity is balanced by SCF(NIPA) during S and G2, which ubiquitinates nuclear cyclin B and prevents the formation of cyclin B-Cdk1 complex, avoiding an early entry in mitosis. However, during late G2, NIPA is phosphorylated and the SCF is disrupted, releasing cyclin B from degradation signal and subsequently accumulating it—promoting its interaction with Cdk1 and entry in mitosis [35] (Fig. 12).

During mitosis, one of the most critical steps is chromatid separation, in order to guarantee each daughter cell will receive the same DNA content. One of the mechanisms used to control correct separation is known as spindle assembly checkpoint (SAC), a protein set that delay mitotic progression until all chromosomes are properly attached to spindle fibers. MAD2 and BubR1, SAC components, bind to Cdc20 and inhibit APC/C action; once all duplicated chromosomes are ready to segregate—and the inhibitor Emi1 is degraded because of SCF(β-TRCP) ubiquitination—inhibitory proteins dissociate from Cdc20 and the APC/C complex is released from inhibition, ubiquitinating checkpoint targets to ensure mitotic progression and exit [35–37]. In parallel, during transition from metaphase to anaphase, SCF(β-TRCP) is also responsible for controlling cycle progression. It targets Bora for degradation, a protein that regulates spindle stability when in contact with kinetochores. Degradation of Bora leads the cell to progress into anaphase and then finishing the division, so the daughter cells can restart the cycle and SCFs as well as APC/Cs can again orchestrate the process [35].

DNA repair

The DNA contains all the necessary information for a functional cell; thus, the cell employs several mechanisms to protect its DNA and also to repair it under any damage. In eukaryotes, DNA is structured in association with many proteins that among several functions, compact it into the cell nucleus. However, when DNA is damaged, those same proteins are targeted for posttranslational modifications that help to indicate there is a need to activate mechanisms of DNA damage repair (DDR) in a certain region of the genome [38]. One of the most destructive types of DNA damage is a double-strand break (DSB)), which causes break in both DNA strands. In order to repair DSBs, the main strategies used by the cells are homologous recombination (HR) and nonhomologous end joining (NHEJ). Both mechanisms, even though different, rely on ubiquitin and ubiquitination to function properly and to repair the DNA [39, 40].

When facing a DSB, the cell activates the ataxia telangiectasia-mutated (ATM) kinase, which is dependent on the action of two E3 ligases: SCF(SKP2) that adds K63-linked ubiquitin chains onto NBS1—responsible for later activating ATM; and UBR5, capable of modifying ATMIN, an interactor of ATM and competitor for NBS1 interaction (Fig. 13). ATM then is able to phosphorylate histone H2A.X generating the γ-H2A.X, which is able to recruit MDC1, a mediator that, after phosphorylated, brings to the region close to the DSB site the UBC13 complex, which contains an E2 (ubiquitin-conjugating enzyme) and the E3 ligase RNF8. This E3 then modifies histone H1, generating a signal for another ligase, RNF168, to be recruited to the damage site, ubiquitinating the histones H2A and γ-H2A.X. RNF168 modification occurs through monoubiquitination, either on lysine 13 or on lysine 15—followed by the action of RNF8, which extends these modifications to K63-linked chains and accumulate RNF168 near to the break site, as both ubiquitin chain and monoubiquitination are able to amplify the signal [39–41].

Ubiquitinated histones are able to recruit factors in the context of DDR, among them BRCA1 e 53BP1, proteins that antagonize each other during the repair process. K63-linked ubiquitin chains recruit BRCA1-A, a multiprotein complex that includes the dimer BRCA1/BARD1—that possess E3 ligase activity—and RAP80, a protein that helps to decide the type of repair will be utilized. BRCA1/BARD1 is able to ubiquitinate H2A at lysines 127/129, which leads to RAD51 attraction and DNA-end resection, a critical step for establishment of HR. The ubiquitination of lysines 127/129 displaces 53BP1 from DSB, which diminishes the NHEJ activity. On the contrary, ubiquitination of H2A/H2A.X at lysine 15 selectively causes the activation of 53BP1 recruiting it to NHEJ. The choice on which mechanism will be used to repair the DNA is also connected to the cell cycle phase active during the repair, as HR is only possible during S/G2, when the other DNA double strand is available to function as a template; in the meanwhile, NHEJ occurs through every phase except for mitosis [39–41].

Finally, deubiquitinating enzymes (DUBs) are also part of these well-orchestrated events, executing crucial actions on the control of DDR. USP7 and USP34 remove ubiquitin from RNF168, stabilizing it and allowing ubiquitination of H2A and H2A.X. Other DUBs, such as DUB3, BAP1, and USP3, act directly on histones, removing ubiquitin from H2A/H2A.X/H2B. They are also parts of BRCA1-A complex, as the BRCC36 DUB is able to regulate which DDR pathway the cell will follow, since its absence may favor the establishment of HR instead of NHEJ [39, 41].

Diseases

E3 ubiquitin ligases do not only confer specificity to the ubiquitin-proteasome system but also play an important role in mediating different cellular processes. Therefore, deregulation in this system underlies the etiology of several diseases (Fig. 14). Mutations in E3 ligase

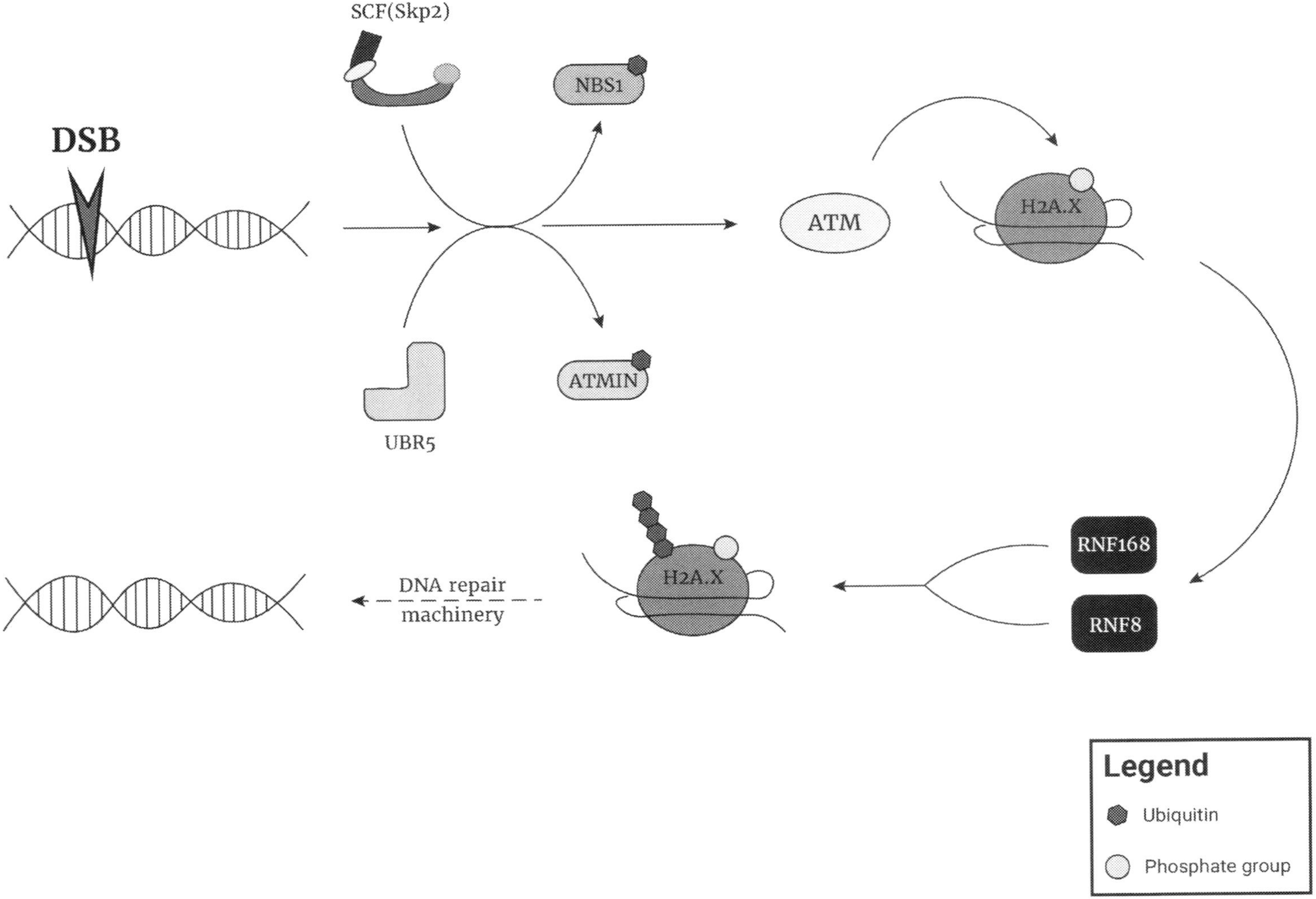

Fig. 13

After facing a double-strand break (DSB), one of the mechanisms the cell employs is the activation of the ATM kinase, which depends on ubiquitination of NBS1 by SCF(SKP2) and ATMIN by UBR5. ATM is then able of phosphorylating H2A.X—which becomes phospho-H2A.X, or γ-H2A.X—resulting in recruitment and activation of RNF8 and RNF168 E3 ligases, which in turn ubiquitinate γ-H2A.X, leading to later recruitment of DNA repair machinery.

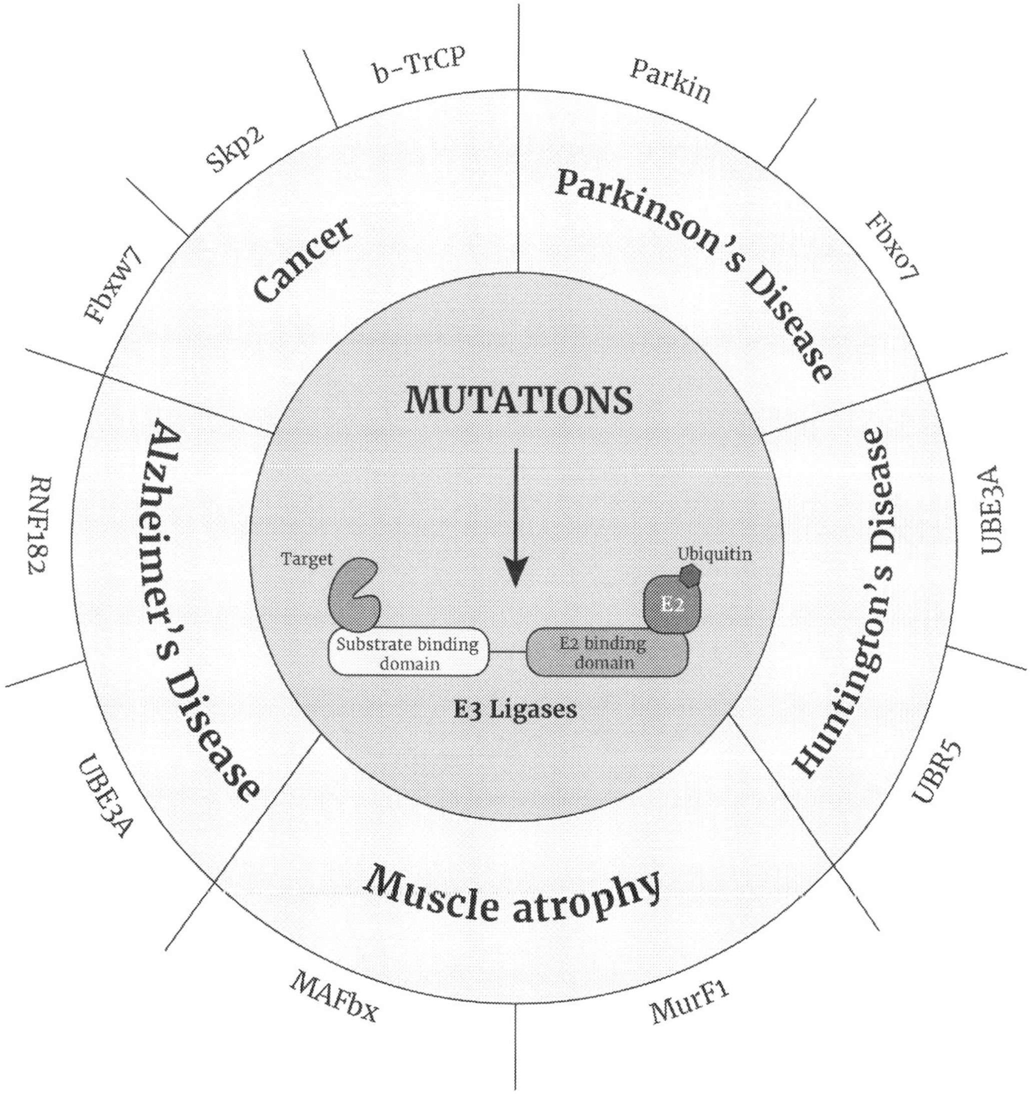

Fig. 14
Mutations found in E3 ubiquitin ligase leading to different types of diseases. The E3 ubiquitin ligase related to each kind of disease is found on the edge of the figure.

complex that do not interfere in vital pathways can generate various types of phenotypes. The pathological state caused by these mutations can be divided into two groups. The first one implicates in the stabilization of several proteins, which is caused by mutations resulting in either loss of function of E3 ligases or compromising its substrate recognition motif and the other group leads to a decrease in protein half-life due to E3 ligase gain of function promoting an acceleration of protein degradation rate [42].

Muscular atrophy

The maintenance of muscle tissue underlies the balance between anabolic and catabolic processes. When protein degradation overcomes, its synthesis generates the loss of muscle mass. Muscular atrophy is a debilitating condition that impacts mobility, whole-body metabolism, disease resistance, and quality of life. There are three major protein degradation pathways present in eukaryotic cells: apoptosis, autophagy-lysosomes, and ubiquitin-proteasome system. Regarding muscular atrophy, it is known that the ubiquitin-proteasome system is crucial for protein degradation and two E3 ubiquitin ligases are essential in this process: muscle atrophy F-box (MAFbx)/atrogin-1 (Fbxo32) and the muscle ring-finger (MuRF-1/Trim63) [43]. These proteins are markers for skeletal muscle atrophy due to its high expression found in this tissue and in many atrophy-inducing conditions [44] (Fig. 15).

Atrogin-1 is an F-box protein found in cardiac and skeletal muscle and has its expression regulated by the family of transcription factors called forkhead, for example, FoxO. Loss of FoxO expression results in inhibition of Atrogin1 and MuRF1 expression. It was discovered that defects in atrogin-1 are also associated with susceptibility to dilated cardiomyopathy, which can result in heart failure and arrhythmia [45]. The other marker for muscle atrophy

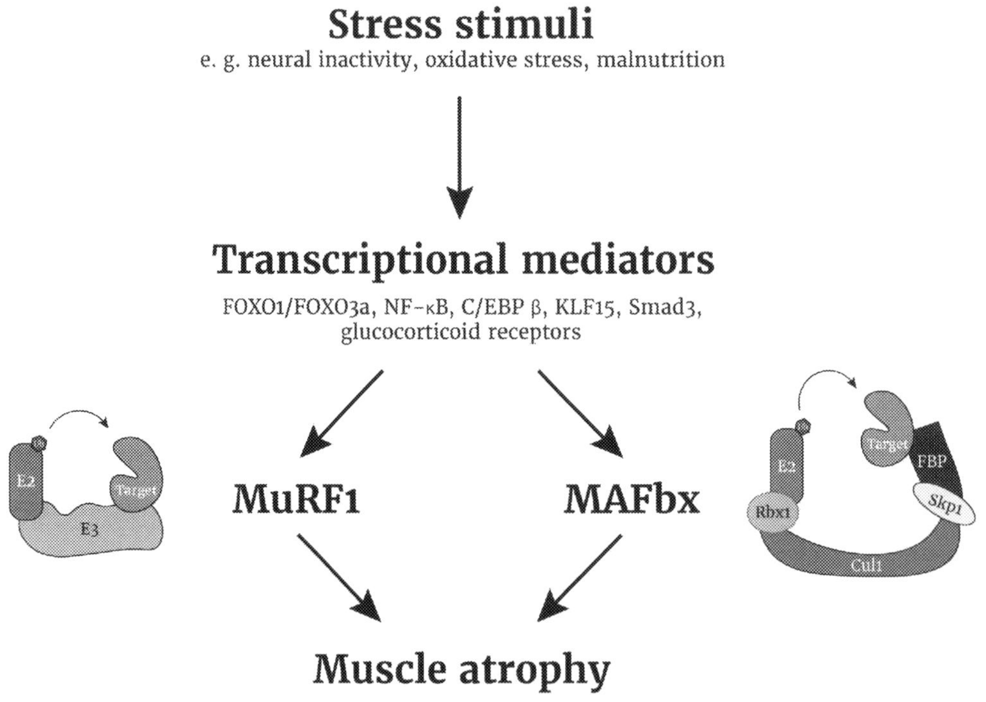

Fig. 15
Muscle RING finger 1 (MuRF1) and muscle atrophy F-box (MAFbx) expression in skeletal muscle under stress conditions regulating muscle atrophy.

MuRF1 localizes in the M-line and Z-disk in the sarcoma and is involved in the mechanical stress sensors via its interaction with titin [46]. Both proteins are induced in response to myostatin/TGFβ signaling [47].

Neurodegenerative diseases

The ubiquitin-proteasome system is important to maintain cell homeostasis; when this system fails, it contributes to generate the formation of protein aggregates, which is one of the most common hallmarks in neurodegenerative diseases. Dysfunction in ubiquitination pathway has been found in some neurodegenerative diseases that are responsible for causing a severe decline in cognitive ability capable of interfering with daily life and that include Alzheimer's disease, Parkinson's disease, and Huntington's disease [48].

Huntington's disease

Huntington's disease (HD) is an autosomal dominant progressive brain disorder, and the earliest neuropathologic modifications in HD patients are found in the striatum and in the cerebral cortex that frequently compromises the movement, cognitive, and psychiatric skills with a broad spectrum of signs and symptoms [49]. HD is caused by the expansion of CAG trinucleotide repeat in exon 1 of the huntingtin gene (*HTT*) coding for poly-glutamine (Poli-Q) in the N-terminus of the Huntington protein (HTT). Usually, in health individuals the *HTT* gene contains up to 18 CAG triplets, however, individuals with 35 CAG repeats is considered critical as a disease-causing allele, and above 40 repeats the disease is pervasive with the CAG repeats length increasing directly correlated to the rate of pathogenesis [50]. The HTT is ubiquitously expressed in eukaryotic cells and is important as an antiapoptotic protein and has importance in embryonic development, vesicle transport, and transcriptional regulation. The expanded poly-Q present in HTT is prone to aggregation that contributes to the development of neurodegeneration due to the accumulation of the mutant HTT fibrils and its intermediate oligomers formed during the aggregation/disaggregation process [51]. HTT accumulation can be toxic to the cell due the interference caused in cell-specific gene expression, posttranslational modifications, and protein-protein interaction.

Recently, it was identified by a genome-wide association analysis of several genetic modifiers for age of onset of HD, and among these data, it was found the E3 ubiquitin ligase UBR5 is a genetic modifier for HD. UBR5 is a component of the HECT class of ubiquitin ligases and it is known to be involved in important cellular pathways including DNA damage response, cell signaling, and apoptosis. The HECT domain shows a structural particularity in the C-lobe that does not have a surface for noncovalent ubiquitin binding; instead, the ubiquitin interacts with UBR5 by the UBA domain [52]. It was observed that UBR5 downregulation is associated with dysregulation of HTT levels resulting in the aggregation of poly-Q-expanded HTT in Huntington's disease immortalized pluripotent stem cell line [51].

Another E3 ligase related to HD is Ube3A, which is also part of the HECT family and it played a crucial role in experience-dependent cortical and neurocortical plasticity. Ube3A is capable of selectively redistributing the nuclear aggregates of mutant HTT. Also, its solubility is reduced in different regions of HD mice brain. Moreover, deficiency in Ube3A culminates in increased levels of HTT aggregates in mice brain [53]. A third E3 ligase component of the HECT family was also discovered regulating the levels of HTT in HD. The WWP1 (WW domain-containing E3 ubiquitin protein ligase 1) was first identified by its WW domain and belongs to the Nedd4-like family. Data showed high levels of expression of WWP1 in mice and in N2 cells expressing mutant HTT [54]. And finally, it was observed in filtration retardation assays that overexpressing SCF(Fbxo25) but not its mutant Fbxo25 without F-box domain of interaction with SKP1, decreases the aggregation of poly-Q-HTT, indicating that ubiquitin ligase activity of SCF(Fbxo25) prevents the aggregation of HTT [55].

Parkinson's disease

Parkinson's disease (PD) is known to be the second most ordinary neurodegenerative disease, which is characterized by the degeneration of dopaminergic neurons in the substantia nigra pars compacta that results in the pathophysiologic changes in downstream basal ganglia circuit, albeit, it is not restricted to this region. The clinical syndrome caused by PD is called "parkinsonism" that includes resting tremor, imbalance, a slow shuffling gait, cogwheel rigidity, and bradykinesia [56]. The major part of PD cases is present in men with an average age of onset of 68 years.

Regarding the genetic factors found in PD patients, mutations in a variety of genes, including a mitochondrial protein kinase *PINK1* and two E3 ubiquitin ligases: Parkin (*PARK2*) and *FBXO7* (F-box only protein 7/*PARK15*), were found to be associated with familial PD [57]. Parkin is an E3 ubiquitin ligase part of the RING finger family and is localized in the cytosol unless mitochondrial depolarization occurs. Once it happens, Parkin is translocated to the mitochondrial outer membrane (MOM) where it ubiquitinates several proteins resulting in the induction of mitophagy. Therefore, Parkin plays an essential role in maintaining mitochondrial quality control and mutation in its gene is one of the major causes of autosomal recessive form of PD [58].

The other protein associated with a familial form of PD is Fbxo7, which is one of the 69 F-box proteins present in humans that is a component of the E3 ubiquitin ligase SCF (Fbxo7) [59]. Fbxo7 interacts with Parkin to regulate mitophagy and point that mutations in *FBXO7* were found in several families with sporadic PD and autosomal recessive early onset of PD [60]. The direct role of Fbxo7 in PD pathology is still unclear, but one hypothesis is that these mutations disable Fbxo7 enzymatic activity triggering toxic accumulation of some unknown protein causing the pathology [61].

Alzheimer's disease

Alzheimer's disease (AD) is the most common brain disorder that causes dementia in older adults. A number of pathologic changes found in AD patients may occur several years before symptom onset, which leaves a reasonable time for implementation of prevention strategies for the earliest stages of AD. The causes of neurodegeneration are attributed to mitochondrial dysfunction, impairment of cerebral glucose metabolism, synaptic dysfunction, formation of neurofibrillary tangles of hyperphosphorylated tau, and toxic amyloid oligomers [62].

Dysfunction in ubiquitin-proteasome pathway plays an important role in AD development. In patients with AD, ubiquitinated proteins are accumulated and this is capable of overwhelming the proteasome in neurons. A reduction of 50% was found in proteasome activity in neurons isolated from amyloid precursor protein (APP) transgenic mice, which is an animal model for AD, compared to neurons isolated from wild-type mice [63, 64].

The accumulation of β-amyloid is involved in the early AD pathogenesis, and it was discovered a ring finger E3 ubiquitin ligase called RNF182 that is overexpressed in postmortem brain tissue of AD patients. In normal conditions, RNF182 is a cytoplasmic protein expressed preferentially in the brain at low levels [65]. RNF182 targets ATP6V0C (C-subunit of the V0 sector of the vacuolar proton ATPase) for degradation by the ubiquitin-proteasome system. ATP6V0C is crucial in the formation of gap junction complexes and neurotransmitter release channels, for instance, it is known that acetylcholine, serotonin, and dopamine are released if ATP6V0C is expressed in cultured cells [66].

Cancer

Cancer is a multifactorial disease where the activity of several E3 ubiquitin ligases is deregulated due to epigenetic and genetic pathways and/or by altered posttranslational modification. Since E3 ubiquitin ligase is involved in several cellular processes, the malfunctioning of this type of enzyme can lead to uncontrolled cell proliferation, DNA damage causing genomic instability contributing to malignant transformation, drug resistance therapy, and tumor progression by modulating the balance between cell growth and survival signals [67].

F-box proteins (FBP) assemble with SKP1, Cullin 1, and Rbx1 forming the SCF-type E3 ubiquitin ligase complex that is a part of the RING-type family. FBPs are known to be associated with a number of types of cancers. For instance, somatic mutations in *FBXW7* (F-box/WD repeat-containing protein 7) were found in a variety of human cancers. This protein acts as a tumor suppressor by promoting the degradation of oncoproteins such as cyclin E, MYC, JUN, mTOR, and NOTCH, which are responsible for regulating crucial cellular processes [68]. Moreover, studies showed that mice with *Fbxw7* depletion developed a predisposition for cancer in a series of tissues [69].

Another FBP protein associated with cancer is SKP2 (S-phase kinase-associated protein 2), which is best known for mediating the degradation of cell cycle regulators via ubiquitin-proteasome system, including p27 and p21. Additionally, it was found that SKP2 is involved in tumorigenesis because it is responsible to ubiquitinate and promote the degradation of tumor suppressor proteins such as p27, p21, p130, and FOXO1 (forkhead box protein O1) [68]. Also, mutated *Skp2* mice models developed breast tumors and the overexpression of SKP2 in a cohort of human cancer supports the idea of SKP2 acting as a proto-oncoprotein [68, 70].

Interestingly, an FBP was discovered acting as both an oncogenic and a tumor suppressor. β-TRCP1 (β-transducin repeat-containing protein 1) and β-TRCP2 (β-transducin repeat-containing protein 2) can execute both functions depending on the cellular context and in a tissue-specific way [68]. In mice, mutation in β-*TRCP1* showed predisposition in developing mammary, ovarian, and uterine carcinomas and mutation in β-*TRCP2* presented a decrease in inflammatory response in skin and hyperplasia, suggesting a possible role of β-TRCP2 in skin cancer [71, 72]. There are several substrates of β-TRCP1 and β-TRCP 2 described, some of those act promoting cell cycle, such as CDC25 and others function in cell cycle arrest and that include Weel e EMI1 [68]. Therefore, to conclude how β-TRCP is going to play a role, we need to consider the whole cellular context.

The fate of a substrate will depend on the topology of the ubiquitin chains added on it. Hence, a more detailed comprehension of the E3 ubiquitin ligase regulation in tumorigenesis can contribute to the identification of prognostic markers enabling the development of anticancer therapies. Inhibition of the proteasome, for example, has been emerging as an interesting method for cancer treatment. Malignant cells are more prompt to suffer from the cytotoxic effects of proteasome inhibition compared to normal cells [73]. One explanation for this effect is the increased metabolism found in tumor cells raising the levels of protein synthesis that would turn these cells more susceptible to the cytotoxic effects of proteasome inhibition.

Bortezomib was the first proteasome inhibitor to be used in clinical practice. It is a peptide boronate inhibitor that acts as a specific inhibitor of the chymotryptic activity of the proteasome that generates cytotoxic effects on tumor cells. Bortezomib is used as a first-line treatment usually combined with other drugs to treat patients with multiple myeloma and malignant lymphoma. It inhibits NF-κB, which is a transcription factor for anti-apoptotic proteins, including Bcl-2, c-IAP2, and surviving, additionally, bortezomib acts stabilizing cell cycle proteins such as p53, p21, and p27 [74]. Understanding the control of checkpoints may be crucial to identify new targets to increase the efficacy of proteasome inhibitor drug development in cancer treatment.

Conclusion

E3 ubiquitin ligases are an abundant family of enzymes composed of more than 600 proteins responsible for the conjugation of ubiquitin to the target substrate regulating numerous cellular responses through proteolytic or nonproteolytic actions in different signaling pathways, such as NF-κB, mitophagy, interferon response, cell cycle, and DNA repair. Because of their pleiotropic actions in cellular regulation, the E3 ligases are therapeutic targets for the treatment of different diseases, such as muscular atrophy, cancer, and neurodegenerative diseases (Parkinson's disease, Alzheimer's disease, and Huntington's disease). Only a small number of the human E3 ubiquitin ligases have been characterized, and expanding the research of this class of enzymes is necessary to understand their role in human health and diseases.

References

[1] Hershko A, Ciechanover A. The ubiquitin system. Ann Rev Biochem 1998;425–79. https://doi.org/10.1146/annurev.biochem.67.1.425.

[2] Goldstein G, Scheid M, Hammerling U, Schlesinger DH, Niall HD, Boyse EA. Isolation of a polypeptide that has lymphocyte-differentiating properties and is probably represented universally in living cells. Proc Natl Acad Sci U S A 1975;72:11–5. https://doi.org/10.1073/pnas.72.1.11.

[3] Metzger MB, Pruneda JN, Klevit RE, Weissman AM. RING-type E3 ligases: master manipulators of E2 ubiquitin-conjugating enzymes and ubiquitination. Biochim Biophys Acta 1843;2014:47–60. https://doi.org/10.1016/j.bbamcr.2013.05.026.

[4] Fouad S, Wells OS, Hill MA, D'Angiolella V. Cullin ring ubiquitin ligases (CRLs) in cancer: responses to ionizing radiation (IR) treatment. Front Physiol 2019;10. https://doi.org/10.3389/fphys.2019.01144.

[5] Zheng N, Schulman BA, Song L, Miller JJ, Jeffrey PD, Wang P, et al. Structure of the Cul1–Rbx1–Skp1–F boxSkp2 SCF ubiquitin ligase complex. Nature 2002;416:703–9.

[6] Kipreos ET, Pagano M. The F-box protein family. Genome Biol 2000;1. reviews3002.1–reviews3002.7.

[7] Nakayama KI, Nakayama K. Ubiquitin ligases: cell-cycle control and cancer. Nat Rev Cancer 2006;6:369–81. https://doi.org/10.1038/nrc1881.

[8] Lafont E, Hartwig T, Walczak H. Paving TRAIL's path with ubiquitin. Trends Biochem Sci 2018;43:44–60. https://doi.org/10.1016/j.tibs.2017.11.002.

[9] Haglund K, Dikic I. Ubiquitylation and cell signaling. EMBO J 2005;24:3353–9. https://doi.org/10.1038/sj.emboj.7600808.

[10] Komander D, Clague MJ, Urbé S. Breaking the chains: structure and function of the deubiquitinases. Nat Rev Mol Cell Biol 2009;10:550–63. https://doi.org/10.1038/nrm2731.

[11] Coux O, Tanaka K, Goldberg AL. Structure and functions of the 20S and 26S proteasomes. Annu Rev Biochem 1996;65:801–47. https://doi.org/10.1146/annurev.bi.65.070196.004101.

[12] Liu S, Chen ZJ. Expanding role of ubiquitination in NF-κB signaling. Cell Res 2011;21:6–21. https://doi.org/10.1038/cr.2010.170.

[13] Xu H, You M, Shi H, Hou Y. Ubiquitin-mediated NFκB degradation pathway. Cell Mol Immunol 2015;12:653–5. https://doi.org/10.1038/cmi.2014.99.

[14] Aksentijevich I, Zhou Q. NF-κB pathway in autoinflammatory diseases: dysregulation of protein modifications by ubiquitin defines a new category of autoinflammatory diseases. Front Immunol 2017;8. https://doi.org/10.3389/fimmu.2017.00399.

[15] Steiner A, Harapas CR, Masters SL, Davidson S. An update on autoinflammatory diseases: relopathies. Curr Rheumatol Rep 2018;20:39. https://doi.org/10.1007/s11926-018-0749-x.

[16] Spinelli JB, Haigis MC. The multifaceted contributions of mitochondria to cellular metabolism. Nat Cell Biol 2018;20:745–54. https://www.nature.com/articles/s41556-018-0124-1. [Accessed 7 January 2021].
[17] Pickles S, Vigié P, Youle RJ. Mitophagy and quality control mechanisms in mitochondrial maintenance. Curr Biol 2018;28:R170–85. https://doi.org/10.1016/j.cub.2018.01.004.
[18] Youle RJ, Narendra DP. Mechanisms of mitophagy. Nat Rev Mol Cell Biol 2011;12:9–14. https://doi.org/10.1038/nrm3028.
[19] Gladkova C, Maslen SL, Skehel JM, Komander D. Mechanism of Parkin activation by PINK1. Nature 2018;559:410–4. https://doi.org/10.1038/s41586-018-0224-x.
[20] Trempe J-F, Sauvé V, Grenier K, Seirafi M, Tang MY, Ménade M, et al. Structure of Parkin reveals mechanisms for ubiquitin ligase activation. Science 2013;340:1451–5. https://doi.org/10.1126/science.1237908.
[21] Chaugule VK, Burchell L, Barber KR, Sidhu A, Leslie SJ, Shaw GS, et al. Autoregulation of Parkin activity through its ubiquitin-like domain. EMBO J 2011;30:2853–67. https://doi.org/10.1038/emboj.2011.204.
[22] Abbas AK, Lichtman AHH, Pillai S. Cellular and molecular immunology. 9th ed; 2017. https://www.elsevier.com/books/cellular-and-molecular-immunology/abbas/978-0-323-47978-3. [Accessed 21 December 2020].
[23] Hu H, Sun S-C. Ubiquitin signaling in immune responses. Cell Res 2016;26:457–83. https://doi.org/10.1038/cr.2016.40.
[24] Zhang Y, Li L-F, Munir M, Qiu H-J. RING-domain E3 ligase-mediated host–virus interactions: orchestrating immune responses by the host and antagonizing immune defense by viruses. Front Immunol 2018;9. https://doi.org/10.3389/fimmu.2018.01083.
[25] Gack MU, Shin YC, Joo C-H, Urano T, Liang C, Sun L, et al. TRIM25 RING-finger E3 ubiquitin ligase is essential for RIG-I-mediated antiviral activity. Nature 2007;446:916–20. https://doi.org/10.1038/nature05732.
[26] Deng L, Wang C, Spencer E, Yang L, Braun A, You J, et al. Activation of the IκB kinase complex by TRAF6 requires a dimeric ubiquitin-conjugating enzyme complex and a unique polyubiquitin chain. Cell 2000;103:351–61. https://doi.org/10.1016/s0092-8674(00)00126-4.
[27] Yoo Y-S, Park Y-Y, Kim J-H, Cho H, Kim S-H, Lee H-S, et al. The mitochondrial ubiquitin ligase MARCH5 resolves MAVS aggregates during antiviral signalling. Nat Commun 2015;6:7910. https://doi.org/10.1038/ncomms8910.
[28] Song G, Liu B, Li Z, Wu H, Wang P, Zhao K, et al. E3 ubiquitin ligase RNF128 promotes innate antiviral immunity through K63-linked ubiquitination of TBK1. Nat Immunol 2016;17:1342–51. https://doi.org/10.1038/ni.3588.
[29] Oshiumi H, Matsumoto M, Hatakeyama S, Seya T. Riplet/RNF135, a RING finger protein, ubiquitinates RIG-I to promote interferon-beta induction during the early phase of viral infection. J Biol Chem 2009;284:807–17. https://doi.org/10.1074/jbc.M804259200.
[30] Liu B, Zhang M, Chu H, Zhang H, Wu H, Song G, et al. The ubiquitin E3 ligase TRIM31 promotes aggregation and activation of the signaling adaptor MAVS through Lys63-linked polyubiquitination. Nat Immunol 2017;18:214–24. https://doi.org/10.1038/ni.3641.
[31] Chen H-W, Yang Y-K, Xu H, Yang W-W, Zhai Z-H, Chen D-Y. Ring finger protein 166 potentiates RNA virus-induced interferon-β production via enhancing the ubiquitination of TRAF3 and TRAF6. Sci Rep 2015;5:14770. https://www.nature.com/articles/srep14770. [Accessed 8 January 2021].
[32] Arimoto K, Takahashi H, Hishiki T, Konishi H, Fujita T, Shimotohno K. Negative regulation of the RIG-I signaling by the ubiquitin ligase RNF125. PNAS 2007;104:7500–5. https://doi.org/10.1073/pnas.0611551104.
[33] Zhong B, Zhang Y, Tan B, Liu T-T, Wang Y-Y, Shu H-B. The E3 ubiquitin ligase RNF5 targets virus-induced signaling adaptor for ubiquitination and degradation. J Immunol 2010;184:6249–55. https://doi.org/10.4049/jimmunol.0903748.
[34] Benanti JA. Coordination of cell growth and division by the ubiquitin–proteasome system. Semin Cell Dev Biol 2012;23:492–8. https://doi.org/10.1016/j.semcdb.2012.04.005.

[35] Bassermann F, Eichner R, Pagano M. The ubiquitin proteasome system—implications for cell cycle control and the targeted treatment of cancer. Biochim Biophys Acta Mol Cell Res 2014;1843:150–62. https://doi.org/10.1016/j.bbamcr.2013.02.028.

[36] Teixeira LK, Reed SI. Ubiquitin ligases and cell cycle control. Annu Rev Biochem 2013;82:387–414. https://doi.org/10.1146/annurev-biochem-060410-105307.

[37] Mocciaro A, Rape M. Emerging regulatory mechanisms in ubiquitin-dependent cell cycle control. J Cell Sci 2012;125(Pt 2):255–63. https://jcs.biologists.org/content/125/2/255. [Accessed 6 January 2021].

[38] Stadler J, Richly H. Regulation of DNA repair mechanisms: how the chromatin environment regulates the DNA damage response. Int J Mol Sci 2017;18:1715. https://doi.org/10.3390/ijms18081715.

[39] Nishi R. Balancing act: to be, or not to be ubiquitylated. Mut Res Fund Mol Mech Mutagen 2017;803–805:43–50. https://doi.org/10.1016/j.mrfmmm.2017.07.006.

[40] Hieu T. Van, Margarida A. Santos, Histone modifications and the DNA double-strand break response. Cell Cycle 2018;17(21–22):2399 https://doi.org/10.1080/15384101.2018.1542899 (accessed December 18, 2020).

[41] Uckelmann M, Sixma TK. Histone ubiquitination in the DNA damage response. DNA Repair 2017;56:92–101. https://doi.org/10.1016/j.dnarep.2017.06.011.

[42] Ciechanover A, Schwartz AL. The ubiquitin system: pathogenesis of human diseases and drug targeting. Biochim Biophys Acta 1695;2004:3–17. https://doi.org/10.1016/j.bbamcr.2004.09.018.

[43] Ogawa T, Furochi H, Mameoka M, Hirasaka K, Onishi Y, Suzue N, et al. Ubiquitin ligase gene expression in healthy volunteers with 20-day bedrest. Muscle Nerve 2006;34:463–9. https://doi.org/10.1002/mus.20611.

[44] Gomes MD, Lecker SH, Jagoe RT, Navon A, Goldberg AL. Atrogin-1, a muscle-specific F-box protein highly expressed during muscle atrophy. Proc Natl Acad Sci U S A 2001;98:14440–5. https://doi.org/10.1073/pnas.251541198.

[45] Al-Hassnan ZN, Shinwari ZM, Wakil SM, Tulbah S, Mohammed S, Rahbeeni Z, et al. A substitution mutation in cardiac ubiquitin ligase, FBXO32, is associated with an autosomal recessive form of dilated cardiomyopathy. BMC Med Genet 2016;17:3. https://doi.org/10.1186/s12881-016-0267-5.

[46] Gumucio JP, Mendias CL. Atrogin-1, MuRF-1, and sarcopenia. Endocrine 2013;43:12–21. https://doi.org/10.1007/s12020-012-9751-7.

[47] Bodine SC, Baehr LM. Skeletal muscle atrophy and the E3 ubiquitin ligases MuRF1 and MAFbx/atrogin-1. Am J Physiol Endocrinol Metab 2014;307:E469–84. https://doi.org/10.1152/ajpendo.00204.2014.

[48] Atkin G, Paulson H. Ubiquitin pathways in neurodegenerative disease. Front Mol Neurosci 2014;7:63. https://doi.org/10.3389/fnmol.2014.00063.

[49] Roos RAC. Huntington's disease: a clinical review. Orphanet J Rare Dis 2010;5:40. https://doi.org/10.1186/1750-1172-5-40.

[50] Zheng Q, Huang T, Zhang L, Zhou Y, Luo H, Xu H, et al. Dysregulation of ubiquitin-proteasome system in neurodegenerative diseases. Front Aging Neurosci 2016;8:303. https://doi.org/10.3389/fnagi.2016.00303.

[51] Koyuncu S, Saez I, Lee HJ, Gutierrez-Garcia R, Pokrzywa W, Fatima A, et al. The ubiquitin ligase UBR5 suppresses proteostasis collapse in pluripotent stem cells from Huntington's disease patients. Nat Commun 2018;9:2886. https://doi.org/10.1038/s41467-018-05320-3.

[52] Shearer RF, Iconomou M, Watts CKW, Saunders DN. Functional roles of the E3 ubiquitin ligase UBR5 in cancer. Mol Cancer Res 2015;13:1523–32. https://doi.org/10.1158/1541-7786.MCR-15-0383.

[53] Maheshwari M, Samanta A, Godavarthi SK, Mukherjee R, Jana NR. Dysfunction of the ubiquitin ligase Ube3a may be associated with synaptic pathophysiology in a mouse model of Huntington disease. J Biol Chem 2012;287:29949–57. https://doi.org/10.1074/jbc.M112.371724.

[54] Lin L, Park JW, Ramachandran S, Zhang Y, Tseng Y-T, Shen S, et al. Transcriptome sequencing reveals aberrant alternative splicing in Huntington's disease. Hum Mol Genet 2016;25:3454–66. https://doi.org/10.1093/hmg/ddw187.

[55] Manfiolli AO, Maragno ALGC, Baqui MMA, Yokoo S, Teixeira FR, Oliveira EB, et al. FBXO25-associated nuclear domains: a novel subnuclear structure. Mol Biol Cell 2008;19:1848–61. https://doi.org/10.1091/mbc.e07-08-0815.

[56] Reich SG, Savitt JM. Parkinson's disease. Med Clin North Am 2019;103:337–50. https://doi.org/10.1016/j.mcna.2018.10.014.
[57] Burchell VS, Nelson DE, Sanchez-Martinez A, Delgado-Camprubi M, Ivatt RM, Pogson JH, et al. The Parkinson's disease-linked proteins Fbxo7 and Parkin interact to mediate mitophagy. Nat Neurosci 2013;16:1257–65. https://doi.org/10.1038/nn.3489.
[58] Seirafi M, Kozlov G, Gehring K. Parkin structure and function. FEBS J 2015;282:2076–88. https://doi.org/10.1111/febs.13249.
[59] Teixeira FR, Randle SJ, Patel SP, Mevissen TET, Zenkeviciute G, Koide T, Komander D, Laman H. Gsk3β and Tomm20 are substrates of the SCFFbxo7/PARK15 ubiquitin ligase associated with Parkinson's disease. Biochem J 2016;473(20):3563–80. https://pubmed.ncbi.nlm.nih.gov/27503909/. [Accessed 18 December 2020].
[60] Stott SR, Randle SJ, Al Rawi S, Rowicka PA, Harris R, Mason B, et al. Loss of FBXO7 results in a Parkinson's-like dopaminergic degeneration via an RPL23-MDM2-TP53 pathway. J Pathol 2019;249:241–54. https://doi.org/10.1002/path.5312.
[61] Randle SJ, Laman H. Structure and function of Fbxo7/PARK15 in Parkinson's disease. Curr Protein Pept Sci 2017;18:715–24. https://doi.org/10.2174/1389203717666160311121433.
[62] Oboudiyat C, Glazer H, Seifan A, Greer C, Isaacson RS. Alzheimer's disease. Semin Neurol 2013;33:313–29. https://doi.org/10.1055/s-0033-1359319.
[63] Almeida CG, Takahashi RH, Gouras GK. Beta-amyloid accumulation impairs multivesicular body sorting by inhibiting the ubiquitin-proteasome system. J Neurosci 2006;26:4277–88. https://doi.org/10.1523/JNEUROSCI.5078-05.2006.
[64] Balducci C, Forloni G. APP transgenic mice: their use and limitations. Neuromolecular Med 2011;13:117–37. https://doi.org/10.1007/s12017-010-8141-7.
[65] Liu QY, Lei JX, Sikorska M, Liu R. A novel brain-enriched E3 ubiquitin ligase RNF182 is up regulated in the brains of Alzheimer's patients and targets ATP6V0C for degradation. Mol Neurodegener 2008;3:4. https://doi.org/10.1186/1750-1326-3-4.
[66] Higashida H, Yokoyama S, Tsuji C, Muramatsu S-I. Neurotransmitter release: vacuolar ATPase V0 sector c-subunits in possible gene or cell therapies for Parkinson's, Alzheimer's, and psychiatric diseases. J Physiol Sci 2017;67:11–7. https://doi.org/10.1007/s12576-016-0462-3.
[67] Senft D, Qi J, Ronai ZA. Ubiquitin ligases in oncogenic transformation and cancer therapy. Nat Rev Cancer 2018;18:69–88. https://doi.org/10.1038/nrc.2017.105.
[68] Wang Z, Liu P, Inuzuka H, Wei W. Roles of F-box proteins in cancer. Nat Rev Cancer 2014;14:233–47. https://doi.org/10.1038/nrc3700.
[69] Tetzlaff MT, Yu W, Li M, Zhang P, Finegold M, Mahon K, et al. Defective cardiovascular development and elevated cyclin E and Notch proteins in mice lacking the Fbw7 F-box protein. Proc Natl Acad Sci U S A 2004;101:3338–45. https://doi.org/10.1073/pnas.0307875101.
[70] Umanskaya K, Radke S, Chander H, Monardo R, Xu X, Pan Z-Q, et al. Skp2B stimulates mammary gland development by inhibiting REA, the repressor of the estrogen receptor. Mol Cell Biol 2007;27:7615–22. https://doi.org/10.1128/MCB.01239-07.
[71] Kudo Y, Guardavaccaro D, Santamaria PG, Koyama-Nasu R, Latres E, Bronson R, et al. Role of F-box protein βTrcp1 in mammary gland development and tumorigenesis. Mol Cell Biol 2004;24:8184–94. https://doi.org/10.1128/MCB.24.18.8184-8194.2004.
[72] Bhatia N, Demmer TA, Sharma AK, Elcheva I, Spiegelman VS. Role of β-TrCP ubiquitin ligase receptor in UVB mediated responses in skin. Arch Biochem Biophys 2011;508:178–84. https://doi.org/10.1016/j.abb.2010.12.023.
[73] Manasanch EE, Orlowski RZ. Proteasome inhibitors in cancer therapy. Nat Rev Clin Oncol 2017;14:417–33. https://doi.org/10.1038/nrclinonc.2016.206.
[74] Strauss SJ, Higginbottom K, Jüliger S, Maharaj L, Allen P, Schenkein D, et al. The proteasome inhibitor bortezomib acts independently of p53 and induces cell death via apoptosis and mitotic catastrophe in B-cell lymphoma cell lines. Cancer Res 2007;67:2783–90. https://doi.org/10.1158/0008-5472.CAN-06-3254.

CHAPTER 3

Lysosomal proteases and their role in signaling pathways

Samuel J. Bose[a], Thamali Ayagama[a], and Rebecca A.B. Burton
Department of Pharmacology, University of Oxford, Oxford, United Kingdom

Introduction

The term “Lysosome” was first used in 1955 by Christian de Duve following the discovery of latent enzyme activity within tissue preparations that were being used to investigate the cellular localization of glucose-6-phosphate [1]. While attempting to separate glucose-6-phosphate from a nonspecific acid phosphatase using the cell fractionation technique developed by Albert Claude and George Emil Palade, de Duve noted that activity of the phosphatase increased after samples had been stored in a refrigerator for 5 days. This observation eventually led to the discovery that the phosphatase was bound within a membrane vesicle distinct from the mitochondria and microsomes [1, 2]. The discovery of an additional four enzymes sharing a similar latency to the original acid phosphatase quickly followed, including β-glucuronidase, ribonuclease, deoxyribonuclease, and a protease that at the time was referred to as “cathepsin” (now known to be cathepsin D) [1, 2]. This initial, and somewhat serendipitous, observation led to the discovery of lysosomes (and additionally peroxisomes), for which de Duve, Claude, and Palade were awarded the 1974 Nobel Prize in Physiology and Medicine. Lysosomes are now recognized as the major location for degradation of both intracellular and extracellular macromolecules and contain more than 50 identified acid hydrolases as well as associated activator proteins [3].

While lysosomes have traditionally been associated primarily with intracellular protein degradation, the roles of these intracellular organelles have now been identified in protein and lipid metabolism, autophagy, antigen presentation, and lysosome-mediated cell death [4]. Lysosomes play key roles in cell signaling, for example, through the endocytosis and degradation of growth factors and their receptors [5], and there is evidence that they may

[a] Joint first authors.

Proteolytic Signaling in Health and Disease. https://doi.org/10.1016/B978-0-323-85696-6.00007-X

play an important role in the regulation of intracellular calcium signaling in excitable cells, including cardiomyocytes [6–8]. Lysosomal dysfunction is directly relevant in the pathogenesis of multiple diseases including cancer [9], heart disease [10], and lysosomal storage disorders (LSDs) [11, 12], which present primarily as degenerative neurological disease but are also linked with disorders in other organs [4]. In this chapter, we will discuss the major families of lysosomal proteases and the roles that these proteases play in cell signaling.

Cathepsins

Since the discovery of those initial five enzymes [2], many more lysosomal hydrolases have been discovered and characterized, including amylases, lipases, nucleases, and proteases, among which are the multiple peptide cleaving proteases of the cathepsin family [13]. The term "cathepsin" was originally derived from the Greek *Kathépsein*, meaning to digest (*kata*) and boil down (*hepsein*), and was applied to the original lysosomal protease identified by de Duve in 1955, although the term was in use prior to de Duve's investigation [2, 5, 14]. Subsequently, the term "cathepsin" has been used more generally to describe multiple, primarily intracellular proteases known to exhibit optimal levels of proteolytic activity within acidic vesicles (including lysosomes, endosomes, multivesicular bodies, and acidic secretory vesicles) [5, 15]. Cathepsins however are now classified according to their catalytic site as either serine (cathepsins A and G), aspartic (cathepsins D and E), or cysteine proteases, the latter group including 11 proteases that have been annotated within the human genome (cathepsins B, C, F, H, K, L, O, S, V, X, and W) [13, 16–18]. A summary of the classification of human cathepsins according to their proteolytic site is shown in Table 1.

As a result of the evolution of the use of the term "cathepsin" over the past century, reference to cathepsins in the literature can appear confusing, with the term occasionally used to refer solely to the cysteine cathepsins and at other times used more generally to apply to all lysosomal proteases. However, not all lysosomal proteases belong to the cathepsin family, e.g., the caspase-type protease legumin, napsin A, and tripeptidyl-peptidase I (TPP1) [11]. Conversely, many proteases traditionally termed as "lysosomal" due to their usual distribution and optimal operating pH may also function outside of the lysosome, exhibiting physiological and pathological functions in the cytosol, nucleus, and mitochondria [5, 50–54]. In addition, some cathepsins may be secreted and function extracellularly, where they can remain functional outside of their optimal operating pH window [13]. Such cathepsins have been shown to play roles in diseases such as arthritis and cancer when acting in the extracellular compartment [4, 5, 55]. Here, we will mainly focus on the largest group of lysosomal proteases, the cysteine cathepsins and their roles in cell signaling and disease; however, we will also discuss other key members of the cathepsin family including the serine and aspartic cathepsins.

Table 1: Classification of human cathepsins.

Cleavage site	Name	Specific functions	Pathology
Serine cathepsins	**Cathepsin A**	Prevents lysosomal degradation of β-galactosidase and neuraminidase [19]. Regulates chaperone-mediated autophagy via lysosomal LAMP-2A degradation [20]	Cardiomyopathies and arterial hypertension [11] Galactosialidosis [11]
	Cathepsin G	Inflammation and immunity—migration of neutrophils, monocytes, and APCs; regulation of autoantigen processing; and activation of lymphocytes [21]	Autoimmune diseases [21]
Aspartic cathepsins	**Cathepsin D**	Neuronal development, brain antigen processing of α-synuclein; tau, amyloid β, and apoE [22–24] Degradation of hormones, proenzymes, and growth factors [25, 26]	Neurodegenerative disorders including Alzheimer's, Parkinson's, and Huntington's diseases [24, 27, 28] Cardiovascular diseases, including ischemic heart disease and sudden cardiac death [29, 30]
	Cathepsin E	Carboxypeptidase A and IgE processing [26]	Atopic dermatitis [31]
Cysteine cathepsins	**Cathepsin B**	Growth factor processing including EGF, IGF-1, and TGF-β [32–34] Degradation of amyloid-β [35] Promotion of viral entry into cells [26]	Neurodegenerative diseases [35] Cancer [36] Acute pancreatitis [26]
	Cathepsin C	Inflammatory responses and activation of serine proteases [26]	Papillon-Lefèvre syndrome and periodontitis [11, 37]
	Cathepsin F	Invariant chain (Ii) processing and MHC II class responses [26]	Lysosomal storage disorders including neuronal ceroid lipofuscinosis (CLN) type 13 [11]
	Cathepsin H	Eye development [11] Immune regulation [11] Prohormone processing [26]	Papillon-Lefèvre syndrome [11] Type 1 diabetes [26]
	Cathepsin K	Bone resorption and extracellular matrix remodeling [38]	Inflammation, including atherosclerosis [38] Periodontitis; pycnodysostosis [11]
	Cathepsin L	Antigen presentation [39] Cardiovascular remodeling [40] Adipogenesis and glucose tolerance [41]	Parkinson's disease Cardiac repair [40] and dilated cardiomyopathy [42] Cancer [9, 43]
	Cathepsin O	Innate immunity [44]	Not known
	Cathepsin S	Major histocompatibility complex class II (MHC II) antigen presentation [38, 45]	Autoimmune diseases [38, 45]
	Cathepsin V	Natural killer cell and CD8 + cytotoxic cell production [26]	Cancer [46]
	Cathepsin X	Immune-cell proliferation, maturation, migration, and adhesion [47, 48]	Cancer [48]
	Cathepsin W	Role in NK cells; endoplasmic reticulum (ER) proteolytic machinery [49]	Not known

Summary of characterized human cathepsins grouped according to catalytic site and key physiological and pathological roles.

Cysteine cathepsins

Cysteine cathepsins belong to the papain family of cysteine proteases (family C1, Clan CA), the most abundant family of cysteine proteases, and are ubiquitously expressed in all living organisms [56]. Cysteine cathepsins have traditionally been associated with lysosomal protein turnover; however, their action is not restricted to the lysosome. Following excretion, they can remain functional in the extracellular space and maintain activity outside of their optimal pH range [57–59]. Cysteine cathepsins are now known to play multiple roles in specific physiological processes including apoptosis, adaptive immunity, prohormone activation, extracellular tissue remodeling, and cell differentiation [13, 16, 38, 56, 60, 61]. In addition, the characterization of inducible cathepsins F, K, S, B, and L has expanded the understanding of these cathepsins to include relevance to multiple diseases, including cancer [4, 55], atherosclerosis [62–64], cardiac repair [15, 65], cardiomyopathy [15, 40, 57] obesity [41, 66], and rheumatoid arthritis [67].

Tissue specificity

The identification of tissue-specific cysteine cathepsins, such as cathepsin K, which is almost exclusively confined to osteoclasts and the ovaries [38, 68], and cathepsin S, which is highly expressed in the spleen, thyroid, and antigen-presenting cells (APCs) [38], provided initial indications that cathepsins may be relevant for more specialized cellular processes [38, 56]. The expression of cathepsin K, for example, may be upregulated in the presence of inflammation, including atherosclerosis [38]. Cathepsin K is the most potent classified mammalian elastase, and despite instability at neutral pH, cathepsin K demonstrates an increased potency at a neutral pH in shorter assays (< 3 h) compared to the more stable and closely related cathepsin S [38]. While instability at neutral pH may be consistent with a primary role in lysosomal degradation, the potency of cathepsin K also makes it an important enzyme for remodeling of the extracellular matrix, where the enzyme acts acutely outside of its normal pH window [38]. This is clearly demonstrated in the case of pycnodysostosis, a rare autosomal recessive disorder characterized by facial hypoplasia, bone malformation, and osteosclerosis, which has been linked to mutations within the cathepsin K gene [69].

Cathepsin S, like cathepsin K, is also a potent elastase, albeit with a lower potency than cathepsin K and demonstrating higher stability and increased enzymatic activity at neutral pH [38]. Cathepsin S is highly expressed in the thyroid, where expression may be regulated, for example, via cytokines including interferon-gamma (IFNγ) and interleukin 1β (IL-1β), and is also highly expressed in APCs including B lymphocytes, macrophages, and dendritic cells [38, 70–72]. Within APCs, cathepsin S plays a vital role in major histocompatibility complex class II (MHC II) antigen presentation [38, 45]. Following synthesis within the ER, MHC II is trafficked to the endosomes where it can bind to antigenic peptides [73]. The resulting MHC II-peptide complexes are presented at the cell surface for recognition by T-cell receptors

(TcR) on T lymphocytes [73]. This trafficking of MHC II is guided by the invariant chain (li), a chaperone that binds to MHC II αβ dimers. A portion of li known as the class II-associated invariant chain peptide (CLIP) binds within the peptide-binding groove of MHC II and prevents the premature association of MHC II with antigenic peptides [39]. Cathepsin S acts through the cleavage of li, leaving just the CLIP portion remaining within the MHC II-peptide binding domain which is subsequently removed by human leukocyte antigen DM (HLA-DM) [38, 39, 74]. This role in antigen presentation can also be carried out in thymic cortical epithelial cells by cathepsin L [39] (for more details on the roles of lysosomal proteolysis in antigen processing and presentation, see Chapter 4).

Serine cathepsins

The serine cathepsins are represented by cathepsins A and G. Cathepsin A is a member of the α/β hydrolase fold family and has been proposed to share a common ancestral relationship with other α/β hydrolase fold enzymes such as cholinesterases [75]. Cathepsin A is predominantly found in lysosomes and stabilizes the protein complex formed by β-galactosidase (β-GAL) and neuromidase-1 (NEU1) to provide protection from lysosomal degradation [11, 19]. Mutations in cathepsin A therefore lead to the LSD galactosialidosis, resulting from NEU1 and β-GAL deficiency [11]. In addition, cathepsin A regulates chaperone-mediated autophagy via initiating degradation of lysosomal activated membrane protein (LAMP) type 2A [20] and is also involved in the regulation of vascular tone and blood pressure [11]. As a result, loss of cathepsin A activity has been linked with the development of cardiomyopathies and arterial hypertensions [11].

Cathepsin G is a member of the family of neutrophil serine proteases and is involved in inflammation and immunity [21]. In neutrophils, cathepsin G stimulates production of cytokines and chemokines and plays important roles in the migration of neutrophils, monocytes, and APCs to pathogen sites and areas of tissue damage [76], while in immune reaction cathepsin G is known to play a role in regulating autoantigen processing and activation of lymphocytes [21]. Cathepsin G is involved in multiple inflammatory diseases including atherosclerosis, neuropathies, and chronic obstructive pulmonary disease (COPD), as well as cancer, either through the promotion or suppression of inflammatory processes depending on the physiological conditions [76].

Aspartic cathepsins

Aspartic cathepsins in human physiology are represented by cathepsin D, which as mentioned above was the first cathepsin identified in lysosomes [1, 2], and cathepsin E. Cathepsin D is the only aspartic cathepsin to be ubiquitously expressed throughout the human body and shows particularly high levels of expression in neuronal tissue [24]. Substrates for cathepsin D include many neuronal proteins that are linked to neurodegenerative diseases, including amyloid precursor protein, tau, lipofuscin, apoE, α-synuclein, and huntingtin, and cathepsin D has been linked to neurodegenerative diseases including Alzheimer's, Huntington's, Parkinson's, and neuronal ceroid lipofuscinosis (CNL) [24, 27, 28]. Cathepsin D has also been implicated in protection against cardiac remodeling and malfunction via the promotion of autophagic flux [30].

Cathepsin E is predominantly expressed within the endosomal compartment of immune-related cells, particularly antigen-presenting cells such as macrophages, dendritic cells, and microglia, and is thought to play a major role in autophagy in these cells [77].

Lysosomal trafficking and cell signaling

Substrates are trafficked to lysosomes via either the endosomal/lysosomal pathway, if entering the cell via endocytosis, or autophagic pathways if substrates are originating from within the cell itself [78]. More detail on the involvement of proteolysis in autophagy is provided in Chapter 5; however, a brief summary is provided below.

The endosomal/lysosomal pathway

Fig. 1 provides an overview of the endosomal/lysosomal pathway via which extracellular components are transferred to the lysosomes for degradation. The main point of entry for extracellular substrates into the pathway is via endocytosis, which facilitates transfer of extracellular substrates that are otherwise unable to cross the plasma membrane into the intracellular environment [43]. The primary mechanism of endocytosis involves the clathrin-dependent formation of primary endocytic vesicles but can also occur via clathrin-independent mechanisms [79, 80]. In clathrin-dependent endocytosis, cytoplasmic domains of plasma membrane proteins interact with clathrin adapter proteins that in turn interact with clathrin to form clathrin-coated vesicles (CCVs) that are then internalized and traffic substrate to early endosomes (EEs) where they are assigned either for lysosomal degradation or recycling [43, 79]. Alternatively, endocytosis may occur via clathrin-independent mechanisms that are in themselves commonly dependent either upon caveolae or actin [81–83]. The EEs themselves are distinguishable by the expression of small guanosine triphosphatase (GTP) Rab5, phosphatidylinositol 3-kinase (PI3K), and the product of PI3K, phosphatidylinositol-3-phosphate (PtdIns3P) [43, 80]. As substrates destined for degradation accumulate, EEs become larger and fewer as they migrate intracellularly. This process is coupled with conversion of Rab5 to Rab7 by the guanine nucleotide exchange factor (GEF) homotypic fusion and protein sorting/Class C vacuolar protein sorting complex (HOPS/Class C Vps), leading to the formation of late endosomes (LEs) [84]. LEs are difficult to differentiate clearly from lysosomes, but are characterized by the expression of Rab7 as well as the RII regulatory subunit of cAMP-dependent protein kinase [43, 85].

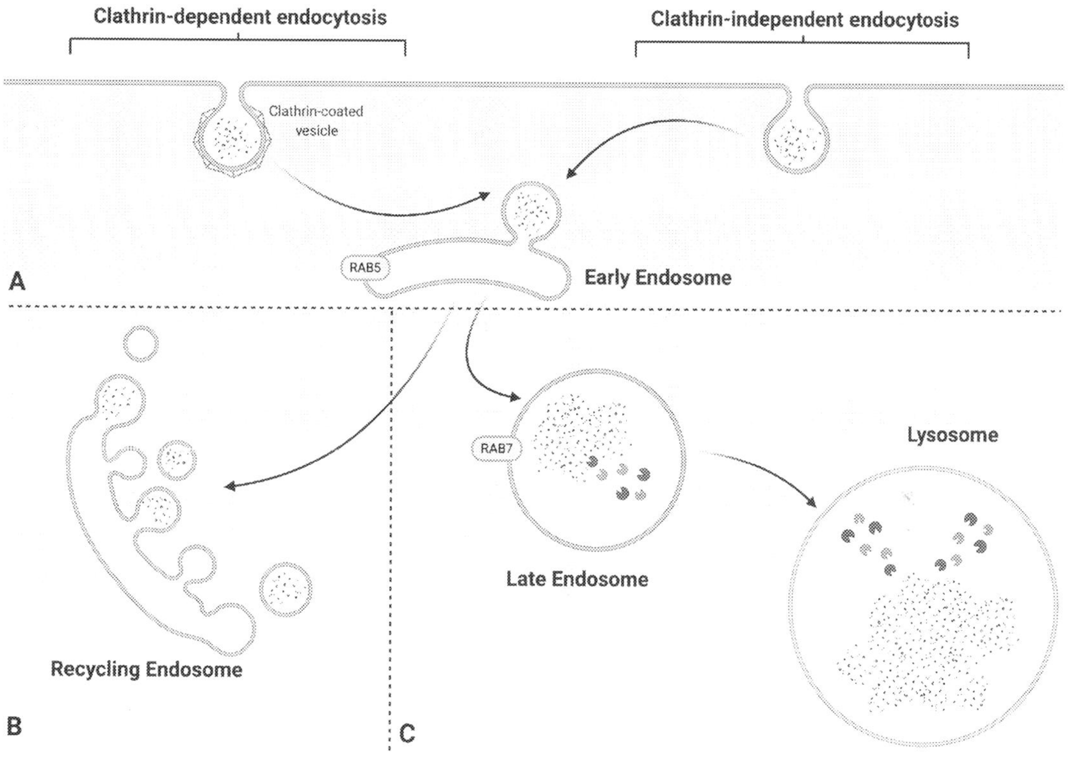

Fig. 1
Summary of the endosomal/lysosomal pathway. Endocytosis occurs via either clathrin-dependent or clathrin-independent formation of vesicles (A). Substrates are then trafficked to early endosomes where they are assigned for either recycling (B) or lysosomal degradation (C).

Autophagy

Autophagy refers to the lysosomal degradation of cytoplasmic components and is distinct from lysosomal degradation of extracellular components via the endosomal/lysosomal pathway [78, 86]. Autophagy can be achieved via three pathways: macroautophagy, chaperone-mediated autophagy (CMA), or microautophagy [78] (Fig. 2), with the general term "autophagy" commonly referring to just macroautophagy. In macroautophagy, a portion of cytoplasm, including organelles, is sequestered by a membrane (phagophore) to form an autophagosome that is then able to fuse with the lysosome [78] (Fig. 2A). Autophagosomes are typically double membraned, with the inner membrane being subject to degradation within the lysosome by lysosomal proteases [78]. CMA occurs via transport of substrates using a chaperone complex and LAMP-2A [87] (Fig. 2B), whereas microautophagy refers to the direct engulfment of substrates by the lysosome itself within invaginations of the

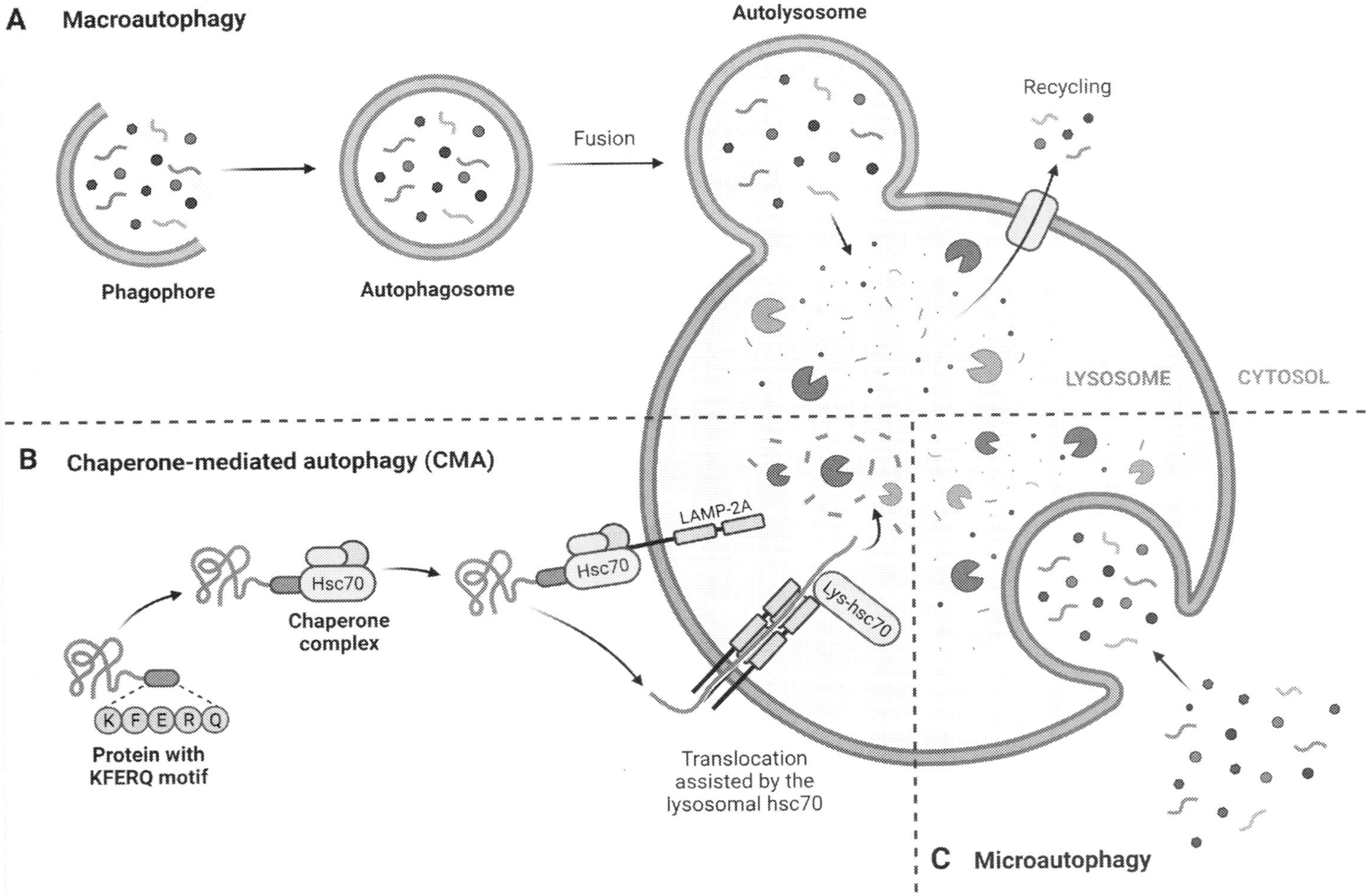

Fig. 2

The three main pathways of autophagy. (A) In macroautophagy, a portion of cytoplasm containing organelles and other substrates is enclosed by a phagophore to form an autophagosome that then fuses with the lysosome. (B) Chaperone-mediated autophagy (CMA) involves the transport of substrates into the lysosome via a chaperone complex, in this case Hsc70, that then interacts with LAMP-2A to facilitate uptake. (C) Microautophagy refers to the direct uptake of substrates by the lysosome.

lysosomal membrane (Fig. 2C). Therefore, while macroautophagy is typically nonspecific, CMA can target substrates carrying specific targeting motifs [87, 88]. For a more detailed account of lysosomal proteolysis and autophagy, we refer readers to Chapter 5.

Targeting and trafficking of lysosomal proteases

Targeting of lysosomal proteases to the lysosome is achieved via recognition of the mannose 6-phosphate (M6P) group, a unique marker that is tagged exclusively to the N-terminal of lysosomal proteases [89]. M6P groups are recognized by mannose 6-phosphate–specific receptors (MPRs), which can be either cation-independent (CI-MPR) or cation-dependent (CD-MPR) [89, 90]. Most MPRs are located on the trans-Golgi network and are responsible for binding proteases on the luminal side of the Golgi membrane. By interacting with adaptins that facilitate assembly of clathrin coats on the cytosolic side, this activation of MPRs leads to the formation of vesicles that enable intracellular transport to endosomes (Fig. 3A). Some CI-MPR may also be expressed on the cell membrane however and can contribute to the uptake of proteases to the endosomes via the endocytic pathway (Fig. 3B) [89,91,92].

Role of cathepsins in growth factor signaling

There is increasing evidence that cysteine cathepsins play crucial roles in the processing of proteins involved in cell signaling, particularly with regard to signaling processes involved in cancer as mentioned earlier in this chapter [43,93]. Specifically, roles have been reported for cysteine cathepsins in the processing of growth factors, such as epidermal growth factors (EGF) [32], transforming growth factor β (TGF-β) [33], and insulin-like growth factors (IGF) [34] as well as their receptors [43]. Cathepsin B, for example, has been linked to degradation of both EGF and the internalized EGF receptor complex in the liver [32], as well as degradation of IGF-1, a known activator of proliferating tumor cells, which shares a common binding site on cathepsin B with EGF [94]. In addition, cathepsin B modulates TGF-β, which may have contradictory effects in tumor development, primarily acting as a suppressor of tumorigenesis, but also acting as a promoter of cell invasion and metastasis [95].

Specific roles of cathepsins in disease

The following sections outline some specific roles for cathepsins that have been identified in disease, as well as observations from cathepsin knockout animal models. Cathepsins are of particular interest as potential therapeutic targets in neurodegenerative disease, cardiovascular disease, cancer, and rheumatoid arthritis, all of which are discussed in more detail below. In addition, cathepsin dysregulation is thought to be involved in the pathogenesis of pulmonary disease; for example, plasma levels of cathepsin S, as well as cystatin C, have been shown to be elevated in patients with chronic obstructive pulmonary disease (COPD) [96] and cathepsin G has also been implicated in the destruction of alveolar tissue in COPD patients [97].

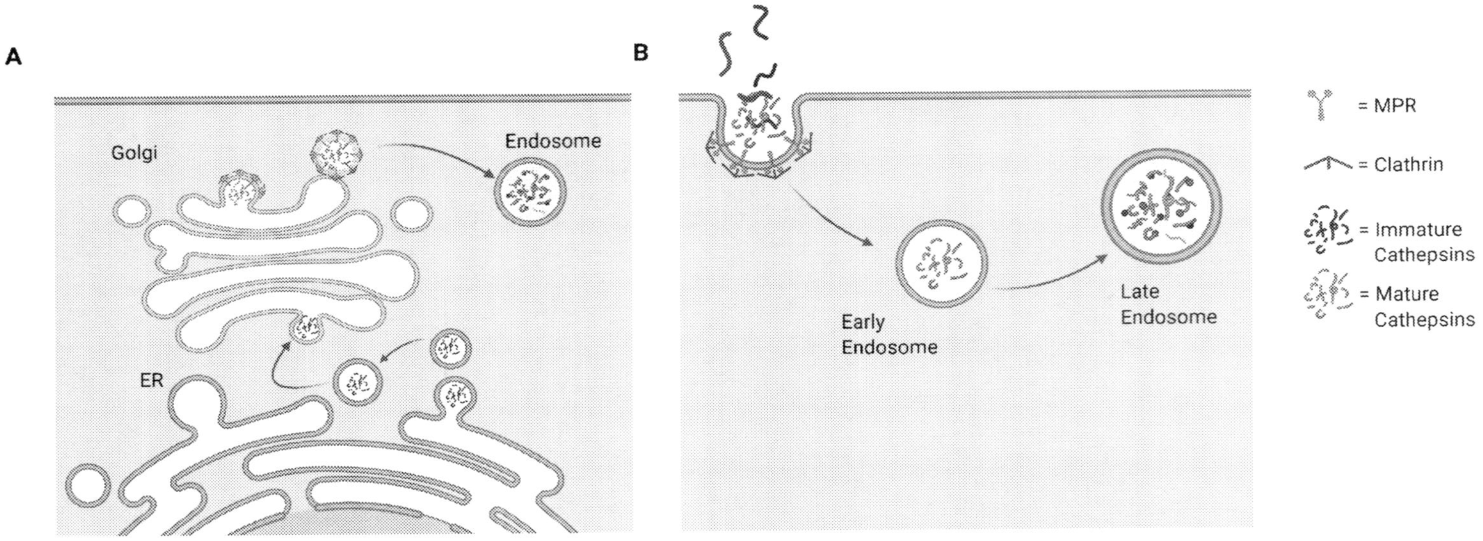

Fig. 3

Trafficking and targeting of lysosomal proteases. (A) Intracellular targeting of lysosomal proteases via the trans-Golgi pathway. (B) Cell surface recognition and targeting of extracellular lysosomal proteases via the endocytic pathway.

Proteases in neurodegenerative disease

Lysosomal failure has been implicated in many neurodegenerative diseases, including both age-related diseases such as Alzheimer's disease (AD) and Parkinson's disease (PD) as well as many inherited LSDs [98]. Alterations in cathepsin expression and intracellular proteolysis have been widely reported in neurodegenerative disease [99,100]; however, the links between neurodegenerative disease and cathepsins are not always clear [99]. Cathepsins D and B, for example, show increased expression in brain tissue with age [28,101], AD [27,102], and PD [22]. In the case of AD, cathepsin D was of initial interest as a potential therapeutic target for the processing of amyloid precursor protein (APP), accumulation of which is linked to the formation of amyloid plaques [99,103]. Cathepsin D was quickly ruled out as being a major APP-processing enzyme however [104], and its role in AD largely appears to be related to its involvement in autophagy [105]. Conversely, cathepsin B has been demonstrated to show both neuroprotective and neurodegenerative effects in AD [23,35,106,107]. Cathepsin B inhibition has been shown to reduce amyloid-β and improve memory deficits in transgenic AD mice [107]; however, cathepsin B may also be neuroprotective for AD via the degradation of amyloid-β [23,106].

Cathepsin D is thought to be the main lysosomal protease involved in degradation of α-synuclein, aggregations of which are linked to PD as well as other neurodegenerative diseases involving Lewy body formation (e.g., Lewy body variants of dementia and AD) [99,108]. Partial inhibition of the cathepsin D encoding gene has been reported to be sufficient to induce a reduction in lysosomal function leading to α-synuclein aggregation [109]; however, other studies have also implicated cathepsins B and L to greater or lesser extents and the dominant pathway is still unclear [99,110,111].

In addition to neurodegenerative disease, lysosomal enzymes are likely to play a role in the natural process of ageing, with cathepsins within the neuronal endosomal/lysosomal system involved in regulating cell death during the natural aging process as well as disease [99,112]. In addition to cathepsin D, cathepsins B, E, and L have been linked to both neuronal degeneration and reactivation of glial cells, with cathepsins B and E also showing age-related increases in expression, while cathepsin L demonstrates a decreased expression [101].

Proteases in cardiovascular disease

Cardiac remodeling occurs as a result of the action of proteolytic enzymes and underlies disease progression in multiple cardiac pathologies, including hypertension, hypertrophic cardiomyopathy, dilated cardiomyopathy, diabetic cardiomyopathy, and myocardial infarction [10,113]. Extracellular proteases are key for the breakdown of cardiac extracellular matrix (ECM) [113]. Matrix metalloproteinases (MMPs) and serine proteases account for the majority of ECM degradation in cardiac tissue [114], but cysteine cathepsins have also been shown to play a role [115]. The expression levels of cathepsins S, K, and L, for example, have been shown to increase in the development of human atherosclerotic or neointimal lesions during vascular remodeling, and cathepsins have been linked to both atherosclerosis and abdominal aortic aneurysms (AAAs) [64,116]. In a heart failure rat model, upregulation of cathepsin S has been demonstrated in response to increased levels of IL-1β, leading to increased proteolytic activity [115]. Indeed, cathepsins have a significant capability for ECM proteolysis via degradation of elastin, fibrillar collagens, and through the activation of pro-MMPs [113]. As such, the use of cathepsin inhibitors has been proposed for the treatment of multiple cardiovascular disease conditions, including hypertension, atherosclerosis, and AAAs [10]. For example, reduced cathepsin S has been shown to attenuate atherosclerotic plaque formation [117] and angiotensin II-induced AAA [118] in vivo. To date however, although multiple cathepsin inhibitors have been developed, the ubiquitous expression of cathepsins and potential for side effects and cytotoxicity have thus far prevented the use of such compounds clinically and tissue specificity will be essential for the future translational benefit of such compounds [15].

In contrast to the potential beneficial effects of reduced cathepsin S expression, reduced levels of cathepsin L have been linked to dilated cardiomyopathy (DCM) [42] and loss of cathepsin L promotes stress-induced cardiac hypertrophy, associated with the accumulation of α-actinin, myosin, connexin-43, and H-cadherin [119]. Cathepsin L-deficient mice (CTSL$^{-/-}$) demonstrate functional and histological cardiac alterations, including interstitial fibrosis, left ventricular hypertrophy, and evidence of pathological impairment of heart rhythm/arrhythmias (including supraventricular tachycardia and atrioventricular heart block) that closely resemble human DCM [42]. CTSL$^{-/-}$ mice develop late-onset DCM characterized by cardiac chamber dilation, fibrosis, and impaired cardiac contraction by 12 months of age, with numerous dysmorphic lysosome-like structures evident within heart muscle from 3 days following birth [120]. Skeletal muscle appears to be unaffected in these CTSL$^{-/-}$ animals [120]. Cardiomyocyte-specific expression of cathepsin L in cathepsin L-deficient mice reverses these changes, albeit without preventing fibrosis [40].

Lysosomal dysfunction is also involved in the pathology of heart failure, with reduced lysosomal degradation implicated in increased autophagy in cardiac tissue [30,87,121]. Cathepsins, principally cathepsin D, are differentially expressed during different phases of the failing heart [29,30,122,123]. Cathepsin D-deficient (CTSD$^{-/-}$) mice develop normally for 2 weeks after birth but die with intestinal necrosis, thromboembolism, and lymphopenia within 4 weeks [25]. The upregulation of cathepsin D is observed during the subacute phase in patients following myocardial infarction (MI) and ischemic heart disease, and this may be followed by downregulation during the chronic phase and final stages of dilated cardiomyopathy [29,30,122,123]. Cathepsin D is also downregulated in sudden cardiac disease (SCD) [29]. Hearts of CTSD$^{-/-}$ mice show restrictive cardiomyopathy along with deposits of myocardial ATP synthase subunit C (ATOPSC) and LC3-II, a marker for autophagosomes, although ejection fraction is maintained [29,30,124,125]. Heterozygous mice (CTSD $^{+/-}$), which do not present with cardiac dysfunction under normal conditions, do not demonstrate upregulation of cathepsin D post-MI levels and show impaired autophagic flux and exacerbated cardiac dysfunction after MI [30]. These data therefore indicate that elevated cathepsin D levels appear to demonstrate a protective role following MI [30].

Proteases in cancer

Cysteine cathepsins are frequently upregulated in cancer and have been implicated in tumorigenic processes such as angiogenesis, cell proliferation, apoptosis, and invasion [9]. Indeed, cysteine cathepsins are known to play roles in a number of signaling pathways closely linked to cancer progression, including growth factor signaling, regulation of apoptosis and autophagy, and epithelial-to-mesenchymal transition (reviewed in Pislar et al. [43]). Increased expression levels of some cathepsins can also serve as prognostic markers. Elevation of cathepsin B, for example, the most abundant and widely expressed cysteine cathepsin [38], has been correlated with poor prognosis in multiple cancers including melanoma, lung, breast, ovarian, brain, and head and neck cancers [9,43,126]. Similarly, increased cathepsin L expression is correlated with decreased survival rates in patients with colorectal, breast, and head and neck cancers [9, 43]. Cathepsin S has also been shown to be highly expressed in malignant tissues and is involved in control of angiogenesis and tumor growth via matrix-derived angiogenic factors [127]. Increased expression of cathepsin S correlates with poor prognosis in colorectal cancer [128]; however in lung cancer, it is decreased levels of cathepsin S which correlate with lower survival rates [129].

Cathepsin B is the most widely studied cysteine cathepsin in relation to tumorigenesis and has been implicated in multiple stages of tumor development including initiation, proliferation, angiogenesis, invasion, inflammation, apoptosis, and metastasis [36]. Cathepsin B has therefore frequently been proposed as a therapeutic target [13,36,130]; however, the precise roles of cysteine cathepsins in cancer are not yet fully understood and there is also strong evidence that the loss of activity of one protease may be compensated for by another [36]. Some preclinical studies however have shown efficacy for the inhibition of cathepsin B as a therapeutic strategy, for example, in the treatment of pancreatic cancer [131,132], and further research into the roles played by cysteine cathepsins in cancer may highlight further breakthroughs in the development of cancer treatment in future.

Proteases in rheumatoid arthritis

Rheumatoid arthritis (RA) is a chronic autoimmune disease triggered by the onset of multiple inflammatory cascades leading to synovial inflammation and eventual joint destruction via the breakdown of bone, cartilage, tendons, and ligaments [133,134]. While multiple inflammatory cascade pathways may eventually lead to RA [134], autophagy is significantly enhanced in RA and is thought to play an important role in RA pathogenesis [135–137]. Cathepsin K inhibition has been shown to reduce immune-cell infiltration, activation of osteoclasts, and erosion of articular cartilage, suggesting that cathepsin K is an important mediator of bone destruction in RA [137]. Cathepsin K is thought to be linked to autophagy and subsequently RA, as well as the linked chronic inflammatory disease periodontitis, via the regulation of the toll-like receptor 9 (TLR9) signaling pathway which is known to be involved in autophagy [136–138]. Inhibition of cathepsin K was found to significantly decrease TLR9-specific cytokine release as well as expression of TLR9-related signaling molecules in vitro [137], and tissue-specific gene therapy targeting cathepsin K has therefore been proposed as a potential future therapeutic strategy for the treatment of RA [136].

Conclusion

Since Christian de Duve's first discovery of lysosomes, our understanding of the role played by lysosomal proteases has greatly expanded, and over 50 lysosomal proteases, as well as their associated activator proteins, have now been characterized [3]. However, it is now clear that the roles of these proteases are not restricted to lysosomal degradation, and indeed, lysosomal proteases are now known to play roles in the regulation of multiple signaling pathways both within and outside of the cell. In addition, many lysosomal proteases may act as diagnostic markers and some have shown initial promise as potential therapeutic targets for diseases, including cancer, neurodegenerative, cardiovascular, and inflammatory diseases. An increased understanding of the involvement of lysosomal proteases and their roles in signaling pathways is therefore likely to yield both new targets for specific therapies, as well as increasing our understanding of the pathology of multiple diseases in the future.

Acknowledgments

R.A.B.B. is funded by the Sir Henry Dale Wellcome Trust and Royal Society Fellowship (109371/Z/15/Z). S.J.B. is funded by the British Heart Foundation, Project Grant Number PG/18/4/33521. T.A. acknowledges the support from The Returning Carers' Fund, Medical Sciences Division, University of Oxford. R.A.B.B. is a Senior Research Fellow of Linacre College, Oxford. All figures were created using Biorender.com. R.A.B.B. acknowledges support from the British Heart Foundation, Centre of Research Excellence, Oxford.

References

[1] de Duve C. The lysosome turns fifty. Nat Cell Biol Sep 2005;7(9):847–9. https://doi.org/10.1038/ncb0905-847.

[2] Duve CD, Pressman BC, Gianetto R, Wattiaux R, Appelmans F. Tissue fractionation studies. 6. Intracellular distribution patterns of enzymes in rat-liver tissue. Biochem J 1955;60(1–4):604–17.

[3] Schröder BA, Wrocklage C, Hasilik A, Saftig P. The proteome of lysosomes. Proteomics Nov 2010;10(22):4053–76. https://doi.org/10.1002/pmic.201000196.

[4] Olson OC, Joyce JA. Cysteine cathepsin proteases: regulators of cancer progression and therapeutic response. Nat Rev Cancer Dec 2015;15(12):712–29. https://doi.org/10.1038/nrc4027.

[5] Muller S, Dennemarker J, Reinheckel T. Specific functions of lysosomal proteases in endocytic and autophagic pathways. Biochim Biophys Acta Jan 2012;1824(1):34–43. https://doi.org/10.1016/j.bbapap.2011.07.003.
[6] Aston D, Capel RA, Ford KL, et al. High resolution structural evidence suggests the sarcoplasmic reticulum forms microdomains with acidic stores (lysosomes) in the heart. Sci Rep Jan 2017. https://doi.org/10.1038/srep40620, 740620.
[7] Capel RA, Bolton EL, Lin WK, et al. Two-pore channels (TPC2s) and nicotinic acid adenine dinucleotide phosphate (NAADP) at lysosomal-sarcoplasmic reticular junctions contribute to acute and chronic β-adrenoceptor signaling in the heart. J Biol Chem Dec 2015;290(50):30087–98. https://doi.org/10.1074/jbc.M115.684076.
[8] Collins TP, Bayliss R, Churchill GC, Galione A, Terrar DA. NAADP influences excitation-contraction coupling by releasing calcium from lysosomes in atrial myocytes. Cell Calcium Nov 2011;50(5):449–58. https://doi.org/10.1016/j.ceca.2011.07.007.
[9] Gocheva V, Joyce JA. Cysteine cathepsins and the cutting edge of cancer invasion. Cell Cycle Jan 2007;6(1):60–4. https://doi.org/10.4161/cc.6.1.3669.
[10] Lutgens SPM, Cleutjens K, Daemen M, Heeneman S. Cathepsin cysteine proteases in cardiovascular disease. FASEB J Oct 2007;21(12):3029–41. https://doi.org/10.1096/fj.06-7924com.
[11] Ketterer S, Gomez-Auli A, Hillebrand LE, Petrera A, Ketscher A, Reinheckel T. Inherited diseases caused by mutations in cathepsin protease genes. FEBS J May 2017;284(10):1437–54. https://doi.org/10.1111/febs.13980.
[12] Platt FM, d'Azzo A, Davidson BL, Neufeld EF, Tifft CJ. Lysosomal storage diseases. Nat Rev Dis Primers Oct 2018;4. https://doi.org/10.1038/s41572-018-0025-4.
[13] Turk V, Stoka V, Vasiljeva O, et al. Cysteine cathepsins: from structure, function and regulation to new frontiers. Biochim Biophys Acta Jan 2012;1824(1):68–88. https://doi.org/10.1016/j.bbapap.2011.10.002.
[14] Willstatter R, Bamann E. The protoease of stomach lining. First essay on the enzymes of leukoctyes. Hoppe-Seylers Zeitschrift Fur Physiologische Chemie Jan 1929;180(1/3):127–43. https://doi.org/10.1515/bchm2.1929.180.1-3.127.
[15] Blondelle J, Lange S, Greenberg BH, Cowling RT. Cathepsins in heart disease-chewing on the heartache? Am J Physiol Heart Circ Physiol May 2015;308(9):H974–6. https://doi.org/10.1152/ajpheart.00125.2015.
[16] Verma S, Dixit R, Pandey KC. Cysteine proteases: modes of activation and future prospects as pharmacological targets. Front Pharmacol Apr 2016;7107. https://doi.org/10.3389/fphar.2016.00107.
[17] Rossi A, Deveraux Q, Turk B, Sali A. Comprehensive search for cysteine cathepsins in the human genome. Biol Chem May 2004;385(5):363–72. https://doi.org/10.1515/bc.2004.040.
[18] Rawlings ND, Barrett AJ, Bateman A. MEROPS: the peptidase database. Nucleic Acids Res Jan 2010;38:D227–33. https://doi.org/10.1093/nar/gkp971.
[19] Potier M, Michaud L, Tranchemontagne J, Thauvette L. Structure of the lysosomal neuraminidase—beta-galactosidase-carboxypeptidase multienzymatic complex. Biochem J Apr 1990;267(1):197–202. https://doi.org/10.1042/bj2670197.
[20] Cuervo AM, Mann L, Bonten EJ, d'Azzo A, Dice JF. Cathepsin A regulates chaperone-mediated autophagy through cleavage of the lysosomal receptor. EMBO J Jan 2003;22(1):47–59. https://doi.org/10.1093/emboj/cdg002.
[21] Gao SM, Zhu HL, Zuo XX, Luo H. Cathepsin G and its role in inflammation and autoimmune diseases. Archiv Rheumatol Dec 2018;33(4):498–504. https://doi.org/10.5606/ArchRheumatol.2018.6595.
[22] Moors T, Paciotti S, Chiasserini D, et al. Lysosomal dysfunction and alpha-synuclein aggregation in parkinson's disease: diagnostic links. Mov Disord Jun 2016;31(6):791–801. https://doi.org/10.1002/mds.26562.
[23] Mueller-Steiner S, Zhou Y, Arai H, et al. Antiamyloidogenic and neuroprotective functions of cathepsin B: implications for Alzheimer's disease. Neuron Sep 2006;51(6):703–14. https://doi.org/10.1016/j.neuron.2006.07.027.
[24] Vidoni C, Follo C, Savino M, Melone MAB, Isidoro C. The role of cathepsin D in the pathogenesis of human neurodegenerative disorders. Med Res Rev Sep 2016;36(5):845–70. https://doi.org/10.1002/med.21394.

[25] Saftig P, Hetman M, Schmahl W, et al. Mice deficient for the lysosomal proteinase cathepsin-D exhibit progressive atrophy of the intestinal-mucosa and profound destruction of lymphoid-cells. EMBO J Aug 1995;14(15):3599–608. https://doi.org/10.1002/j.1460-2075.1995.tb00029.x.
[26] Yadati T, Houben T, Bitorina A, Shiri-Sverdlov R. The ins and outs of cathepsins: physiological function and role in disease management. Cell 2020;9(7):1679. https://doi.org/10.3390/cells9071679.
[27] Haas U, Sparks DL. Cortical cathepsin D activity and immunolocalization in Alzheimer disease, critical coronary artery disease, and aging. Mol Chem Neuropathol Sep 1996;29(1):1–14. https://doi.org/10.1007/bf02815189.
[28] Banayschwartz M, Deguzman T, Kenessey A, Palkovits M, Lajtha A. The distribution of cathepsin-D activity in adult and aging human brain-regions. J Neurochem Jun 1992;58(6):2207–11. https://doi.org/10.1111/j.1471-4159.1992.tb10965.x.
[29] Kakimoto Y, Sasaki A, Niioka M, Kawabe N, Osawa M. Myocardial cathepsin D is downregulated in sudden cardiac death. PLoS One 2020;15(3). https://doi.org/10.1371/journal.pone.0230375, e0230375.
[30] Wu P, Yuan X, Li F, et al. Myocardial upregulation of cathepsin D by ischemic heart disease promotes autophagic flux and protects against cardiac remodeling and heart failure. Circ Heart Fail 2017;10(7). https://doi.org/10.1161/CIRCHEARTFAILURE.117.004044.
[31] Tsukuba T, Okamoto K, Yamamoto K. Cathepsin E and atopic dermatitis. J Pharmacol Sci 2003;91:13P.
[32] Authier F, Metioui M, Bell AW, Mort JS. Negative regulation of epidermal growth factor signaling by selective proteolytic mechanisms in the endosome mediated by cathepsin B. J Biol Chem Nov 1999;274(47):33723–31. https://doi.org/10.1074/jbc.274.47.33723.
[33] Yin M, Soikkeli J, Jahkola T, Virolainen S, Saksela O, Holtta E. TGF-beta signaling, activated stromal fibroblasts, and cysteine cathepsins B and L drive the invasive growth of human melanoma cells. Am J Pathol Dec 2012;181(6):2202–16. https://doi.org/10.1016/j.ajpath.2012.08.027.
[34] Navab R, Chevet E, Authier F, Di Guglielmo GM, Bergeron JJM, Brodt P. Inhibition of endosomal insulin-like growth factor-I processing by cysteine proteinase inhibitors blocks receptor-mediated functions. J Biol Chem Apr 2001;276(17):13644–9. https://doi.org/10.1074/jbc.M100019200.
[35] Hook V, Yoon M, Mosier C, et al. Cathepsin B in neurodegeneration of Alzheimer's disease, traumatic brain injury, and related brain disorders. Biochim Biophys Acta Aug 2020;1868(8). https://doi.org/10.1016/j.bbapap.2020.140428, 140428.
[36] Aggarwal N, Sloane BF. Cathepsin B: multiple roles in cancer. Proteomics Clin Appl Jun 2014;8(5–6):427–37. https://doi.org/10.1002/prca.201300105.
[37] Hart TC, Hart PS, Michalec MD, et al. Haim-Munk syndrome and Papillon-Lefevre syndrome are allelic mutations in cathepsin C. J Med Genet Feb 2000;37(2):88–94. https://doi.org/10.1136/jmg.37.2.88.
[38] Chapman HA, Riese RJ, Shi GP. Emerging roles for cysteine proteases in human biology. Annu Rev Physiol 1997;59:63–88. https://doi.org/10.1146/annurev.physiol.59.1.63.
[39] Beers C, Burich A, Kleijmeer MJ, Griffith JM, Wong P, Rudensky AY. Cathepsin S controls MHC class II-mediated antigen presentation by epithelial cells in vivo. J Immunol Feb 2005;174(3):1205–12. https://doi.org/10.4049/jimmunol.174.3.1205.
[40] Spira D, Stypmann J, Tobin DJ, et al. Cell type-specific functions of the lysosomal protease cathepsin L in the heart. J Biol Chem Dec 2007;282(51):37045–52. https://doi.org/10.1074/jbc.M703447200.
[41] Yang M, Zhang Y, Pan J, et al. Cathepsin L activity controls adipogenesis and glucose tolerance. Nat Cell Biol Aug 2007;9(8):970–U130.
[42] Stypmann J, Glaser K, Roth W, et al. Dilated cardiomyopathy in mice deficient for the lysosomal cysteine peptidase cathepsin L. Proc Natl Acad Sci U S A Apr 2002;99(9):6234–9. https://doi.org/10.1073/pnas.092637699.
[43] Pislar A, Nanut MP, Kos J. Lysosomal cysteine peptidases—molecules signaling tumor cell death and survival. Semin Cancer Biol Dec 2015;35:168–79. https://doi.org/10.1016/j.semcancer.2015.08.001.
[44] Sun YX, Zhu BJ, Tang L, et al. Cathepsin o is involved in the innate immune response and metamorphosis of Antheraea pernyi. J Invertebr Pathol Nov 2017;150:6–14. https://doi.org/10.1016/j.jip.2017.08.015.

[45] Villadangos JA, Riese RJ, Peters C, Chapman HA, Ploegh HL. Degradation of mouse invariant chain: roles of cathepsins S and D and the influence of major histocompatibility complex polymorphism. J Exp Med Aug 1997;186(4):549–60. https://doi.org/10.1084/jem.186.4.549.
[46] Al-Hashimi A, Venugopalan V, Sereesongsaeng N, et al. Significance of nuclear cathepsin V in normal thyroid epithelial and carcinoma cells. BBA-Mol Cell Res Dec 2020;1867(12). https://doi.org/10.1016/j.bbamcr.2020.118846, 118846.
[47] Kos J, Jevnikar Z, Obermajer N. The role of cathepsin X in cell signaling. Cell Adh Migr Apr-Jun 2009;3(2):164–6. https://doi.org/10.4161/cam.3.2.7403.
[48] Kos J, Vizin T, Fonovic UP, Pislar A. Intracellular signaling by cathepsin X: molecular mechanisms and diagnostic and therapeutic opportunities in cancer. Semin Cancer Biol Apr 2015;31:76–83. https://doi.org/10.1016/j.semcancer.2014.05.001.
[49] Wex T, Buhling F, Wex H, et al. Human cathepsin W, a cysteine protease predominantly expressed in NK cells, is mainly localized in the endoplasmic reticulum. J Immunol Aug 2001;167(4):2172–8. https://doi.org/10.4049/jimmunol.167.4.2172.
[50] Goulet B, Baruch A, Moon NS, et al. A cathepsin L isoform that is devoid of a signal peptide localizes to the nucleus in S phase and processes the CDP/Cux transcription factor. Mol Cell Apr 2004;14(2):207–19. https://doi.org/10.1016/s1097-2765(04)00209-6.
[51] Sever S, Altintas MM, Nankoe SR, et al. Proteolytic processing of dynamin by cytoplasmic cathepsin L is a mechanism for proteinuric kidney disease. J Clin Investig Aug 2007;117(8):2095–104. https://doi.org/10.1172/jci32022.
[52] Duncan EM, Muratore-Schroeder TL, Cook RG, et al. Cathepsin L proteolytically processes histone H3 during mouse embryonic stem cell differentiation. Cell Oct 2008;135(2):284–94. https://doi.org/10.1016/j.cell.2008.09.055.
[53] Reiser J, Adair B, Reinheckel T. Specialized roles for cysteine cathepsins in health and disease. J Clin Investig Oct 2010;120(10):3421–31. https://doi.org/10.1172/jci42918.
[54] Muntener K, Zwicky R, Csucs G, Rohrer J, Baici A. Exon skipping of cathepsin B-mitochondrial targeting of a lysosomal peptidase provokes cell death. J Biol Chem Sep 2004;279(39):41012–7. https://doi.org/10.1074/jbc.M405333200.
[55] Mohamed MM, Sloane BF. Cysteine cathepsins: multifunctional enzymes in cancer. Nat Rev Cancer Oct 2006;6(10):764–75. https://doi.org/10.1038/nrc1949.
[56] Turk B, Turk D, Turk V. Lysosomal cysteine proteases: more than scavengers. BBA-Protein Struct M Mar 2000;1477(1–2):98–111. https://doi.org/10.1016/s0167-4838(99)00263-0.
[57] Cheng XW, Shi GP, Kuzuya M, Sasaki T, Okumura K, Murohara T. Role for cysteine protease cathepsins in heart disease focus on biology and mechanisms with clinical implication. Circulation Mar 2012;125(12):1551–62. https://doi.org/10.1161/circulationaha.111.066712.
[58] Haka AS, Grosheva I, Chiang E, et al. Macrophages create an acidic extracellular hydrolytic compartment to digest aggregated lipoproteins. Mol Biol Cell Dec 2009;20(23):4932–40. https://doi.org/10.1091/mbc.E09-07-0559.
[59] Vaes G. On mechanisms of bone resorption—action of parathyroid hormone on excretion and synthesis of lysosomal enzymes and on extracellular release of acid by bone cells. J Cell Biol 1968;39(3):676. https://doi.org/10.1083/jcb.39.3.676.
[60] Turk V, Turk B, Guncar G, Turk D, Kos J. Lysosomal cathepsins: structure, role in antigen processing and presentation, and cancer. Adv Enzyme Regul Proc 2002;42:285–303. Pii: s0065-2571(01)00034-6. doi: 10.1016/s0065-2571(01)00034-6.
[61] Vasiljeva O, Reinheckel T, Peters C, Turk D, Turk V, Turk B. Emerging roles of cysteine cathepsins in disease and their potential as drug targets. Curr Pharm Des 2007;13(4):387–403. https://doi.org/10.2174/138161207780162962.
[62] Sasaki T, Kuzuya M, Nakamura K, et al. AT1 blockade attenuates atherosclerotic plaque destabilization accompanied by the suppression of cathepsin S activity in apoE-deficient mice. Atherosclerosis Jun 2010;210(2):430–7. https://doi.org/10.1016/j.atherosclerosis.2009.12.031.

[63] Liu J, Sukhova GK, Yang JT, et al. Cathepsin L expression and regulation in human abdominal aortic aneurysm, atherosclerosis, and vascular cells. Atherosclerosis Feb 2006;184(2):302–11. https://doi.org/10.1016/j.atherosclerosis.2005.05.012.
[64] Sukhova GK, Shi GP, Simon DI, Chapman HA, Libby P. Expression of the elastolytic cathepsins S and K in human atheroma and regulation of their production in smooth muscle cells. J Clin Investig Aug 1998;102(3):576–83. https://doi.org/10.1172/jci181.
[65] Sun M, Chen MY, Liu YA, et al. Cathepsin-L contributes to cardiac repair and remodelling post-infarction. Cardiovasc Res Feb 2011;89(2):374–83. https://doi.org/10.1093/cvr/cvq328.
[66] Taleb S, Cancello R, Poitou C, et al. Weight loss reduces adipose tissue cathepsin S and its circulating levels in morbidly obese women. J Clin Endocrinol Metab Mar 2006;91(3):1042–7. https://doi.org/10.1210/jc.2005-1601.
[67] Hou WS, Li WJ, Keyszer G, et al. Comparison of cathepsins K and S expression within the rheumatoid and osteoarthritic synovium. Arthritis Rheum Mar 2002;46(3):663–74. https://doi.org/10.1002/art.10114.
[68] Bromme D, Okamoto K, Wang BB, Biroc S. Human cathepsin O-2, a matrix protein-degrading cysteine protease expressed in osteoclasts - Functional expression of human cathepsin O-2 in Spodoptera frugiperda and characterization of the enzyme. J Biol Chem Jan 1996;271(4):2126–32. https://doi.org/10.1074/jbc.271.4.2126.
[69] Gelb BD, Shi GP, Chapman HA, Desnick RJ. Pycnodysostosis, a lysosomal disease caused by cathepsin K deficiency. Science Aug 1996;273(5279):1236–8. https://doi.org/10.1126/science.273.5279.1236.
[70] Shi GP, Munger JS, Meara JP, Rich DH, Chapman HA. Molecular-cloning and expression of human alveolar macrophage cathepsin-S, an elastinolytic cysteine protease. J Biol Chem Apr 1992;267(11):7258–62.
[71] Shi GP, Webb AC, Foster KE, et al. Human cathepsin-S—chromosomal localization, gene structure, and tissue distribution. J Biol Chem Apr 1994;269(15):11530–6.
[72] Morton PA, Zacheis ML, Giacoletto KS, Manning JA, Schwartz BD. Delivery of nascent MHC class II-invariant chain complexes to lysosomal compartments and proteolysis of invariant chain by cysteine proteases precedes peptide binding in b-lymphoblastoid cells. J Immunol Jan 1995;154(1):137–50.
[73] Landsverk OJB, Bakke O, Gregers TF. MHC II and the endocytic pathway: regulation by invariant chain. Scand J Immunol Sep 2009;70(3):184–93. https://doi.org/10.1111/j.1365-3083.2009.02301.x.
[74] Ghosh P, Amaya M, Mellins E, Wiley DC. The structure of an intermediate in class-II MHC maturation—clip bound to HLA-DR3. Nature Nov 1995;378(6556):457–62. https://doi.org/10.1038/378457a0.
[75] Hiraiwa M. Cathepsin A/protective protein: an unusual lysosomal multifunctional protein. Cell Mol Life Sci Dec 1999;56(11–12):894–907. https://doi.org/10.1007/s000180050482.
[76] Zamolodchikova TS, Tolpygo SM, Svirshchevskaya EV. Cathepsin G-not only inflammation: the immune protease can regulate normal physiological processes. Front Immunol Mar 2020. https://doi.org/10.3389/fimmu.2020.00411, 11411.
[77] Tsukuba T, Yanagawa M, Kadowaki T, et al. Cathepsin E deficiency impairs autophagic proteolysis in macrophages. PLoS One Dec 2013;8(12). https://doi.org/10.1371/journal.pone.0082415, e82415.
[78] Mizushima N. Autophagy: process and function. Genes Dev Nov 2007;21(22):2861–73. https://doi.org/10.1101/gad.1599207.
[79] Doherty GJ, McMahon HT. Mechanisms of endocytosis. Annu Rev Biochem 2009;78:857–902. https://doi.org/10.1146/annurev.biochem.78.081307.110540.
[80] Grant BD, Donaldson JG. Pathways and mechanisms of endocytic recycling. Nat Rev Mol Cell Biol Sep 2009;10(9):597–608. https://doi.org/10.1038/nrm2755.
[81] Kiss AL. Caveolae and the regulation of endocytosis. In: Caveolins and caveolae: roles in signaling and disease mechanisms, 729; 2012. p. 14–28.
[82] Donaldson JG, Porat-Shliom N, Cohen LA. Clathrin-independent endocytosis: a unique platform for cell signaling and PM remodeling. Cell Signal Jan 2009;21(1):1–6. https://doi.org/10.1016/j.cellsig.2008.06.020.
[83] Engqvist-Goldstein AEY, Drubin DG. Actin assembly and endocytosis: from yeast to mammals. Annu Rev Cell Dev Biol 2003;19:287–332. https://doi.org/10.1146/annurev.cellbio.19.111401.093127.

[84] Rink J, Ghigo E, Kalaidzidis Y, Zerial M. Rab conversion as a mechanism of progression from early to late endosomes. Cell Sep 2005;122(5):735–49. https://doi.org/10.1016/j.cell.2005.06.043.
[85] Repnik U, Cesen MH, Turk B. The endolysosomal system in cell death and survival. Cold Spring Harb Perspect Biol Jan 2013;5(1). https://doi.org/10.1101/cshperspect.a008755, a008755.
[86] Levine B, Klionsky DJ. Development by self-digestion: molecular mechanisms and biological functions of autophagy. Dev Cell Apr 2004;6(4):463–77. https://doi.org/10.1016/s1534-5807(04)00099-1.
[87] Lavandero S, Chiang M, Rothermel BA, Hill JA. Autophagy in cardiovascular biology. J Clin Investig Jan 2015;125(1):55–64. https://doi.org/10.1172/jci73943.
[88] Pedrozo Z, Torrealba N, Fernandez C, et al. Cardiomyocyte ryanodine receptor degradation by chaperone-mediated autophagy. Cardiovasc Res May 2013;98(2):277–85. https://doi.org/10.1093/cvr/cvt029.
[89] Coutinho MF, Prata MJ, Alves S. Mannose-6-phosphate pathway: a review on its role in lysosomal function and dysfunction. Mol Genet Metab Apr 2012;105(4):542–50. https://doi.org/10.1016/j.ymgme.2011.12.012.
[90] Pohlmann R, Nagel G, Hille A, et al. Mannose 6-phosphate specific receptors—structure and function. Biochem Soc Trans Feb 1989;17(1):15–6. https://doi.org/10.1042/bst0170015.
[91] Vonfigura K, Hasilik A. Lysosomal-enzymes and their receptors. Annu Rev Biochem 1986;55:167–93. https://doi.org/10.1146/annurev.biochem.55.1.167.
[92] Brix K. Host cell proteases: cathepsins. In: Activation of Viruses by Host Proteases, vol. 16. Cham: Springer; 2018. p. 249–76. https://doi.org/10.1007/978-3-319-75474-1_10.
[93] Brix K, Dunkhorst A, Mayer K, Jordans S. Cysteine cathepsins: cellular roadmap to different functions. Biochimie Feb 2008;90(2):194–207. https://doi.org/10.1016/j.biochi.2007.07.024.
[94] Authier F, Kouach M, Briand G. Endosomal proteolysis of insulin-like growth factor-I at its C-terminal D-domain by cathepsin B. FEBS Lett Aug 2005;579(20):4309–16. https://doi.org/10.1016/j.febslet.2005.06.066.
[95] Massague J. TGF beta in cancer. Cell Jul 2008;134(2):215–30. https://doi.org/10.1016/j.cell.2008.07.001.
[96] Nakajima T, Nakamura H, Owen CA, et al. Plasma cathepsin S and cathepsin S/cystatin C ratios are potential biomarkers for COPD. Dis Markers 2016. https://doi.org/10.1155/2016/4093870, 20164093870.
[97] Pandey KC, De S, Mishra PK. Role of proteases in chronic obstructive pulmonary disease. Front Pharmacol Aug 2017. https://doi.org/10.3389/fphar.2017.00512, 8512.
[98] Nixon RA. The aging lysosome: an essential catalyst for late-onset neurodegenerative diseases. Biochim Biophys Acta Sep 2020;1868(9). https://doi.org/10.1016/j.bbapap.2020.140443, 140443.
[99] Stoka V, Turk V, Turk B. Lysosomal cathepsins and their regulation in aging and neurodegeneration. Ageing Res Rev Dec 2016;32:22–37. https://doi.org/10.1016/j.arr.2016.04.010.
[100] Kurz DJ, Decary S, Hong Y, Erusalimsky JD. Senescence-associated beta-galactosidase reflects an increase in lysosomal mass during replicative ageing of human endothelial cells. J Cell Sci Oct 2000;113(20):3613–22.
[101] Nakanishi H, Tominaga K, Amano T, Hirotsu I, Inoue T, Yamamoto K. Age-related-changes in activities and localizations of cathepsin-D, cathepsin-E, cathepsin-B, and cathepsin-L in the rat-brain tissues. Exp Neurol Mar 1994;126(1):119–28. https://doi.org/10.1006/exnr.1994.1048.
[102] Nakanishi H. Neuronal and microglial cathepsins in aging and age-related diseases. Ageing Res Rev Oct 2003;2(4):367–81. https://doi.org/10.1016/s1568-1637(03)00027-8.
[103] Cataldo AM, Nixon RA. Enzymatically active lysosomal proteases are associated with amyloid deposits in Alzheimer brain. Proc Natl Acad Sci U S A May 1990;87(10):3861–5. https://doi.org/10.1073/pnas.87.10.3861.
[104] Saftig P, Peters C, von Figura K, Craessaerts K, Van Leuven F, De Strooper B. Amyloidogenic processing of human amyloid precursor protein in hippocampal neurons devoid of cathepsin D. J Biol Chem Nov 1996;271(44):27241–4. https://doi.org/10.1074/jbc.271.44.27241.
[105] Nixon RA. The role of autophagy in neurodegenerative disease. Nat Med Aug 2013;19(8):983–97. https://doi.org/10.1038/nm.3232.
[106] Sun BG, Zhou YG, Halabisky B, et al. Cystatin C-cathepsin B axis regulates amyloid beta levels and associated neuronal deficits in an animal model of Alzheimer's disease. Neuron Oct 2008;60(2):247–57. https://doi.org/10.1016/j.neuron.2008.10.001.

[107] Hook G, Hook V, Kindy M. The cysteine protease inhibitor, E64d, reduces brain amyloid-beta and improves memory deficits in Alzheimer's disease animal models by inhibiting cathepsin B, but not BACE1, beta-secretase activity. J Alzheimer Dis 2011;26(2):387–408. https://doi.org/10.3233/jad-2011-110101.
[108] Sevlever D, Jiang PZ, Yen SHC. Cathepsin D is the main lysosomal enzyme involved in the degradation of alpha-synuclein and generation of its carboxy-terminally truncated species. Biochemistry Sep 2008;47(36):9678–87. https://doi.org/10.1021/bi800699v.
[109] Bae EJ, Yang NY, Lee C, Kim S, Lee HJ, Lee SJ. Haploinsufficiency of cathepsin D leads to lysosomal dysfunction and promotes cell-to-cell transmission of alpha-synuclein aggregates. Cell Death Dis Oct 2015;6e1901. https://doi.org/10.1038/cddis.2015.283.
[110] McGlinchey RP, Lee JC. Cysteine cathepsins are essential in lysosomal degradation of alpha-synuclein. Proc Natl Acad Sci U S A Jul 2015;112(30):9322–7. https://doi.org/10.1073/pnas.1500937112.
[111] Tsujimura A, Taguchi K, Watanabe Y, et al. Lysosomal enzyme cathepsin B enhances the aggregate forming activity of exogenous alpha-synuclein fibrils. Neurobiol Dis Jan 2015;73:244–53. https://doi.org/10.1016/j.nbd.2014.10.011.
[112] Nixon RA, Cataldo AM, Mathews PM. The endosomal-lysosomal system of neurons in Alzheimer's disease pathogenesis: a review. Neurochem Res Oct 2000;25(9–10):1161–72. https://doi.org/10.1023/a:1007675508413.
[113] Muller AL, Dhalla NS. Role of various proteases in cardiac remodeling and progression of heart failure. Heart Fail Rev May 2012;17(3):395–409. https://doi.org/10.1007/s10741-011-9269-8.
[114] Spinale FG. Bioactive peptide signaling within the myocardial interstitium and the matrix metalloproteinases. Circ Res Dec 2002;91(12):1082–4. https://doi.org/10.1161/01.res.0000047874.80576.5a.
[115] Cheng XW, Obata K, Kuzuya M, et al. Elastolytic cathepsin induction/activation system exists in myocardium and is upregulated in hypertensive heart failure. Hypertension Nov 2006;48(5):979–87. https://doi.org/10.1161/01.HYP.0000242331.99369.2f.
[116] Cheng XW, Kuzuya M, Sasaki T, et al. Increased expression of elastolytic cysteine proteases, cathepsins S and K, in the neointima of balloon-injured rat carotid arteries. Am J Pathol Jan 2004;164(1):243–51. https://doi.org/10.1016/s0002-9440(10)63114-8.
[117] Sukhova GK, Zhang Y, Pan JH, et al. Deficiency of cathepsin S reduces atherosclerosis in LDL receptor-deficient mice. J Clin Investig Mar 2003;111(6):897–906. https://doi.org/10.1172/jci200314915.
[118] Qin YW, Cao X, Guo J, et al. Deficiency of cathepsin S attenuates angiotensin II-induced abdominal aortic aneurysm formation in apolipoprotein E-deficient mice. Cardiovasc Res Dec 2012;96(3):401–10. https://doi.org/10.1093/cvr/cvs263.
[119] Sun M, Ouzounian M, de Couto G, et al. Cathepsin-L ameliorates cardiac hypertrophy through activation of the autophagy-lysosomal dependent protein processing pathways. J Am Heart Assoc Apr 2013;2(2). https://doi.org/10.1161/jaha.113.000191, e000191.
[120] Petermann I, Mayer C, Stypmann J, et al. Lysosomal, cytoskeletal, and metabolic alterations in cardiomyopathy of cathepsin L knockout mice. FASEB J Jun 2006;20(8):1266. https://doi.org/10.1096/fj.05-5517fje.
[121] Wang XJ, Robbins J. Proteasomal and lysosomal protein degradation and heart disease. J Mol Cell Cardiol Jun 2014;71:16–24. https://doi.org/10.1016/j.yjmcc.2013.11.006.
[122] Kostin S, Pool L, Elsasser A, et al. Myocytes die by multiple mechanisms in failing human hearts. Circ Res Apr 2003;92(7):715–24. https://doi.org/10.1161/01.res.0000067471.95890.5c.
[123] Kanamori H, Takemura G, Goto K, et al. The role of autophagy emerging in postinfarction cardiac remodelling. Cardiovasc Res Jul 2011;91(2):330–9. https://doi.org/10.1093/cvr/cvr073.
[124] Koike M, Nakanishi H, Saftig P, et al. Cathepsin D deficiency induces lysosomal storage with ceroid lipofuscin in mouse CNS neurons. J Neurosci Sep 2000;20(18):6898–906. https://doi.org/10.1523/jneurosci.20-18-06898.2000.
[125] Koike M, Shibata M, Waguri S, et al. Participation of autophagy in storage of lysosomes in neurons from mouse models of neuronal ceroid-lipofuscinoses (Batten disease). Am J Pathol Dec 2005;167(6):1713–28. https://doi.org/10.1016/s0002-9440(10)61253-9.

[126] Jedeszko C, Sloane BF. Cysteine cathepsins in human cancer. Biol Chem Nov 2004;385(11):1017–27. https://doi.org/10.1515/bc.2004.132.
[127] Wang B, Sun JS, Kitamoto S, et al. Cathepsin S controls angiogenesis and tumor growth via matrix-derived angiogenic factors. J Biol Chem Mar 2006;281(9):6020–9. https://doi.org/10.1074/jbc.M509134200.
[128] Gormley JA, Hegarty SM, O'Grady A, et al. The role of cathepsin S as a marker of prognosis and predictor of chemotherapy benefit in adjuvant CRC: a pilot study. Br J Cancer Nov 2011;105(10):1487–94. https://doi.org/10.1038/bjc.2011.408.
[129] Kos J, Sekirnik A, Kopitar G, et al. Cathepsin S in tumours, regional lymph nodes and sera of patients with lung cancer: relation to prognosis. Br J Cancer Oct 2001;85(8):1193–200. https://doi.org/10.1054/bjoc.2001.2057.
[130] Frosch BA, Berquin I, Emmert-Buck MR, Moin K, Sloane BF. Molecular regulation, membrane association and secretion of tumor cathepsin B. Apmis Jan 1999;107(1):28–37. https://doi.org/10.1111/j.1699-0463.1999.tb01523.x.
[131] Elie BT, Gocheva V, Shree T, Dalrymple SA, Holsinger LJ, Joyce JA. Identification and pre-clinical testing of a reversible cathepsin protease inhibitor reveals anti-tumor efficacy in a pancreatic cancer model. Biochimie Nov 2010;92(11):1618–24. https://doi.org/10.1016/j.biochi.2010.04.023.
[132] Bell-McGuinn KM, Garfall AL, Bogyo M, Hanahan D, Joyce JA. Inhibition of cysteine cathepsin protease activity enhances chemotherapy regimens by decreasing tumor growth and invasiveness in a mouse model of multistage cancer. Cancer Res Aug 2007;67(15):7378–85. https://doi.org/10.1158/0008-5472.can-07-0602.
[133] Lee DM, Weinblatt ME. Rheumatoid arthritis. Lancet Sep 2001;358(9285):903–11. https://doi.org/10.1016/s0140-6736(01)06075-5.
[134] Scott DL, Wolfe F, Huizinga TWJ. Rheumatoid arthritis. Lancet Sep-Oct 2010;376(9746):1094–108. https://doi.org/10.1016/s0140-6736(10)60826-4.
[135] Vomero M, Barbati C, Colasanti T, et al. Autophagy and rheumatoid arthritis: current knowledges and future perspectives. Front Immunol Jul 2018. https://doi.org/10.3389/fimmu.2018.01577, 91577.
[136] Yue Y, Yin WW, Yang Q, et al. Inhibition of cathepsin K alleviates autophagy-related inflammation in periodontitis-aggravating arthritis. Infect Immun Dec 2020;88(12):e00498–520. https://doi.org/10.1128/iai.00498-20.
[137] Pan WY, Yin WW, Yang L, et al. Inhibition of Ctsk alleviates periodontitis and comorbid rheumatoid arthritis via downregulation of the TLR9 signalling pathway. J Clin Periodontol Mar 2019;46(3):286–96. https://doi.org/10.1111/jcpe.13060.
[138] Lim JS, Kim HS, Nguyen KCT, Cho KA. The role of TLR9 in stress-dependent autophagy formation. Biochem Biophys Res Commun Dec 2016;481(3–4):219–26. https://doi.org/10.1016/j.bbrc.2016.10.105.

CHAPTER 4

Antigen processing and presentation through MHC molecules

Tâmisa Seeko Bandeira Honda[a,*], Barbara Nunes Padovani[a,*], and Niels Olsen Saraiva Câmara[a,b]

[a]*Department of Immunology, Institute of Biomedical Sciences, University of São Paulo, São Paulo, Brazil,* [b]*Nephrology Division, Department of Medicine, Federal University of São Paulo, São Paulo, Brazil*

Introduction

The adaptive immune response can be triggered by the recognition of protein antigens from transformed cells, pathogens, and in some cases autoantigens [1, 2]. The proteolytic cleavage of these antigens let them to be loaded into their respective class I and II major histocompatibility complex (MHC), triggering specific responses of $CD8^+$ and $CD4^+$ T cells [3, 4], respectively.

The existence of MHC-like proteins in some cell types (B cells, dendritic cells, and thymic cells) allows the anchoring of molecules with hydrophobic characteristics inside of the biding groove [5], which assume a narrow and deep pockets' conformation [6]. The processing and presentation of lipids also take place inside the acidic compartments, where glycolytic enzymes, lipases (e.g., lysosomal phospholipase A2), and also the CD1-presenting molecules are present [7].

This chapter will focus on the pathways involved in the processing and presentation of protein antigens. The class I and II MHC have similar structural characteristics, which will be described in more detail in the course of this chapter. In addition, it is through the recognition of the MHC-peptide complex (pMHC) that naive T cells become effector cells, differentiate, and proliferate. However, the activation of T cell requires a second signal, such as recognition of costimulatory molecules and cytokine secretion by antigen-presenting cells (APCs) [8, 9].

At the time of presentation, the density of T-cell receptors (TCR) and the aggregation of costimulatory molecules inside the lipid rafts let the formation of immunologic synapse and, consequently, the activation of tyrosine-based immunoreceptors (ITAMs), which are present in the structural domains ζ of the chains of the CD3 coreceptor [10–12]. The activation of

[*] All the authors contribute equally.

Proteolytic Signaling in Health and Disease. https://doi.org/10.1016/B978-0-323-85696-6.00005-6

ITAMs triggers a signal transduction, resulting in the transcription of genes involved in T-cell proliferation, cytokine secretion, cytoskeleton rearrangement, and metabolic changes [13, 14].

The presentation process ends with the activation of several cellular proteases that cleave intracellular and extracellular proteins into peptides, making them eligible of being transported to the MHC groove [15]. The presence of these peptidases in different cell compartments (cytosol, endoplasmic reticulum, endosome, lysosomes, and phagosome) and their different cleavage patterns guarantees the diversity in the generation of epitopes presented to T cells [16–18]. Moreover, these different antigenic epitopes contribute to the generation of different clones of T cells.

In addition to the processing of peptides, the presentation of antigens depends on the structural characteristics of the MHC binding groove, determined by the genetic polymorphisms observed in the genes encoding class I and class II MHC [19].

The chromosomal region responsible for encoding information related to the MHC's immune role is located at 6p21. This super locus extends for about 4 Mb, being responsible for ~ 0.13% of whole genome [20]. The magnitude of this genetic region and its codominant pattern contribute to the high genetic variability observed in different individuals.

According to IPD-IMGT/HLA (IMmunoGeneTics HLA, accessed in January 2021), there are about 28,000 different alleles at the HLA locus, of which 20,597 are related to MHC I and 7,723 to MHC II. The most polymorphic gene associated with HLA class I in ascending order is HLA-B with 7.5×10^3 different alleles, followed by HLA-A (6.29×10^3), HLA-C (6.22×10^3), HLA-E (256), HLA-G (82), and HLA-F (45). The same occurs with HLA class II in which the most polymorphic gene related is HLA-DRB (3.5×10^3), HLA-DQB1 (1.9×10^3), HLA-DPB1 (1.6×10^3), and HLA-DQA1 (264). These variations and the pattern of expression can be associated with different outcomes for viral infections and transplantation [20].

Therefore, the factors that contribute to greater efficiency in the antigen processing and presentation will be discussed in more detail throughout this chapter. In addition, the main proteases and their cleavage mechanisms will be also emphasized, as well as clinical defects arising from changes in the processing pathways associated with class I and II MHC.

Processing and presentation through class I MHC pathway

Class I MHC is a protein complex expressed on the surface of all nucleated cells containing a heavy chain α [21], which is divided into three domains (α1, α2, and α3). The α1–α2 domains are organized in structures of α helices overlaying an eight antiparallel β-sheet strands, forming a groove [22]. Such conformation permits the formation of a binding groove, which is closed by two tyrosine residues, thus allowing the binding of only small peptides with 8–10 amino acids (aa) [23]. The α3 domain, together with α2, is able to bind to CD8 coreceptor and activate T cells [23, 24].

The presence of the immunoglobulin-like domain (Ig) allows the formation of a noncovalent association of beta-2 microglobulin (β2m) with the heavy chain. This complex is anchored in the plasma membrane by the presence of α transmembrane helices. The presence of β2m is crucial for the structural maintenance of the class I MHC complex [4].

The binding and anchoring of intracellular peptides in class I MHC molecules occur only after proteolytic cleavage and correct conformational stabilization. The first step in the correct processing of cytoplasmic antigens is the cleavage by the proteasome [25]. Structurally, the 26S proteasome is formed by two regulatory portions (RP) 19S, which are located at the ends of the 20S cleavage core (CP), thus generating a cylindrical structure (20S) with two caps (19S), responsible for the peptide preprocessing and generation of long oligopeptides [26].

The 19S RP is responsible for the recognition of polyubiquitinated proteins. The presence of three deubiquitylating enzymes [27], USP14, UCHL5, and PSMD14 [28], allows the proteasome to remove ubiquitin (Ub) residues, making the peptides unfolded and suitable for proteolytic cleavage by the core 20S. However, DUBs have an ambiguous function: remove the Ub to address the peptides for processing, whereas make substrates more stable through the removal of Ub from Lys residues, promoting its accumulation in the cell [29, 30]. As a consequence, several studies have shown that DUBs are pharmacological targets for the treatment of certain types of cancer (e.g., esophagus and lung cancer [31, 32]) or infectious diseases (e.g., HIV [27]).

After the removal of Ub, the peptides are oriented and directed into the core 20S, where they will be cleaved, generating oligopeptides. The core 20S consists of fourteen subunits, seven of which are α and seven β subunits [33–35]. The presence of 3β catalytic subunits (β1, β2, and β5), located inside the CP, allows the peptides to be hydrolyzed in a proteolytic pattern similar to that of caspase, trypsin, and chymotrypsin [36, 37]. Inflammatory stimuli, such as IFNγ and TNF-α, have been described for inducing an immune adaptation of 20S [38, 39], stimulating the expression of the iβ1, iβ2, and iβ5 subunits and the assembly of the immunoproteasome [38, 40]. On the other hand, it was observed that chronic inflammation, observed in murine EAE models, reduces the catalytic activity of the β1 and β5 subunits resulting in the accumulation of carbonylated proteins [37].

After being released by the proteasome, part of the peptides still has the N-terminal or the wrong length, which make it impossible binding to class I MHC groove [41]. Thus, many of these oligopeptides are trimmered by cytosolic peptidases [42], even before they are translocated to the endoplasmic reticulum [43, 44]. However, it is highlighted that during trimming several peptides are completely destroyed, being used as a source of aa recycling.

Leucine aminopeptidase (LAP) is a cytosolic peptidase that cleaves leucine residues at the N-terminal of peptides from the proteasome. Several studies indicate that LAP is regulated by IFN-γ, so that exposure to this cytokine increases peptide trimming [45]. However, it was noted that the absence of LAP did not reduce the generation of mature peptides, suggesting that other aminopeptidases, such as puromycin-sensitive aminopeptidases (PSA) and bleomycin hydrolase (BH) [46], can play redundant functions. In addition, changes in the expression of PSA and BH have been shown to regulate the expression of class I MHC on cell surface as well as the quality of peptides bound to it [47]. Moreover, in the cytoplasm, some peptides can be processed independently of the proteasome [43, 48, 49]. The main protagonist of this process is tripeptidyl peptidase II (TPPII) [50], found strategically close to the proteasome [51]. Functionally, TPPII seems to prefer long peptides, in which it performs two distinct catalytic functions: (1) cleavage of tripeptides close to the N-terminal and (2) generation of long peptides with nine or more amino acids (aa) [52].

Similarly, to TPPII, caspases 5 and 10 are considered to be capable of cleaving proteins up to 19 aa, independently of the proteasome [53, 54]. Insulin-degrading enzyme (IDE) is another cytosolic peptidase able to generate peptides of different composition (sequence of aa) than those produced by the proteasome, as well as participating in protein turnover during cellular metabolism [55]. Equivalently, other proteases (*thimet oligopeptidase* and *nardilysin*) act in parallel to proteasome [56] in order to generate a great diversity of epitopes, of the same protein, presented by class I MHC.

After the cytoplasmic proteolytic cascade is completed, the preprocessed peptides interact with TAP [57] and are translocated to the ER. However, TAP is part of a multiprotein complex, the PLC [58], which is formed by TAP, tapasin, ERp57, and the canexin-calnexin system [59]. In this complex, empty class I MHC heterodimers are recruited by the calnexin–calreticulin system and maintained in the correct conformation by the tapasin chaperone and the ERp57 thiol oxidoreductase [60].

When assembled, this complex allows peptides to pass through the ER lumen, thus being able to interact with endopeptidases [59], aminopeptidase associated with antigen processing (ERAAP), or its analogous in human ER-associated aminopeptidase 1 (ERAP1) and ER-associated aminopeptidase 2 (ERAP2). However, some studies show that ERAAP can edit peptides already linked to class I MCH [61, 62]. Functionally, the ER's aminopeptidases trimmer the N-terminal end of peptides, generating peptides with 8–10 aa able to fit and stabilize the class I MHC [63, 64]. The importance of peptidases found in the ER was demonstrated by a series of studies that showed that the knockdown for ER proteases reduces up to 75% the presentation of cytoplasmic antigens [64–67].

Thus, a large number of proteases capable of cleaving the same substrate can be seen as an essential mechanism for maintaining immune surveillance, so that the redundancy of proteases is necessary for the generation of a great diversity of epitopes from the same peptide.

Class II MHC-associated pathway

Unlike class I MHC, which is expressed in all nucleated cells, class II MHC is expressed basically in professional APCs, such as macrophages, dendritic cells, and B cells, and can also be found in the thymic epithelial cells. Occasionally, in the face of inflammatory stimuli, endothelial and epithelial cells can also express class II MHC [68, 69].

Structurally, class II MHC is formed by two heavy chains that are divided into two α domains (α1 and α2) and two β domains (β1 and β2). The three-dimensional structures that form the binding groove are similar to those observed in the class I MHC, but are organized in an open way, which allows the binding and anchoring of peptides with 13–25 aa residues. The addition of invariant chain (Ii) by class II-associated invariant chain peptide (CLIP) stabilizes the class II MHC, in order to prevent the αβ formation, newly formed in the ER, from being occupied by ER peptides. When the αβIi complex is formed, it is directed to the endocytic pathway, where CLIP remotion is catalyzed by HLA-DM and by endosomal proteases (cathepsins S and L) giving space for the binding and anchoring of extracellular antigenic peptides [70]. The presence of HLA-DM, which is negatively regulated by HLA-DO [71], is related to its ability to stabilize empty class II MHC molecules and also to favor the binding of high-affinity peptides [72].

Class II MHC presents extracellular antigens, which were picked up mainly through pinocytosis, macropinocytosis, or receptor-mediated endocytosis. Once inside the cells, vesicles with external content are processed by acid proteases present in the early endosomes (pH 5.5–6.8). The preprocessed content in the early endosomes is transferred to the late endosome (pH 5.0–5.5), as the products from phagocytosis [73].

Among the enzymes that participate in the endosome pathway are Cys (AEP, asparaginyl endopeptidases, and cathepsins B, F, H, L, S, and Z), aspartate (cathepsins D and E), and serine proteases (cathepsins A and G). However, due to redundant activities [74], different tissue locations, asymmetric distribution among APCs, and preference for certain protein antigens, it is difficult to determine the exact role of these endosomal hydrolases.

Among the group of Cys proteases, the role of CatS [75] stands out, because of its action in the cleavage of Ii in the αβIi complex [76, 77]. The cleavage of Ii is carried out in several stages [78], resulting in the formation of the intermediate lip22, which is cleaved at position 115–125, generating its degradation product Iip10 and, finally, CLIP. The removal of the Iip10 portion of the MHC-Ii complex is a function of CatS [79, 80], as such in the absence of CatS is observed an accumulation of lip22 and αβ-Iip10 dimers is observed [81].

However, in a study developed by Bania et al. [77] it was observed that even in the absence of CatS, Iip10 was generated, reiterating the existence of other proteases capable of performing the same function.

In addition to the classic roles played by EAP Cys proteases, recent studies show that during neurodegenerative processes there is a change in the proteolytic and spatial pattern of EAP [79, 80], which translocates from the lysosomal compartment to the cytoplasm [82], where cleaves the tau protein and α-synuclein, contributing to the maintenance of the degenerative state [83].

The aspartate proteases represented by CatD and CatE are proteases that have a high degree of homology but occupy different cell compartments. CatE is located in the early endosomes [84], where it appears to play a role in the proteolytic regulation of lysosomal glycoproteins LAMP-1 and LAMP-2. Therefore, the absence of CatE in the macrophages caused an increase in the expression of these glycoproteins and lysosomal pH (6.4 ± 0.3) [85], which influences the ability to degrade products of bacterial origin [86]. In addition, the same group showed that the knockout of CatE is related to lesser infiltration of macrophages in adipose tissue and, consequently, less accumulation of it [87].

Unlike CatE, CatD is found in late endosomes. However, its role in endosomal proteolysis is not well understood. Interestingly, Saftig et al. [88] showed that knockout for CatD causes defects during development stage (such as atrophy of the ileum mucosa that progresses to intestinal necrosis). Such observations may be due to the nonregulation of proteolysis of certain growth factors, necessary for tissue homeostasis, due to the absence of CatD in the endosomal compartments.

CatA is a carboxypeptidase of the serine protease family that, in addition to being located in lysosomes, can be secreted or found in the plasma membrane. Several studies, using a cardiac injury model, have shown that this protease participates in a tissue remodeling process [89, 90]. When located in lysosomes, CatA is able to cleave bioactive peptides such as angiotensin I, endothelin, and bradykinin [91]. Through the use of mice with CatA catalytic inactivity, less degradation of endothelin-1 was observed, which resulted in an increase in blood pressure [92].

The other member of serine proteases is CatG reported to be upregulated in PBMCs of patients with type 1 diabetes (T1D), when compared to the same cells of control subjects. Thus, CatG appears to act in the proteolytic degradation of proinsulin in T1D, activation of $CD4^+$ T cells, and secretion of IFN-γ, IL-17, and IL-22 [93]. In addition, recent studies show that CatG plays an important role in the recognition and inhibition of tumor growth [94], as well as seems to contribute to the worsening of lesions associated with autoimmune diseases [95, 96].

Last but not least, GILT (gamma-interferon-inducible lysosomal thiol reductase) is a protease located in endosomes, lysosomes, and phagosomes. Its main function is to keep the disulfide bridges in the captured proteins reduced, in order to make them accessible for processing by the proteases described above [97, 98].

As mentioned so far, proteases that participate in the processing of extracellular antigens have their functions restricted to compartments with controlled acid pH. Due to the sharing of restriction sites, these proteases play redundant roles, which make it difficult to understand their isolated functions, but on the other hand, show the need for the processes performed by them.

Cross presentation

The cross presentation consists of the process of presenting proteins external to the cell by class I MHC pathway. However, until now, this process is restricted only to subtypes of macrophages and dendritic cells [99]. This process allows the immune system to be able to generate $CD8^+$ T cells against intracellular bacteria or pathogens that remain in phagocytes, in the same way that it allows the generation of a $CD8^+$ T-dependent immune response after vaccination [97, 98].

In order for the peptides captured in the phagosomes to have access to the class I MHC binding groove, the peptides must follow two distinct pathways, which involve the participation of different proteases.

The phagosome-cytosol pathway begins when the peptides are found in endosomes or phagosomes. In this compartment, the pH will be high, in order to avoid the activation of proteases and the destruction of the peptides. Then, they are transported via endosomal TAP to the proteasome, where they will be cleaved and later translocated to the ER, as previously described, or returned to the endosomes. If they go to the ER, these peptides will be trimmed by the ERPA, described above. But, if they return to the endosomes, they will be processed by the endosomal aminopeptidase, IRAP [100], which has the same function as the ERAP. Still in endosomes, these peptides can be loaded to class I MHC, newly synthesized in the ER and transported to the endosome, or class I MHC already loaded that have been endocytosed and will have their peptides processed in this vesicle.

The other possible cross-presentation pathway for extracellular peptides is through vacuolar processing mediated by CatS. Studies show that other Cys proteases, such as CatB and CatL, are not involved in this processing pathway [101]. The exact reason why only CatS operates in vacuole-dependent cross presentation is not fully known, but it is believed that the ability to remain active at neutral pH is one of the reasons [101, 102], since acidic microenvironments do not allow the correct assembly of the class I MHC [103].

Thus, the signaling pathways associated with the processing and presentation of antigens, as well as the signals that trigger the expression of class I and II MHC, will be discussed in more detail in the course of this chapter (Fig. 1).

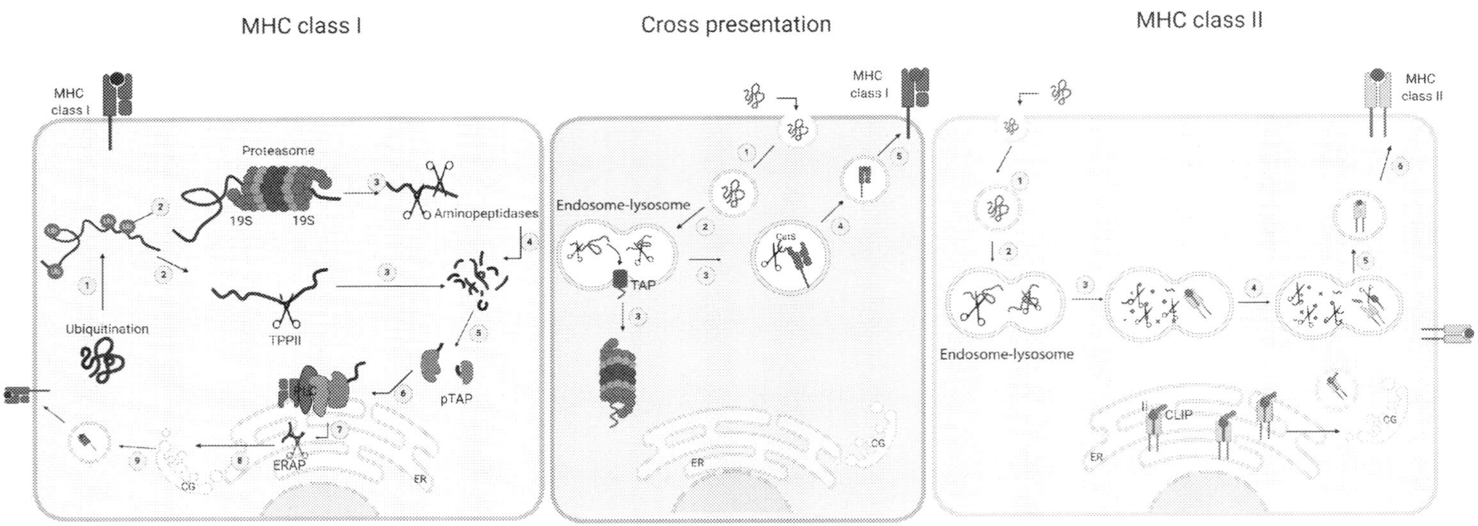

Fig. 1 Pathways related to class I MHC, class II MHC, and cross presentation. Class I MHC pathway: The endogenous antigen is polyubiquitinated (1), processed by the proteasome and aminopeptidases (2/3), and translocated to the endoplasmic reticulum (ER) by TAP (4/5), where it binds to the free class I MHC (6/7/8). Cross-presentation pathway: When the exogenous antigen is endocitized (1/2), it can escape the endosome and be processed by the proteasome (3), following the pathway related to class I MHC (3), or it can follow the vacuolar pathway being bound to recycled class I MHC (4/5). Class II MHC: The exogenous antigen is endocitized (1/2), degraded in the phagolysosome by Cys proteases (3/4). After removing the CLIP (5), the class II MHC bind groove. Created with BioRender.com.

Protease dysregulation during disease

From the information addressed so far, we have been able to understand the importance of MHC molecules and proteases during the processing and antigen presentation. In addition, it has been shown that these play an important role in the regulation of various cellular processes, among which are cell cycle control, transcription, and aging [104–106]. However, it is worth mentioning that incorrect regulation of these signaling pathways may result in disruption of cellular homeostasis and contribute to many pathologies, such as in the progression of inflammatory diseases, various types of cancers, cardiovascular diseases, and neurodegenerative disorders [107–109].

In humans, a defective regulation of HLA molecules can result in several clinical manifestations, one of which is the immunodeficiency disorder known as bare lymphocyte syndrome (BLS). This rare syndrome is characterized by the overall decrease of HLA molecules, implying a faulty self-recognition and nonself-recognition, and consequently an inefficient immune response [110, 111]. Depending on which molecule is being negatively regulated, HLA-I or HLA-II, the BLS can be categorized as type I or II [110].

The development of type I BLS can occur from mutations in the subunits that make up TAP (TAP 1/TAP 2) since the functions performed by this complex are essences for HLA-I processing and expression of the antigen-HLA complex on the cell surface [57]. Usually in TAP1-deficient patients, this syndrome is started from the generation of a premature stop codon [112]. However, Yabe et al. [58] showed that genetic changes in the tapasin molecule can also lead to a diagnosis of BLS. On the other hand, type II BLS also known as MHC class II deficiency is a primary autosomal recessive immunodeficiency disorder which may be caused by defects at least four different trans-acting regulatory genes (CIITA, RFX5, RFXAP, and RFXANK) that are required for transcription of MHC class II [58]. In both cases, the absence of MHC molecules results in greater susceptibility to several infections, especially in the respiratory tract [112, 113].

Neurodegenerative diseases

The proteases, in special the cathepsins, participate in different physiological processes, and one of them is MHC class II-mediated antigen presentation. However, their functions need to be highly regulated since proteases irreversibly cleave peptide bonds. Studies show that active cathepsin when erroneously localized in other cell compartments may result in the progression of neurodegenerative disorders [114–116]. In addition, a lysosomal proteolytic capacity decline is associated with some proteinopathies such as Alzheimer's, Parkinson's, and Huntington's diseases. López-Otín et al. included loss of proteostasis among the nine hallmarks of aging. In fact, the impaired proteostasis is related to aging and the development of diseases linked to this process [117, 118].

Cathepsin B has been related to both Alzheimer's disease and traumatic brain injury given that lysosomal leakage into the cytosol is related to the inflammation, cell death, and injury reported in the neurodegeneration of these diseases [119, 120]. Because of this, CatB has become a drug target candidate since studies show that the use of chemical inhibitors of this lysosomal cysteine protease, such as E64d, CA-074Me, and E64c, may reduce the effects of neuropathology [119, 121, 122].

On the other hand, the cathepsin D is highly expressed in the brain and plays an important role to neuronal cell homeostasis through in the degradation of defective and unfolded proteins [123]. It has been shown that high levels of CatD are connected to decreased α-synuclein aggregation and toxicity in animal models of Parkinson's disease [107, 124]. A study developed by Siintola et al. observed that individuals with the rare disease, congenital neuronal ceroid-lipofuscinosis disease, have little or almost no expression of CatD in brain tissues. In addition, this same study related the presence of lysosomal storage to the homozygous mutation in the CatD gene [125].

Cardiovascular alterations

Studies have also shown the relationship between proteases and cardiovascular diseases [126, 127]. Patients with chronic heart failure had an increase in CatK in the circulation [128]. This collagenase can act on cardiac remodeling, and Hua et al. showed that CatK knockout attenuates pressure overload-induced cardiac hypertrophy and contractile dysfunction [129].

LAMP-2 is important for MHC II-mediated presentation of exogenous antigens; however, mutations in the coding sequence of LAMP-2 may result in Danon disease, a multisystem glycogen-storage disease characterized by hypertrophic cardiomyopathy, myopathy, and intellectual disability [130–132]. Tanaka et al. showed that LAMP-2-deficient mice have reduced the contractile function of heart muscle; in addition, accumulation of autophagic vacuoles in liver, kidney, pancreas, and cardiac and skeletal muscle was also observed [127].

Cancer

The reduction in the expression of class I MHC molecules on the surface of cancer cells allows them to escape surveillance by the immune system [133, 134]. Indeed, this maneuver can be used as predictive for the identification of tumor resistant to immunotherapies [133, 135].

Several studies have shown that during the development of malignant tumors there is a reduction in the expression of HLA-DM [136, 137], which suggest a greater expression of MHC class II-CLIP complex, instead of pMHC II. The presence of CLIP in mature class II MHC molecules had already been associated with poor cancer prognosis [138].

Another route that has been extensively studied with a possible form of intervention for the treatment of cancer focuses on the role of the proteasome [139]. Seminal studies have observed, in cancer cell lines, the overexpression of the proteasome or its subunits can be considered a mechanism of a drug resistance [140–142]. This observation can be explained by the "addiction" of cancer cell by a proteotoxic situation and increase in the protein homeostasis burden [143]. Based on this, a recent study showed the increased expression of the 26S proteasome in a cancer cell line. However, this increase was accompanied by a greater sensitivity to apoptosis associated with 26S inhibition [144].

Other pathological processes

Deregulation of cathepsins is not only linked to neurodegenerative disorders. The cysteine cathepsins also play an important role in the pathophysiology of lung diseases, e.g., patients with cystic fibrosis showed elevated levels of cathepsins B, S, and L in the lung [145]. CatS is an essential enzyme in the processing and presentation of MHC class II antigens. However, it also has the role of cleavage of important proteins such as surfactant protein A, lactoferrin, LL-37 and β-defensins that help in the defense against pathogens [75, 146, 147]. Therefore, the high levels of this protease make individuals with lung diseases more susceptible to infections. Proteases may also be involved in the development of cancers as they may be associated with tumor growth [148]. The cathepsin K possibly plays a role in bone metastasis formation since it is a critical enzyme in bone resorption by osteoclasts. Le Gall et al. show that the use of cathepsin K inhibitors reduces skeletal tumor burden and osteolysis in animal models of breast cancer bone metastasis [149].

All these data indicate that the deregulation of proteases can cause several pathologies. Based on this, many of them, especially cathepsins, have become targets for therapeutic strategies for their role in maintaining homeostasis.

Conclusions

The adaptive immune response initiated by T cells can only be performed after they recognize the antigens presented. The antigen presentation depends on their mechanism of cell entrance. In the case of antigen from transformed cells, the participation of the 20S proteasome and cytoplasmic proteases, as well as those associated with ER, stands out. In other hand, in the case of extracellular antigens, there are the participation of endosomal and lysosomal proteases, mainly cathepsin and its isoforms.

Currently, several groups have shown that proteases involved in antigen processing can also be associated with the aging-related process, such as neurodegenerative and cardiovascular diseases. One reason for that correlations is the reduction of protease expression during the aging. In this way, the understanding of the mechanism of proteases involved in antigenic processing can emerge as a way of predicting the mechanisms of cell edition, as well as manipulating the use of macromolecules for therapeutic purposes.

Abbreviations used in this chapter

MHC (major histocompatibility complex), pMHC (MHC-peptide complex), ITAMs (tyrosine-based immunoreceptors), HLA (human leukocyte antigen), HC (heavy chain), RP (regulatory portion), CP (cleavage core), DUBs (deubiquitylating enzymes), LAP (leucine aminopeptidases), PSA (puromycin-sensitive aminopeptidases), BH (bleomycin hydrolase), TPPII (tripeptidyl peptidase II), IDE (insulin-degrading enzyme), ER (endoplasmic reticulum), TAP (tapasin), PLC (phospholipase C), ERAP1/ERAP2 (ER-associated aminopeptidase 1/2), ERAAP (aminopeptidase associated with antigen processing), APC (antigen-presenting cell), CLIP (class II-associated invariant chain peptide), Ii (invariant chain), Cat (cathepsin), AEP (asparaginyl endopeptidases), Cys (cysteine), LAMP-1/2 (lysosomal-associated membrane protein 1/2), T1D (type 1 diabetes), GILT (gamma-interferon-inducible lysosomal thiol reductase), IRAP (insulin-regulated aminopeptidase).

References

[1] Lukasch B, Westerdahl H, Strandh M, et al. Genes of the major histocompatibility complex highlight interactions of the innate and adaptive immune system. PeerJ 2017;5, e3679.

[2] Thakur A, Mikkelsen H, Jungersen G. Intracellular pathogens: host immunity and microbial persistence strategies. J Immunol Res 2019;2019:1356540.

[3] Holling TM, Schooten E, van Den Elsen PJ. Function and regulation of MHC class II molecules in T-lymphocytes: of mice and men. Hum Immunol 2004;65(4):282–90.

[4] Wieczorek M, Abualrous ET, Sticht J, et al. Major histocompatibility complex (MHC) class I and MHC class II proteins: conformational plasticity in antigen presentation. Front Immunol 2017;8:292.

[5] Vartabedian VF, Savage PB, Teyton L. The processing and presentation of lipids and glycolipids to the immune system. Immunol Rev 2016;272(1):109–19.

[6] Zeng Z, Castaño AR, Segelke BW, Stura EA, Peterson PA, Wilson IA. Crystal structure of mouse CD1: an MHC-like fold with a large hydrophobic binding groove. Science (New York, NY) 1997;277(5324):339–45.

[7] Briken V, Jackman RM, Watts GF, Rogers RA, Porcelli SA. Human CD1b and CD1c isoforms survey different intracellular compartments for the presentation of microbial lipid antigens. J Exp Med 2000;192(2):281–8.

[8] Li XC, Raghavan M. Structure and function of major histocompatibility complex class I antigens. Curr Opin Organ Transplant 2010;15(4):499–504.

[9] Min B, Spontaneous T. Cell proliferation: a physiologic process to create and maintain homeostatic balance and diversity of the immune system. Front Immunol 2018;9:547.

[10] Love PE, Hayes SM. ITAM-mediated signaling by the T-cell antigen receptor. Cold Spring Harb Perspect Biol 2010;2(6):a002485.

[11] Courtney AH, Lo WL, Weiss A. TCR signaling: mechanisms of initiation and propagation. Trends Biochem Sci 2018;43(2):108–23.

[12] Pennock ND, White JT, Cross EW, Cheney EE, Tamburini BA, Kedl RM. T cell responses: naive to memory and everything in between. Adv Physiol Educ 2013;37(4):273–83.

[13] Hwang S, Palin AC, Li L, et al. TCR ITAM multiplicity is required for the generation of follicular helper T-cells. Nat Commun 2015;6:6982.

[14] Cochran JR, Cameron TO, Stern LJ. The relationship of MHC-peptide binding and T cell activation probed using chemically defined MHC class II oligomers. Immunity 2000;12(3):241–50.
[15] Manoury B. Proteases: essential actors in processing antigens and intracellular toll-like receptors. Front Immunol 2013;4:299.
[16] Riese RJ, Wolf PR, Brömme D, et al. Essential role for cathepsin S in MHC class II-associated invariant chain processing and peptide loading. Immunity 1996;4(4):357–66.
[17] Weimershaus M, Evnouchidou I, Saveanu L, van Endert P. Peptidases trimming MHC class I ligands. Curr Opin Immunol 2013;25(1):90–6.
[18] Saveanu L, Fruci D, van Endert P. Beyond the proteasome: trimming, degradation and generation of MHC class I ligands by auxiliary proteases. Mol Immunol 2002;39(3–4):203–15.
[19] Beck S, Trowsdale J. The human major histocompatability complex: lessons from the DNA sequence. Annu Rev Genomics Hum Genet 2000;1:117–37.
[20] Shiina T, Hosomichi K, Inoko H, Kulski JK. The HLA genomic loci map: expression, interaction, diversity and disease. J Hum Genet 2009;54(1):15–39.
[21] Noble S, Godoy R, Affaticati P, Ekker M. Transgenic zebrafish expressing mCherry in the mitochondria of dopaminergic neurons. Zebrafish 2015;12(5):349–56.
[22] Zacharias M, Springer S. Conformational flexibility of the MHC class I alpha1-alpha2 domain in peptide bound and free states: a molecular dynamics simulation study. Biophys J 2004;87(4):2203–14.
[23] Collins EJ, Garboczi DN, Karpusas MN, Wiley DC. The three-dimensional structure of a class I major histocompatibility complex molecule missing the alpha 3 domain of the heavy chain. Proc Natl Acad Sci U S A 1995;92(4):1218–21.
[24] Pittet MJ, Rubio-Godoy V, Bioley G, et al. Alpha 3 domain mutants of peptide/MHC class I multimers allow the selective isolation of high avidity tumor-reactive CD8 T cells. J Immunol 2003;171(4):1844–9.
[25] Eggers M, Boes-Fabian B, Ruppert T, Kloetzel PM, Koszinowski UH. The cleavage preference of the proteasome governs the yield of antigenic peptides. J Exp Med 1995;182(6):1865–70.
[26] Livneh I, Cohen-Kaplan V, Cohen-Rosenzweig C, Avni N, Ciechanover A. The life cycle of the 26S proteasome: from birth, through regulation and function, and onto its death. Cell Res 2016;26(8):869–85.
[27] Setz C, Friedrich M, Rauch P, et al. Inhibitors of deubiquitinating enzymes block HIV-1 replication and augment the presentation of Gag-derived MHC-I epitopes. Viruses 2017;9(8):1–22.
[28] Komander D, Clague MJ, Urbé S. Breaking the chains: structure and function of the deubiquitinases. Nat Rev Mol Cell Biol 2009;10(8):550–63.
[29] Li J, Yakushi T, Parlati F, et al. Capzimin is a potent and specific inhibitor of proteasome isopeptidase Rpn11. Nat Chem Biol 2017;13(5):486–93.
[30] Liao Y, Liu N, Hua X, et al. Proteasome-associated deubiquitinase ubiquitin-specific protease 14 regulates prostate cancer proliferation by deubiquitinating and stabilizing androgen receptor. Cell Death Dis 2017;8(2), e2585.
[31] Wei R, Liu X, Yu W, et al. Deubiquitinases in cancer. Oncotarget 2015;6(15):12872–89.
[32] Young MJ, Hsu KC, Lin TE, Chang WC, Hung JJ. The role of ubiquitin-specific peptidases in cancer progression. J Biomed Sci 2019;26(1):42.
[33] Groll M, Ditzel L, Löwe J, et al. Structure of 20S proteasome from yeast at 2.4 A resolution. Nature 1997;386(6624):463–71.
[34] Latham MP, Sekhar A, Kay LE. Understanding the mechanism of proteasome 20S core particle gating. Proc Natl Acad Sci U S A 2014;111(15):5532–7.
[35] Löwe J, Stock D, Jap B, Zwickl P, Baumeister W, Huber R. Crystal structure of the 20S proteasome from the archaeon T. acidophilum at 3.4 A resolution. Science (New York, NY) 1995;268(5210):533–9.
[36] Coux O, Tanaka K, Goldberg AL. Structure and functions of the 20S and 26S proteasomes. Annu Rev Biochem 1996;65:801–47.
[37] Zheng J, Bizzozero OA. Reduced proteasomal activity contributes to the accumulation of carbonylated proteins in chronic experimental autoimmune encephalomyelitis. J Neurochem 2010;115(6):1556–67.
[38] Heink S, Ludwig D, Kloetzel PM, Krüger E. IFN-gamma-induced immune adaptation of the proteasome system is an accelerated and transient response. Proc Natl Acad Sci U S A 2005;102(26):9241–6.

[39] Zheng J, Dasgupta A, Bizzozero OA. Changes in 20S subunit composition are largely responsible for altered proteasomal activities in experimental autoimmune encephalomyelitis. J Neurochem 2012;121(3):486–94.
[40] Moritz KE, McCormack NM, Abera MB, et al. The role of the immunoproteasome in interferon-γ-mediated microglial activation. Sci Rep 2017;7(1):9365.
[41] Kisselev AF, Akopian TN, Woo KM, Goldberg AL. The sizes of peptides generated from protein by mammalian 26 and 20 S proteasomes. Implications for understanding the degradative mechanism and antigen presentation. J Biol Chem 1999;274(6):3363–71.
[42] Rock KL, York IA, Goldberg AL. Post-proteasomal antigen processing for major histocompatibility complex class I presentation. Nat Immunol 2004;5(7):670–7.
[43] Oliveira CC, van Hall T. Alternative antigen processing for MHC class I: multiple roads lead to Rome. Front Immunol 2015;6:298.
[44] York IA, Mo AX, Lemerise K, et al. The cytosolic endopeptidase, thimet oligopeptidase, destroys antigenic peptides and limits the extent of MHC class I antigen presentation. Immunity 2003;18(3):429–40.
[45] Beninga J, Rock KL, Goldberg AL. Interferon-gamma can stimulate post-proteasomal trimming of the N terminus of an antigenic peptide by inducing leucine aminopeptidase. J Biol Chem 1998;273(30):18734–42.
[46] Towne CF, York IA, Neijssen J, et al. Leucine aminopeptidase is not essential for trimming peptides in the cytosol or generating epitopes for MHC class I antigen presentation. J Immunol 2005;175(10):6605–14.
[47] Kim E, Kwak H, Ahn K. Cytosolic aminopeptidases influence MHC class I-mediated antigen presentation in an allele-dependent manner. J Immunol 2009;183(11):7379–87.
[48] Glas R, Bogyo M, McMaster JS, Gaczynska M, Ploegh HL. A proteolytic system that compensates for loss of proteasome function. Nature 1998;392(6676):618–22.
[49] Seifert U, Marañón C, Shmueli A, et al. An essential role for tripeptidyl peptidase in the generation of an MHC class I epitope. Nat Immunol 2003;4(4):375–9.
[50] Guil S, Rodríguez-Castro M, Aguilar F, Villasevil EM, Antón LC, Del Val M. Need for tripeptidyl-peptidase II in major histocompatibility complex class I viral antigen processing when proteasomes are detrimental. J Biol Chem 2006;281(52):39925–34.
[51] Fukuda Y, Beck F, Plitzko JM, Baumeister W. In situ structural studies of tripeptidyl peptidase II (TPPII) reveal spatial association with proteasomes. Proc Natl Acad Sci U S A 2017;114(17):4412–7.
[52] Reits E, Neijssen J, Herberts C, et al. A major role for TPPII in trimming proteasomal degradation products for MHC class I antigen presentation. Immunity 2004;20(4):495–506.
[53] López D, García-Calvo M, Smith GL, Del Val M. Caspases in virus-infected cells contribute to recognition by CD8+ T lymphocytes. J Immunol 2010;184(9):5193–9.
[54] López D, Jiménez M, García-Calvo M, Del Val M. Concerted antigen processing of a short viral antigen by human caspase-5 and -10. J Biol Chem 2011;286(19):16910–3.
[55] Parmentier N, Stroobant V, Colau D, et al. Production of an antigenic peptide by insulin-degrading enzyme. Nat Immunol 2010;11(5):449–54.
[56] Kessler JH, Khan S, Seifert U, et al. Antigen processing by nardilysin and thimet oligopeptidase generates cytotoxic T cell epitopes. Nat Immunol 2011;12(1):45–53.
[57] Law-Ping-Man S, Toutain F, Rieux-Laucat F, et al. Chronic granulomatous skin lesions leading to a diagnosis of TAP1 deficiency syndrome. Pediatr Dermatol 2018;35(6):e375–7.
[58] Yabe T, Kawamura S, Sato M, et al. A subject with a novel type I bare lymphocyte syndrome has tapasin deficiency due to deletion of 4 exons by Alu-mediated recombination. Blood 2002;100(4):1496–8.
[59] Blees A, Januliene D, Hofmann T, et al. Structure of the human MHC-I peptide-loading complex. Nature 2017;551(7681):525–8.
[60] Fisette O, Schröder GF, Schäfer LV. Atomistic structure and dynamics of the human MHC-I peptide-loading complex. Proc Natl Acad Sci U S A 2020;117(34):20597–606.
[61] Chen H, Li L, Weimershaus M, Evnouchidou I, van Endert P, Bouvier M. ERAP1-ERAP2 dimers trim MHC I-bound precursor peptides; implications for understanding peptide editing. Sci Rep 2016;6:28902.

[62] Infantes S, Samino Y, Lorente E, et al. H-2Ld class I molecule protects an HIV N-extended epitope from in vitro trimming by endoplasmic reticulum aminopeptidase associated with antigen processing. J Immunol 2010;184(7):3351–5.
[63] Hammer GE, Gonzalez F, Champsaur M, Cado D, Shastri N. The aminopeptidase ERAAP shapes the peptide repertoire displayed by major histocompatibility complex class I molecules. Nat Immunol 2006;7(1):103–12.
[64] Serwold T, Gonzalez F, Kim J, Jacob R, Shastri N. ERAAP customizes peptides for MHC class I molecules in the endoplasmic reticulum. Nature 2002;419(6906):480–3.
[65] Hammer GE, Gonzalez F, James E, Nolla H, Shastri N. In the absence of aminopeptidase ERAAP, MHC class I molecules present many unstable and highly immunogenic peptides. Nat Immunol 2007;8(1):101–8.
[66] Saveanu L, Carroll O, Lindo V, et al. Concerted peptide trimming by human ERAP1 and ERAP2 aminopeptidase complexes in the endoplasmic reticulum. Nat Immunol 2005;6(7):689–97.
[67] Blanchard N, Kanaseki T, Escobar H, et al. Endoplasmic reticulum aminopeptidase associated with antigen processing defines the composition and structure of MHC class I peptide repertoire in normal and virus-infected cells. J Immunol 2010;184(6):3033–42.
[68] Koyama M, Mukhopadhyay P, Schuster IS, et al. MHC class II antigen presentation by the intestinal epithelium initiates graft-versus-host disease and is influenced by the microbiota. Immunity 2019;51(5):885–98. e887.
[69] Scott NA, Zhao Y, Krishnamurthy B, Mannering SI, Kay TWH, Thomas HE. IFNγ-induced MHC class II expression on islet endothelial cells is an early marker of insulitis but is not required for diabetogenic CD4(+) T cell migration. Front Immunol 2018;9:2800.
[70] Roche PA, Cresswell P. Invariant chain association with HLA-DR molecules inhibits immunogenic peptide binding. Nature 1990;345(6276):615–8.
[71] Jiang W, Strohman MJ, Somasundaram S, et al. pH-susceptibility of HLA-DO tunes DO/DM ratios to regulate HLA-DM catalytic activity. Sci Rep 2015;5:17333.
[72] Reyes-Vargas E, Barker AP, Zhou Z, He X, Jensen PE. HLA-DM catalytically enhances peptide dissociation by sensing peptide-MHC class II interactions throughout the peptide-binding cleft. J Biol Chem 2020;295(10):2959–73.
[73] Tjelle TE, Saigal B, Froystad M, Berg T. Degradation of phagosomal components in late endocytic organelles. J Cell Sci 1998;111(Pt 1):141–8.
[74] Deussing J, Roth W, Saftig P, Peters C, Ploegh HL, Villadangos JA. Cathepsins B and D are dispensable for major histocompatibility complex class II-mediated antigen presentation. Proc Natl Acad Sci U S A 1998;95(8):4516–21.
[75] Small DM, Brown RR, Doherty DF, et al. Targeting of cathepsin S reduces cystic fibrosis-like lung disease. Eur Respir J 2019;53(3):1–11.
[76] Kim H, Mazumdar B, Bose SK, et al. Hepatitis C virus-mediated inhibition of cathepsin S increases invariant-chain expression on hepatocyte surface. J Virol 2012;86(18):9919–28.
[77] Bania J, Gatti E, Lelouard H, et al. Human cathepsin S, but not cathepsin L, degrades efficiently MHC class II-associated invariant chain in nonprofessional APCs. Proc Natl Acad Sci U S A 2003;100(11):6664–9.
[78] Villadangos JA, Ploegh HL. Proteolysis in MHC class II antigen presentation: who's in charge? Immunity 2000;12(3):233–9.
[79] Nakagawa TY, Brissette WH, Lira PD, et al. Impaired invariant chain degradation and antigen presentation and diminished collagen-induced arthritis in cathepsin S null mice. Immunity 1999;10(2):207–17.
[80] Shi GP, Villadangos JA, Dranoff G, et al. Cathepsin S required for normal MHC class II peptide loading and germinal center development. Immunity 1999;10(2):197–206.
[81] Costantino CM, Ploegh HL, Hafler DA. Cathepsin S regulates class II MHC processing in human CD4+ HLA-DR+ T cells. J Immunol 2009;183(2):945–52.
[82] Wang ZH, Wu W, Kang SS, et al. BDNF inhibits neurodegenerative disease-associated asparaginyl endopeptidase activity via phosphorylation by AKT. JCI Insight 2018;3(16):1–21.

[83] Zhang Z, Kang SS, Liu X, et al. Asparagine endopeptidase cleaves α-synuclein and mediates pathologic activities in Parkinson's disease. Nat Struct Mol Biol 2017;24(8):632–42.
[84] Chain BM, Free P, Medd P, Swetman C, Tabor AB, Terrazzini N. The expression and function of cathepsin E in dendritic cells. J Immunol 2005;174(4):1791–800.
[85] Yanagawa M, Tsukuba T, Nishioku T, et al. Cathepsin E deficiency induces a novel form of lysosomal storage disorder showing the accumulation of lysosomal membrane sialoglycoproteins and the elevation of lysosomal pH in macrophages. J Biol Chem 2007;282(3):1851–62.
[86] Tsukuba T, Yamamoto S, Yanagawa M, et al. Cathepsin E-deficient mice show increased susceptibility to bacterial infection associated with the decreased expression of multiple cell surface Toll-like receptors. J Biochem 2006;140(1):57–66.
[87] Kadowaki T, Kido MA, Hatakeyama J, Okamoto K, Tsukuba T, Yamamoto K. Defective adipose tissue development associated with hepatomegaly in cathepsin E-deficient mice fed a high-fat diet. Biochem Biophys Res Commun 2014;446(1):212–7.
[88] Saftig P, Hetman M, Schmahl W, et al. Mice deficient for the lysosomal proteinase cathepsin D exhibit progressive atrophy of the intestinal mucosa and profound destruction of lymphoid cells. EMBO J 1995;14(15):3599–608.
[89] Hohl M, Erb K, Lang L, et al. Cathepsin A mediates ventricular remote remodeling and atrial cardiomyopathy in rats with ventricular ischemia/reperfusion. JACC Basic Transl Sci 2019;4(3):332–44.
[90] Petrera A, Gassenhuber J, Ruf S, et al. Cathepsin A inhibition attenuates myocardial infarction-induced heart failure on the functional and proteomic levels. J Transl Med 2016;14(1):153.
[91] Timur ZK, Akyildiz Demir S, Seyrantepe V. Lysosomal cathepsin A plays a significant role in the processing of endogenous bioactive peptides. Front Mol Biosci 2016;3:68.
[92] Seyrantepe V, Hinek A, Peng J, et al. Enzymatic activity of lysosomal carboxypeptidase (cathepsin) A is required for proper elastic fiber formation and inactivation of endothelin-1. Circulation 2008;117(15):1973–81.
[93] Zou F, Schäfer N, Palesch D, et al. Regulation of cathepsin G reduces the activation of proinsulin-reactive T cells from type 1 diabetes patients. PLoS One 2011;6(8), e22815.
[94] Sionov RV, Fainsod-Levi T, Zelter T, Polyansky L, Pham CT, Granot Z. Neutrophil cathepsin G and tumor cell RAGE facilitate neutrophil anti-tumor cytotoxicity. Oncoimmunology 2019;8(9), e1624129.
[95] Guo J, Tu J, Hu Y, Song G, Yin Z. Cathepsin G cleaves and activates IL-36γ and promotes the inflammation of psoriasis. Drug Des Devel Ther 2019;13:581–8.
[96] Huang S, Thomsson KA, Jin C, et al. Cathepsin g degrades both glycosylated and unglycosylated regions of lubricin, a synovial mucin. Sci Rep 2020;10(1):4215.
[97] Embgenbroich M, Burgdorf S. Current concepts of antigen cross-presentation. Front Immunol 2018;9:1643.
[98] Rock KL, Farfán-Arribas DJ, Shen L. Proteases in MHC class I presentation and cross-presentation. J Immunol 2010;184(1):9–15.
[99] Rock KL, Shen L. Cross-presentation: underlying mechanisms and role in immune surveillance. Immunol Rev 2005;207:166–83.
[100] Saveanu L, Carroll O, Weimershaus M, et al. IRAP identifies an endosomal compartment required for MHC class I cross-presentation. Science (New York, NY) 2009;325(5937):213–7.
[101] Shen L, Sigal LJ, Boes M, Rock KL. Important role of cathepsin S in generating peptides for TAP-independent MHC class I crosspresentation in vivo. Immunity 2004;21(2):155–65.
[102] Vasiljeva O, Dolinar M, Pungercar JR, Turk V, Turk B. Recombinant human procathepsin S is capable of autocatalytic processing at neutral pH in the presence of glycosaminoglycans. FEBS Lett 2005;579(5):1285–90.
[103] Reich Z, Altman JD, Boniface JJ, et al. Stability of empty and peptide-loaded class II major histocompatibility complex molecules at neutral and endosomal pH: comparison to class I proteins. Proc Natl Acad Sci U S A 1997;94(6):2495–500.
[104] Kito Y, Matsumoto M, Hatano A, et al. Cell cycle-dependent localization of the proteasome to chromatin. Sci Rep 2020;10(1):5801.

[105] Chapman HA. Cathepsins as transcriptional activators? Dev Cell 2004;6(5):610–1.
[106] Chondrogianni N, Gonos ES. Proteasome function determines cellular homeostasis and the rate of aging. Adv Exp Med Biol 2010;694:38–46.
[107] Stoka V, Turk V, Turk B. Lysosomal cathepsins and their regulation in aging and neurodegeneration. Ageing Res Rev 2016;32:22–37.
[108] Hua Y, Nair S. Proteases in cardiometabolic diseases: pathophysiology, molecular mechanisms and clinical applications. Biochim Biophys Acta 2015;1852(2):195–208.
[109] Eatemadi A, Aiyelabegan HT, Negahdari B, et al. Role of protease and protease inhibitors in cancer pathogenesis and treatment. Biomed Pharmacother 2017;86:221–31.
[110] Shrestha D, Szöllosi J, Jenei A. Bare lymphocyte syndrome: an opportunity to discover our immune system. Immunol Lett 2012;141(2):147–57.
[111] von Boehmer H, Kisielow P. Self-nonself discrimination by T cells. Science 1990;248(4961):1369–73.
[112] Zimmer J, Andrès E, Donato L, Hanau D, Hentges F, de la Salle H. Clinical and immunological aspects of HLA class I deficiency. QJM 2005;98(10):719–27.
[113] Hanna S, Etzioni A. MHC class I and II deficiencies. J Allergy Clin Immunol 2014;134(2):269–75.
[114] Aufschnaiter A, Kohler V, Büttner S. Taking out the garbage: cathepsin D and calcineurin in neurodegeneration. Neural Regen Res 2017;12(11):1776–9.
[115] Santos-Rosa H, Kirmizis A, Nelson C, et al. Histone H3 tail clipping regulates gene expression. Nat Struct Mol Biol 2009;16(1):17–22.
[116] Duncan EM, Muratore-Schroeder TL, Cook RG, et al. Cathepsin L proteolytically processes histone H3 during mouse embryonic stem cell differentiation. Cell 2008;135(2):284–94.
[117] López-Otín C, Blasco MA, Partridge L, Serrano M, Kroemer G. The hallmarks of aging. Cell 2013;153(6):1194–217.
[118] Koga H, Kaushik S, Cuervo AM. Protein homeostasis and aging: the importance of exquisite quality control. Ageing Res Rev 2011;10(2):205–15.
[119] Hook V, Yoon M, Mosier C, et al. Cathepsin B in neurodegeneration of Alzheimer's disease, traumatic brain injury, and related brain disorders. Biochim Biophys Acta Proteins Proteom 1868;2020(8):140428.
[120] Sun Y, Rong X, Lu W, et al. Translational study of Alzheimer's disease (AD) biomarkers from brain tissues in AβPP/PS1 mice and serum of AD patients. J Alzheimers Dis 2015;45(1):269–82.
[121] Hook GR, Yu J, Sipes N, Pierschbacher MD, Hook V, Kindy MS. The cysteine protease cathepsin B is a key drug target and cysteine protease inhibitors are potential therapeutics for traumatic brain injury. J Neurotrauma 2014;31(5):515–29.
[122] Luo CL, Chen XP, Yang R, et al. Cathepsin B contributes to traumatic brain injury-induced cell death through a mitochondria-mediated apoptotic pathway. J Neurosci Res 2010;88(13):2847–58.
[123] Vidoni C, Follo C, Savino M, Melone MA, Isidoro C. The role of cathepsin D in the pathogenesis of human neurodegenerative disorders. Med Res Rev 2016;36(5):845–70.
[124] Sevlever D, Jiang P, Yen SH. Cathepsin D is the main lysosomal enzyme involved in the degradation of alpha-synuclein and generation of its carboxy-terminally truncated species. Biochemistry 2008;47(36):9678–87.
[125] Siintola E, Partanen S, Strömme P, et al. Cathepsin D deficiency underlies congenital human neuronal ceroid-lipofuscinosis. Brain 2006;129(Pt 6):1438–45.
[126] Saftig P, Tanaka Y, Lüllmann-Rauch R, von Figura K. Disease model: LAMP-2 enlightens Danon disease. Trends Mol Med 2001;7(1):37–9.
[127] Tanaka Y, Guhde G, Suter A, et al. Accumulation of autophagic vacuoles and cardiomyopathy in LAMP-2-deficient mice. Nature 2000;406(6798):902–6.
[128] Zhao G, Li Y, Cui L, et al. Increased circulating cathepsin K in patients with chronic heart failure. PLoS One 2015;10(8), e0136093.
[129] Hua Y, Xu X, Shi GP, Chicco AJ, Ren J, Nair S. Cathepsin K knockout alleviates pressure overload-induced cardiac hypertrophy. Hypertension 2013;61(6):1184–92.

[130] Crotzer VL, Glosson N, Zhou D, Nishino I, Blum JS. LAMP-2-deficient human B cells exhibit altered MHC class II presentation of exogenous antigens. Immunology 2010;131(3):318–30.
[131] Endo Y, Furuta A, Nishino I. Danon disease: a phenotypic expression of LAMP-2 deficiency. Acta Neuropathol 2015;129(3):391–8.
[132] Samad F, Jain R, Jan MF, et al. Malignant cardiac phenotypic expression of Danon disease (LAMP2 cardiomyopathy). Int J Cardiol 2017;245:201–6.
[133] Cornel AM, Mimpen IL, Nierkens S. MHC class I downregulation in cancer: underlying mechanisms and potential targets for cancer immunotherapy. Cancers 2020;12(7):1–31.
[134] Morrison BJ, Steel JC, Morris JC. Reduction of MHC-I expression limits T-lymphocyte-mediated killing of cancer-initiating cells. BMC Cancer 2018;18(1):469.
[135] Gettinger S, Choi J, Hastings K, et al. Impaired HLA class I antigen processing and presentation as a mechanism of acquired resistance to immune checkpoint inhibitors in lung cancer. Cancer Discov 2017;7(12):1420–35.
[136] Callahan MJ, Nagymanyoki Z, Bonome T, et al. Increased HLA-DMB expression in the tumor epithelium is associated with increased CTL infiltration and improved prognosis in advanced-stage serous ovarian cancer. Clin Cancer Res 2008;14(23):7667–73.
[137] God JM, Cameron C, Figueroa J, et al. Elevation of c-MYC disrupts HLA class II-mediated immune recognition of human B cell tumors. J Immunol 2015;194(4):1434–45.
[138] Chamuleau ME, Souwer Y, Van Ham SM, et al. Class II-associated invariant chain peptide expression on myeloid leukemic blasts predicts poor clinical outcome. Cancer Res 2004;64(16):5546–50.
[139] Manasanch EE, Orlowski RZ. Proteasome inhibitors in cancer therapy. Nat Rev Clin Oncol 2017;14(7):417–33.
[140] Fuchs D, Berges C, Opelz G, Daniel V, Naujokat C. Increased expression and altered subunit composition of proteasomes induced by continuous proteasome inhibition establish apoptosis resistance and hyperproliferation of Burkitt lymphoma cells. J Cell Biochem 2008;103(1):270–83.
[141] Kuhn DJ, Berkova Z, Jones RJ, et al. Targeting the insulin-like growth factor-1 receptor to overcome bortezomib resistance in preclinical models of multiple myeloma. Blood 2012;120(16):3260–70.
[142] Rückrich T, Kraus M, Gogel J, et al. Characterization of the ubiquitin-proteasome system in bortezomib-adapted cells. Leukemia 2009;23(6):1098–105.
[143] Deshaies RJ. Proteotoxic crisis, the ubiquitin-proteasome system, and cancer therapy. BMC Biol 2014;12:94.
[144] Tsvetkov P, Adler J, Myers N, Biran A, Reuven N, Shaul Y. Oncogenic addiction to high 26S proteasome level. Cell Death Dis 2018;9(7):773.
[145] Taggart C, Mall MA, Lalmanach G, et al. Protean proteases: at the cutting edge of lung diseases. Eur Respir J 2017;49(2):1–12.
[146] Andrault PM, Samsonov SA, Weber G, et al. Antimicrobial peptide LL-37 is both a substrate of cathepsins S and K and a selective inhibitor of cathepsin L. Biochemistry 2015;54(17):2785–98.
[147] Lecaille F, Naudin C, Sage J, et al. Specific cleavage of the lung surfactant protein A by human cathepsin S may impair its antibacterial properties. Int J Biochem Cell Biol 2013;45(8):1701–9.
[148] Mason SD, Joyce JA. Proteolytic networks in cancer. Trends Cell Biol 2011;21(4):228–37.
[149] Le Gall C, Bellahcène A, Bonnelye E, et al. A cathepsin K inhibitor reduces breast cancer induced osteolysis and skeletal tumor burden. Cancer Res 2007;67(20):9894–902.
[150] Sivitz WI, Yorek MA. Mitochondrial dysfunction in diabetes: from molecular mechanisms to functional significance and therapeutic opportunities. Antioxid Redox Signal 2010;12(4):537–77.

CHAPTER 5

Proteolytic processing in autophagy

João Agostinho Machado-Neto[a] and Andrei Leitão[b]

[a]Department of Pharmacology, Institute of Biomedical Sciences, University of São Paulo, São Paulo, Brazil, [b]Medicinal & Biological Chemistry Group, Institute of Chemistry of São Carlos, University of São Paulo, Brazil

Autophagy in health and disease

Autophagy is a well-conserved catabolic process that regulates cell homeostasis by eliminating damaged whole organelles or macromolecules, such as proteins and lipids. The success of the autophagic flow (molecular processes from the beginning of the autophagy to the degradation of the cargo) allows for the recycling of nutrients, making it possible to obtain basic building blocks (i.e., amino acids, lipids, sugars, and nucleotides). This cellular process was initially identified in yeast as an adaptive response to nutrient deprivation, hence its name derived from the Greek and which means "self-eating." Currently, it is well-established that autophagy is conserved in higher eukaryotes and that it has a much larger repertoire of cellular and physiological functions than those initially identified in yeast [1, 2]. The importance of autophagy has been evidenced through the analysis of the development of murine knockout models for the core autophagy-related genes, which reduces the ability to maintain the autophagic flow and results in impairment in the development of those animals [3]. In fact, in humans, the dysregulation of the autophagic flow has been associated with several diseases such as cancer, metabolic diseases, lung diseases, neurodegenerative diseases, aging, vascular diseases, and infectious diseases [4, 5].

According to the current knowledge, there are three main autophagic pathways: macroautophagy, microautophagy, and chaperone-mediated autophagy (CMA), which differ in the way in which the cargo is transferred to the lysosome. Macroautophagy comprises the formation of a double-membrane vesicular structure (autophagosome), which will fuse with the lysosome (autophagolysosome). Microautophagy results from invaginations in the lysosome membrane to deliver the cargo, whereas, in CMA, the cargo (proteins that contain the recognition sequence KFERQ or KFERQ-like motif) is delivered by chaperones to specific receptors in the lysosomes. Besides, other determinations for macroautophagy of specific organelles/structures have been described in the literature to designate the degradation of these elements, including mitophagy (mitochondria), ER-phagy

Proteolytic Signaling in Health and Disease. https://doi.org/10.1016/B978-0-323-85696-6.00011-1

(endoplasmic reticulum), lipophagy (lipid droplets), aggrephagy (protein aggregates), and xenophagy (invading microorganisms) [1, 4, 5].

Focusing on macroautophagy (hereinafter simply referred to as autophagy), this process is finely orchestrated by multiple proteins and a cell signaling apparatus that begins with the recognition of stressful stimuli such as nutrient deprivation, oxidative stress, hypoxia, protein aggregates, or toxic molecules and culminates in the cargo degradation. Stressful stimuli inactivate the mammalian (or mechanistic) target of rapamycin (mTOR) which consequently activates the UNC-51-like kinase 1 and 2 (ULK1/2, hereafter referred as ULK) allowing the formation of the ATG13/ULK/FIP200 complex, which in turn will participate in the recruitment of other ATG proteins and in the formation of the autophagosome. Phagophore formation is regulated by the beclin 1-class III PI3K complex (beclin 1/VPS34/ATG14L/p150), which can be negatively regulated by the antiapoptotic proteins of the B-cell lymphoma 2 (BCL2) family (i.e., BCL2 and BCL-XL). Then, the ATG5-ATG12 conjugation system is complexed with ATG16L1, which leads to the elongation of the phagophore. Finally, the LC3 conjugation system results in the conversion of the pro-LC3 (also called MAP1LC3), into LC3I and then into LC3II allowing the maturation of the phagophore in an autophagosome. During this process, other proteins, such as sequestosome 1 (SQSTM1/p62), participate in the sequestration of the cargo into the autophagosomes under formation [4, 6].

Initially, autophagy was described as a nonselective process for protein turnover, but currently, it has been shown that protein aggregates formed from the failure of degradation via the proteasome can be selectively degraded by autophagy [7].

Inducing-autophagy proteases

The proteolytic process is involved in the entire autophagic flow. In the early stages of autophagy, the ATG4 cysteine protease family (also known as autophagins) plays a central role in the processes of ATG8 mammalian orthologues (subfamilies LC3, GABARAP, and GATE16) and LC3 conjugation system, allowing the elongation of the phagophore under formation [8, 9]. LC3B is synthesized as a soluble and inactive protein (also named pro-LC3B), and through post-translational processing and the binding of a phosphatidylethanolamine (PE) molecule (also called lipidation), this protein (LC3BII) is able to be inserted into the autophagosome membrane under formation and to exercise its function in the biosynthesis of this complex structure. ATG4 is the protease responsible for cleaving a C-terminal region of LC3B and exposing the glycine residue that will be conjugated to PE. The lipidation of LC3B is reversible and also depends on ATG4, which is able to cleave the binding between LC3B and PE, providing the necessary dynamics for activation and recovery of LC3B during autophagic flow [6, 10].

In mammals, four orthologous proteases were identified to yeast's ATG4 (ATG4A, ATG4B, ATGA4C, and ATGA4D) that may differ in terms of substrate specificity, expression in different tissues, and location in cell compartments, which could provide spatiotemporal specificity, but yet little is known. On the other hand, an apparent functional redundancy of ATG4 proteins may have made it possible to generate knockout murine models for the genes that encode the different members of these autophagins [10].

Deregulation of ATG4 has been linked to human diseases, such as inflammatory diseases, infectious diseases, and cancer. For instance, mutations and genetic variants in the ATG4 genes have already been associated with inflammatory bowel diseases and granuloma formation in Crohn's disease [11, 12]. In addition, the increase in ATG4 activity was associated with the progression of infectious diseases such as HIV [13], hepatitis C [14], Chagas disease [15], and leishmaniasis [16]. The role of ATG4, as well as its potential as a therapeutic target, is still an active field of study and debate in cancer. ATG4B has been identified as a promising anticancer target, since this protease is highly expressed in different types of cancer and that preclinical study data initiate that ATG4B inhibition or the expression of its dominant-negative form ($ATG4B^{C74A}$) robustly inhibits the autophagic flow and viability of tumor cells [8, 17–20]. Similarly, some studies have indicated the participation of ATG4A, ATG4C, and ATG4B in the development, progression, or therapeutic response in several types of cancer [21–26].

In later stages of autophagy, the efficiency of the lysosome function is determinant for the effectiveness of the autophagic flow. Lysosomes are membrane-bound cytoplasmic organelles that contain a plethora of hydrolases in their interior, including peptidases, phosphatases, nucleases, glycosidases, proteases, and lipases, which digest the macromolecules and then provide basic unit blocks back to the cytoplasm through specific transporters [27]. Indeed, lysosomes are responsible for the degradation of long lived proteins and play a key role in proteostasis (i.e., regulation of cellular concentration, folding, interactions, and localization of the proteome) [28]. Focusing on the proteolytic process of lysosomes in the context of autophagy, cathepsins are essential for the degradation of the cargo. Lysosomal cathepsins are subdivisions into cysteine (B, C, F, H, L, K, O, S, W, and Z), aspartic (D and E), and serine (A and G) cathepsins, being cathepsins B, L, and D the most abundant and participating in the completion of the autophagic flow [29, 30]. Similar to what occurs with murine knockout models for some genes related to the early autophagy regulation, the knockout for cathepsins L and B results in abnormalities in the development of these animals in the postnatal period, and the ultrastructural analysis indicates that a reduction in autophagic flow is involved [31–35]. Another important aspect for cathepsins to be activated and to efficiently degrade their substrates is the acidification of lysosomes, so drugs that inhibit vacuolar-type H + ATPase (V-ATPase) (e.g., bafilomycin A1) also inhibit autophagic flow mediated by cathepsins [36, 37]. Aberrant cathepsin activity has been associated with a range of human diseases, including cancer, immune deregulation, osteoporosis, arthritis, cystic fibrosis, neurodegenerative diseases, and cardiovascular diseases [38, 39].

Cysteine cathepsin inhibitors and the autophagic process

Based on these premises, compounds that could modulate autophagy have been described in the literature against many targets, revised in He et al. [40]. Much less is known about cysteine protease inhibitors in the autophagy regulation process, but there is growing evidence supporting their utility and drug discovery potential.

Cysteine proteases present similar structural features, having a shallow binding pocket that is divided into subsites following the recognition of the side chains of the peptide-like substrate (Fig. 1A) [41]. Hence, substrate-like compounds known as peptidomimetics are designed to work as enzyme inhibitors. The affinity toward the cysteine proteases is improved when a reactive group (warhead) is added to the chemical structure, which can react with the catalytic cysteine into the binding pocket. Despite the potential druggability of cysteine proteases, cross-class activity is often a problem that must be faced when novel inhibitors are studied [45]. Therefore, selectivity is usually analyzed using a panel of cysteine proteases, even though it is hardly covered in the scientific literature. Here, some of the most potent compounds are addressed.

It is known that cysteine cathepsin inhibitors could trigger the autophagic response in the target cell. Hence, compounds were studied regarding the linkage of cathepsin inhibition and neuroblastoma autophagy using in vitro and in vivo models. Five known cysteine protease inhibitors were tested, and Fmoc-Tyr-Ala-CHN_2 (FYAD) was chosen for further assays (Fig. 1B). FYAD is a selective cathepsin L and B inhibitor, where the diazonium forms a stable covalent bond with the cysteine thiolate into the binding pocket working as an irreversible inhibitor [41]. Its cytotoxicity was evaluated against SK-N-SH neuroblastoma cell lines, with $IC_{50}=10\,\mu M$ after 72 h of incubation. LC-3-II level was increased in cells for all inhibitors, indicating that autophagy may be involved in the cell death process [46]. The in vivo xenograft subcutaneous mouse model with the SK-N-SH cells showed that FYAD was active, arresting tumor growth. Moreover, FYAD could inhibit the cathepsin activity up to 100% in vitro and 70% in the in vivo model.

Five known cysteine protease inhibitors were tested, and Fmoc-Tyr-Ala-CHN_2 (FYAD) was chosen for further assays (Fig. 1B). FYAD is a selective cathepsin L and B inhibitor, where the diazonium forms a stable covalent bond with the cysteine thiolate into the binding pocket working as an irreversible inhibitor [41]. Its cytotoxicity was evaluated against SK-N-SH neuroblastoma cell lines, with $IC_{50}=10\,\mu M$ after 72 h incubation. LC-3-II level was increased in cells for all inhibitors, indicating that autophagy may be involved in the cell death process [46]. The in vivo xenograft subcutaneous mouse model with the SK-N-SH cells showed that FYAD was active, arresting tumor growth. Moreover, FYAD could inhibit the cathepsin activity up to 100% in vitro and 70% in the in vivo model.

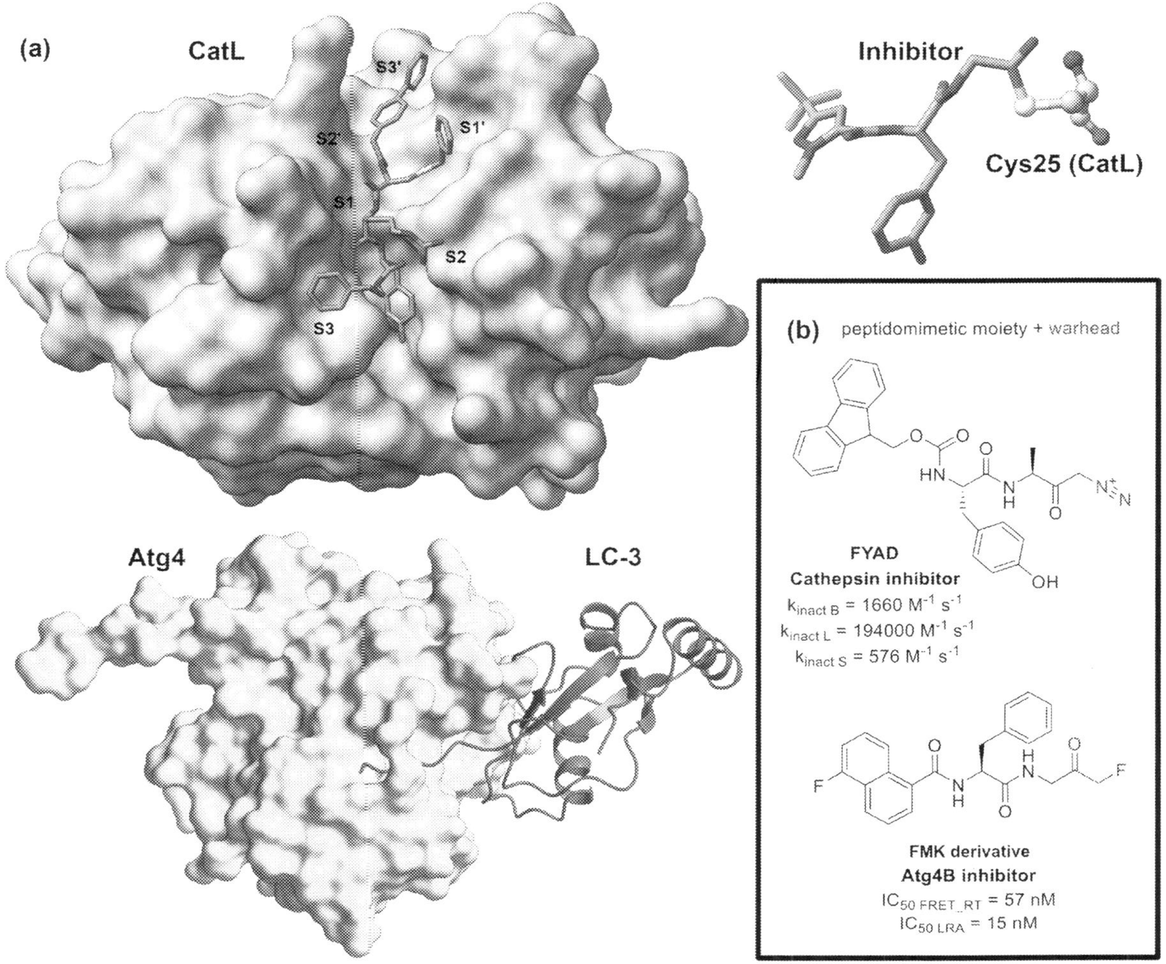

Fig. 1

Representation of the cysteine proteases (A) and chemical structures and potency of cysteine protease inhibitors (B) that modulate autophagy. (A, *Upper left*) Structure of a representative human cathepsin L (CatL) with a peptidomimetic ligand showing the subsites. (*Lower left*) LC-3 protein (*blue cartoon* (gray in the print version)) interacting with the human Atg4B protease (*green surface* (light in the print version), the catalytic site is shown in *yellow* (light in the print version)). (*Upper right*) Example of a nitrile-bound inhibitor (*stick*) with the catalytic cysteine residue (*ball and stick*) from the active site of CatL. (B) Two examples of highly potent CatL and Atg4 inhibitors. Data retrieved from Xiang et al. [41] and PDB codes for all structures: cathepsin L (3BC3 [42] and 3HHA) [43] and Atg4 (2ZZP) [44]. Protein crystallographic structures were analyzed in DeepViewer v. 4.1.0, and images were generated using UCSF ChimeraX v. 1.1. Ligands were drawn using ChemDraw v. 19.1. (For interpretation of the references to color in this figure legend, the reader is referred to the web version of this article.)

ATG4 inhibitors have also been described in the literature, and some compounds could reach nanomolar potency in biochemical assays, like the fluoromethylketone (FMK) derivatives (Fig. 1B). These ATG4 compounds also bind the active site of the enzyme impairing the LC-3 substrate to reach the catalytic site (Fig. 1A) in a similar way to the cathepsin inhibitors.

Inhibiting-autophagy proteases

Similar to the proteases that initiate and sustain the autophagic flow, there are proteases that may inhibit key stages of autophagy; for example, several ATG proteins are cleaved by caspases (cysteine aspartate proteases) [47]. It has been reported that the initiator (caspases 8, 9, and 10) or executioner (caspases 3, 6, and 7) caspases are able to cleave ATG3, ATG7, beclin 1, PI3KC3, ATG4D, AMBRA1, and SQSTM1/p62 [29, 48]. The proteolytic inactivation of autophagy-related proteins by caspases may generate a switch from autophagy to apoptosis. The cleavage of beclin 1 by caspase 3 generates a truncated beclin 1 fragment that causes mitochondrial outer membrane permeabilization and releases proapoptotic factors (i.e., cytochrome *c*) [49]. Besides, z-VAD-fmk-mediated caspases' inhibition not only reduces apoptosis but also restores autophagy in several experimental models [50].

Another group of proteases that act in the inhibition of autophagy is the calcium-dependent non-lysosomal cysteine protease, calpain 1 and calpain 2. Calpain 1 regulates the levels of AGT5, which impacts the formation of ATG5-ATG12 conjugates, and calpain 1 inhibition is enough to trigger autophagy [51]. Calpain 2 degrades ATG7 and beclin 1, which deregulates mitophagy and increases mitochondrial permeability [52]. Similar to what happens with caspases, the inhibition of autophagy mediated by calpains also may translocate the balance from autophagy to apoptosis [53].

Conclusion remarks and perspectives

Knowledge about the molecular mechanisms that govern the autophagy has been expanded considerably in recent years, as well as a better understanding of the role of proteases in this context. An overview of autophagy and the main proteolytic processes that induce and inhibit autophagic flow are illustrated in Fig. 2 and Table 1. The deregulation of autophagy is involved in human diseases, and several preclinical studies and clinical observations have been showing aberrant expression or activity of the involved proteases in the autophagic flow, which may create opportunities for therapeutic intervention. ATG4 inhibitors have been developed, such as N-ethylmaleimide, hypericin, aurin tricarboxylic acid, ZL-Phe-chloromethyl ketone, and fluoromethylketone (FMK) 9a, but there is still a lot of work to increase their selectivity, since these compounds may act on other cysteine proteases [54]. A better understanding of the distinct functions of the ATG4 isoforms may provide the identification of compounds with greater selectivity in inhibiting autophagy in specific

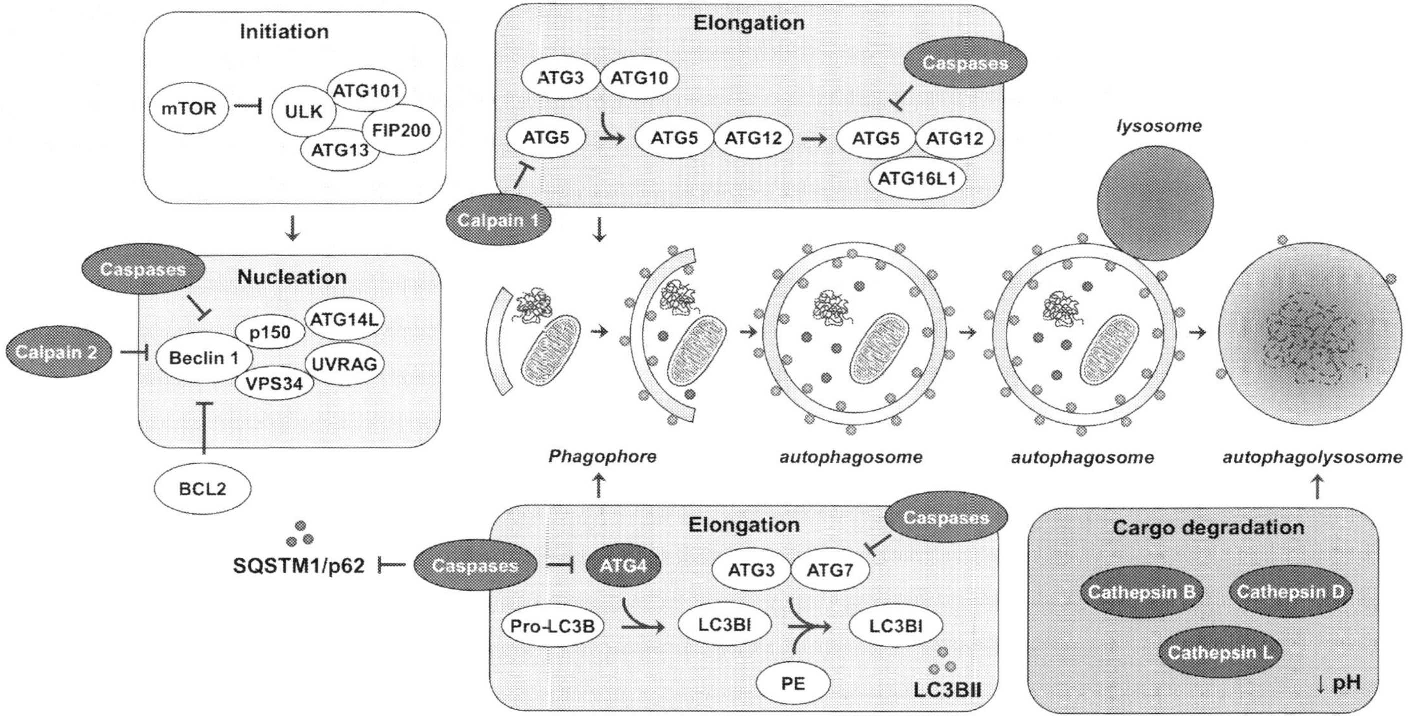

Fig. 2

Overview of proteolytic processing in autophagy. The mTORC1 complex (represented by mTOR) acts as a sensor and with a favorable balance of nutrients inhibits the ULK complex (ULK/ATG13/FIP200/ATG101). In the presence of stressors (e.g., nutrient deprivation, oxidative stress, hypoxia, protein aggregates, and toxic molecules), mTOR is inactivated and the ULK complex triggers the autophagic flow (initiation), phosphorylating targets that allow the release between beclin 1 and the antiapoptotic proteins of the BCL2 family (i.e., BCL2, BCL-XL). Then, beclin 1 forms a complex with p150 and VPS34 (class III PI3K [PI3K III] complex) and recruits the proteins ATG14L, UVRAG, and AMBRA1. The PI3K III complex participates in the regulation of proteins that will nucleate the phagophore (nucleation). The elongation process is mediated on two fronts, one through the formation of the ATG12/ATG5/ATG16L1 complex and the other through the generation of LC3BII from proteolytic processing of pro-LCB3 and lipidation of LCB3I. During the maturation process, the SQSTM1/p62 protein acts in the sequestration of the cargo into the closing phagophore. The complete autophagosome merges with the lysosome, and the cargo is degraded by lysosomal hydrolases. Among the main proteases involved in cargo degradation are cathepsins B, D, and L. The activity of caspases and calpain may negatively regulate the autophagic flow. For instance, caspases may cleave ATG3, ATG5, ATG7, beclin 1, PI3KC3, ATG4D, AMBRA1, and SQSTM1/p62, whereas calpain 1 cleaves ATG5 and calpain 2 cleaves beclin 1. Proteases that induce and inhibit the autophagic flow are highlighted in *green* (dark gray in the print version) and *red* (gray in the print version), respectively. (For interpretation of the references to color in this figure legend, the reader is referred to the web version of this article.)

Table 1: Effects of proteases in autophagy.

Protease	Role in autophagy	Substrate	Effects
ATG4	Induction	LC3B	Generation of LC3BI and LCB3BII recycling by delipidation [10, 29]
Lysosomal cathepsins[a]	Induction	Cargo in the autophagolysosome	Degradation of the cargo [29]
Caspase 8	Inhibition	ATG3, ATG5, beclin 1	Cleavage of substrates. Truncated beclin 1 fragment may induce MOMP [49, 61]
Caspase 9	Inhibition	ATG7, ATG5, beclin 1	Cleavage of substrates. Truncated beclin 1 fragment may induce MOMP [62, 63]
Caspase 10	Inhibition	ATG5, beclin 1	Cleavage of substrates. Truncated beclin 1 fragment may induce MOMP [62, 63]
Caspase 3	Inhibition	ATG5, beclin 1, AMBRA-1	Cleavage of substrates. AMBRA1 degradation reduces beclin 1/BCL2 dissociation. Truncated beclin 1 fragment may induce MOMP [49, 63–65]
Caspase 6	Inhibition	ATG3, ATG5, beclin 1, SQSTM1/p62	Cleavage of substrates. Truncated beclin 1 fragment may induce MOMP [47, 63, 66]
Caspase 7	Inhibition	ATG16L1, beclin 1	Cleavage of substrates. Truncated beclin 1 fragment may induce MOMP [49, 67]
Calpain 1	Inhibition	AGT5	Cleavage of the substrates [51]
Calpain 2	Inhibition	ATG7, beclin 1	Cleavage of substrates. Truncated beclin 1 fragment may induce MOMP [52]

Abbreviations: *ATG4*, autophagy-related 4 cysteine peptidase family; *MOMP*, mitochondrial outer membrane permeabilization; *ATG*, autophagy-related gene; *SQSTM1/p62*, sequestosome 1.
[a]Mainly cathepsins B, L, and D.

contexts or tissues. On the other hand, the selective pharmacological inhibitors of cathepsins are better elucidated and have already been tested in preclinical models [55–58]. In fact, inhibitors of lysosomal acidification, such as bafilomycin A1, which impacts the activity of cathepsins have proved to be a valuable experimental tool in the study of autophagy [59, 60].

In conclusion, a better understanding of how proteases that act under the autophagic flow impact the development and progression of human pathologies is an active and expanding research topic and the control of the proteolytic process in this context should be considered as a strategy therapy in the future studies.

References

[1] Mizushima N. A brief history of autophagy from cell biology to physiology and disease. Nat Cell Biol 2018 May;20(5):521–7.

[2] Levy JMM, Towers CG, Thorburn A. Targeting autophagy in cancer. Nat Rev Cancer 2017 Sep;17(9):528–42.

[3] Kuma A, Komatsu M, Mizushima N. Autophagy-monitoring and autophagy-deficient mice. Autophagy 2017 Oct 3;13(10):1619–28.

[4] Choi AM, Ryter SW, Levine B. Autophagy in human health and disease. N Engl J Med 2013 Feb 14;368(7):651–62.

[5] Saha S, Panigrahi DP, Patil S, Bhutia SK. Autophagy in health and disease: a comprehensive review. Biomed Pharmacother 2018 Aug;104:485–95.

[6] Folkerts H, Hilgendorf S, Vellenga E, Bremer E, Wiersma VR. The multifaceted role of autophagy in cancer and the microenvironment. Med Res Rev 2019 Mar;39(2):517–60.

[7] Khaminets A, Behl C, Dikic I. Ubiquitin-dependent and independent signals in selective autophagy. Trends Cell Biol 2016 Jan;26(1):6–16.

[8] Akin D, Wang SK, Habibzadegah-Tari P, Law B, Ostrov D, Li M, et al. A novel ATG4B antagonist inhibits autophagy and has a negative impact on osteosarcoma tumors. Autophagy 2014;10(11):2021–35.

[9] Shpilka T, Weidberg H, Pietrokovski S, Elazar Z. Atg8: an autophagy-related ubiquitin-like protein family. Genome Biol 2011 Jul 27;12(7):226.

[10] Fernandez AF, Lopez-Otin C. The functional and pathologic relevance of autophagy proteases. J Clin Invest 2015 Jan;125(1):33–41.

[11] Cho JH. The genetics and immunopathogenesis of inflammatory bowel disease. Nat Rev Immunol 2008 Jun;8(6):458–66.

[12] Brinar M, Vermeire S, Cleynen I, Lemmens B, Sagaert X, Henckaerts L, et al. Genetic variants in autophagy-related genes and granuloma formation in a cohort of surgically treated Crohn's disease patients. J Crohns Colitis 2012 Feb;6(1):43–50.

[13] Wang X, Gao Y, Tan J, Devadas K, Ragupathy V, Takeda K, et al. HIV-1 and HIV-2 infections induce autophagy in Jurkat and CD4+ T cells. Cell Signal 2012 Jul;24(7):1414–9.

[14] Dreux M, Gastaminza P, Wieland SF, Chisari FV. The autophagy machinery is required to initiate hepatitis C virus replication. Proc Natl Acad Sci U S A 2009 Aug 18;106(33):14046–51.

[15] Alvarez VE, Niemirowicz GT, Cazzulo JJ. The peptidases of Trypanosoma cruzi: digestive enzymes, virulence factors, and mediators of autophagy and programmed cell death. Biochim Biophys Acta 2012 Jan;1824(1):195–206.

[16] Williams RA, Mottram JC, Coombs GH. Distinct roles in autophagy and importance in infectivity of the two ATG4 cysteine peptidases of Leishmania major. J Biol Chem 2013 Feb 1;288(5):3678–90.

[17] Fu Y, Hong L, Xu J, Zhong G, Gu Q, Guan Y, et al. Discovery of a small molecule targeting autophagy via ATG4B inhibition and cell death of colorectal cancer cells in vitro and in vivo. Autophagy 2019 Feb;15(2):295–311.

[18] Fujita N, Hayashi-Nishino M, Fukumoto H, Omori H, Yamamoto A, Noda T, et al. An Atg4B mutant hampers the lipidation of LC3 paralogues and causes defects in autophagosome closure. Mol Biol Cell 2008 Nov;19(11):4651–9.

[19] Rothe K, Lin H, Lin KB, Leung A, Wang HM, Malekesmaeili M, et al. The core autophagy protein ATG4B is a potential biomarker and therapeutic target in CML stem/progenitor cells. Blood 2014 Jun 5;123(23):3622–34.

[20] Tran E, Chow A, Goda T, Wong A, Blakely K, Rocha M, et al. Context-dependent role of ATG4B as target for autophagy inhibition in prostate cancer therapy. Biochem Biophys Res Commun 2013 Nov 29;441(4):726–31.

[21] Wolf J, Dewi DL, Fredebohm J, Muller-Decker K, Flechtenmacher C, Hoheisel JD, et al. A mammosphere formation RNAi screen reveals that ATG4A promotes a breast cancer stem-like phenotype. Breast Cancer Res 2013 Nov 14;15(6):R109.

[22] Mao JJ, Wu LX, Wang W, Ye YY, Yang J, Chen H, et al. Nucleotide variation in ATG4A and susceptibility to cervical cancer in southwestern Chinese women. Oncol Lett 2018 Mar;15(3):2992–3000.

[23] He Q, Lu Y, Hu S, Huang Q, Li S, Huang Y, et al. An intron SNP rs807185 in ATG4A decreases the risk of lung cancer in a southwest Chinese population. Eur J Cancer Prev 2016 Jul;25(4):255–8.

[24] Antonelli M, Strappazzon F, Arisi I, Brandi R, D'Onofrio M, Sambucci M, et al. ATM kinase sustains breast cancer stem-like cells by promoting ATG4C expression and autophagy. Oncotarget 2017 Mar 28;8(13):21692–709.

[25] Betin VM, Lane JD. Caspase cleavage of Atg4D stimulates GABARAP-L1 processing and triggers mitochondrial targeting and apoptosis. J Cell Sci 2009 Jul 15;122(Pt 14):2554–66.

[26] Gil J, Ramsey D, Pawlowski P, Szmida E, Leszczynski P, Bebenek M, et al. The influence of tumor microenvironment on ATG4D gene expression in colorectal cancer patients. Med Oncol 2018 Oct 29;35(12):159.

[27] Trivedi PC, Bartlett JJ, Pulinilkunnil T. Lysosomal biology and function: modern view of cellular debris bin. Cell 2020 May;4:9(5).
[28] Jackson MP, Hewitt EW. Cellular proteostasis: degradation of misfolded proteins by lysosomes. Essays Biochem 2016 Oct 15;60(2):173–80.
[29] Kaminskyy V, Zhivotovsky B. Proteases in autophagy. Biochim Biophys Acta 2012 Jan;1824(1):44–50.
[30] Uchiyama Y. Autophagic cell death and its execution by lysosomal cathepsins. Arch Histol Cytol 2001 Aug;64(3):233–46.
[31] Saftig P, Hetman M, Schmahl W, Weber K, Heine L, Mossmann H, et al. Mice deficient for the lysosomal proteinase cathepsin D exhibit progressive atrophy of the intestinal mucosa and profound destruction of lymphoid cells. EMBO J 1995 Aug 1;14(15):3599–608.
[32] Koike M, Nakanishi H, Saftig P, Ezaki J, Isahara K, Ohsawa Y, et al. Cathepsin D deficiency induces lysosomal storage with ceroid lipofuscin in mouse CNS neurons. J Neurosci 2000 Sep 15;20(18):6898–906.
[33] Koike M, Shibata M, Waguri S, Yoshimura K, Tanida I, Kominami E, et al. Participation of autophagy in storage of lysosomes in neurons from mouse models of neuronal ceroid-lipofuscinoses (batten disease). Am J Pathol 2005 Dec;167(6):1713–28.
[34] Petermann I, Mayer C, Stypmann J, Biniossek ML, Tobin DJ, Engelen MA, et al. Lysosomal, cytoskeletal, and metabolic alterations in cardiomyopathy of cathepsin L knockout mice. FASEB J 2006 Jun;20(8):1266–8.
[35] Muller S, Dennemarker J, Reinheckel T. Specific functions of lysosomal proteases in endocytic and autophagic pathways. Biochim Biophys Acta 2012 Jan;1824(1):34–43.
[36] Yamamoto A, Tagawa Y, Yoshimori T, Moriyama Y, Masaki R, Tashiro Y. Bafilomycin A1 prevents maturation of autophagic vacuoles by inhibiting fusion between autophagosomes and lysosomes in rat hepatoma cell line, H-4-II-E cells. Cell Struct Funct 1998 Feb;23(1):33–42.
[37] Machado-Neto JA, Coelho-Silva JL, Santos FPS, Scheucher PS, Campregher PV, Hamerschlak N, et al. Autophagy inhibition potentiates ruxolitinib-induced apoptosis in JAK2(V617F) cells. Invest New Drugs 2020 Jun;38(3):733–45.
[38] Gocheva V, Joyce JA. Cysteine cathepsins and the cutting edge of cancer invasion. Cell Cycle 2007 Jan 1;6(1):60–4.
[39] Turk V, Stoka V, Vasiljeva O, Renko M, Sun T, Turk B, et al. Cysteine cathepsins: from structure, function and regulation to new frontiers. Biochim Biophys Acta 2012 Jan;1824(1):68–88.
[40] He S, Li Q, Jiang X, Lu X, Feng F, Qu W, et al. Design of small molecule autophagy modulators: a promising druggable strategy. J Med Chem 2018 Jun 14;61(11):4656–87.
[41] Xing R, Addington AK, Mason RW. Quantification of cathepsins B and L in cells. Biochem J 1998 Jun 1;332(Pt 2):499–505.
[42] Chowdhury SF, Joseph L, Kumar S, Tulsidas SR, Bhat S, Ziomek E, et al. Exploring inhibitor binding at the S' subsites of cathepsin L. J Med Chem 2008 Mar 13;51(5):1361–8.
[43] Asaad N, Bethel PA, Coulson MD, Dawson JE, Ford SJ, Gerhardt S, et al. Dipeptidyl nitrile inhibitors of cathepsin L. Bioorg Med Chem Lett 2009 Aug 1;19(15):4280–3.
[44] Satoo K, Noda NN, Kumeta H, Fujioka Y, Mizushima N, Ohsumi Y, et al. The structure of Atg4B-LC3 complex reveals the mechanism of LC3 processing and delipidation during autophagy. EMBO J 2009 May 6;28(9):1341–50.
[45] Cianni L, Feldmann CW, Gilberg E, Gutschow M, Juliano L, Leitao A, et al. Can cysteine protease cross-class inhibitors achieve selectivity? J Med Chem 2019 Dec 12;62(23):10497–525.
[46] Cartledge DM, Colella R, Glazewski L, Lu G, Mason RW. Inhibitors of cathepsins B and L induce autophagy and cell death in neuroblastoma cells. Invest New Drugs 2013 Feb;31(1):20–9.
[47] Norman JM, Cohen GM, Bampton ET. The in vitro cleavage of the hAtg proteins by cell death proteases. Autophagy 2010 Nov;6(8):1042–56.
[48] Tsapras P, Nezis IP. Caspase involvement in autophagy. Cell Death Differ 2017 Aug;24(8):1369–79.
[49] Wirawan E, Vande Walle L, Kersse K, Cornelis S, Claerhout S, Vanoverberghe I, et al. Caspase-mediated cleavage of Beclin-1 inactivates Beclin-1-induced autophagy and enhances apoptosis by promoting the release of proapoptotic factors from mitochondria. Cell Death Dis 2010;1, e18.

[50] Ojha R, Ishaq M, Singh SK. Caspase-mediated crosstalk between autophagy and apoptosis: mutual adjustment or matter of dominance. J Cancer Res Ther 2015 Jul-Sep;11(3):514–24.
[51] Xia HG, Zhang L, Chen G, Zhang T, Liu J, Jin M, et al. Control of basal autophagy by calpain1 mediated cleavage of ATG5. Autophagy 2010 Jan;6(1):61–6.
[52] Kim JS, Nitta T, Mohuczy D, O'Malley KA, Moldawer LL, Dunn Jr WA, et al. Impaired autophagy: a mechanism of mitochondrial dysfunction in anoxic rat hepatocytes. Hepatology 2008 May;47(5):1725–36.
[53] Yousefi S, Perozzo R, Schmid I, Ziemiecki A, Schaffner T, Scapozza L, et al. Calpain-mediated cleavage of Atg5 switches autophagy to apoptosis. Nat Cell Biol 2006 Oct;8(10):1124–32.
[54] Maruyama T, Noda NN. Autophagy-regulating protease Atg4: structure, function, regulation and inhibition. J Antibiot (Tokyo) 2017 Sep 13.
[55] Dana D, Pathak SK. A review of small molecule inhibitors and functional probes of human Cathepsin L. Molecules 2020 Feb 6;25(3).
[56] Schmitz J, Gilberg E, Loser R, Bajorath J, Bartz U, Gutschow M. Cathepsin B: active site mapping with peptidic substrates and inhibitors. Bioorg Med Chem 2019 Jan 1;27(1):1–15.
[57] Li YY, Fang J, Ao GZ. Cathepsin B and L inhibitors: a patent review (2010-present). Expert Opin Ther Pat 2017 Jun;27(6):643–56.
[58] Khaket TP, Kwon TK, Kang SC. Cathepsins: potent regulators in carcinogenesis. Pharmacol Ther 2019 Jun;198:1–19.
[59] Oda K, Nishimura Y, Ikehara Y, Kato K. Bafilomycin A1 inhibits the targeting of lysosomal acid hydrolases in cultured hepatocytes. Biochem Biophys Res Commun 1991 Jul 15;178(1):369–77.
[60] Mauvezin C, Neufeld TP. Bafilomycin A1 disrupts autophagic flux by inhibiting both V-ATPase-dependent acidification and Ca-P60A/SERCA-dependent autophagosome-lysosome fusion. Autophagy 2015;11(8):1437–8.
[61] Oral O, Oz-Arslan D, Itah Z, Naghavi A, Deveci R, Karacali S, et al. Cleavage of Atg3 protein by caspase-8 regulates autophagy during receptor-activated cell death. Apoptosis 2012 Aug;17(8):810–20.
[62] Kang R, Zeh HJ, Lotze MT, Tang D. The Beclin 1 network regulates autophagy and apoptosis. Cell Death Differ 2011 Apr;18(4):571–80.
[63] You M, Savaraj N, Kuo MT, Wangpaichitr M, Varona-Santos J, Wu C, et al. TRAIL induces autophagic protein cleavage through caspase activation in melanoma cell lines under arginine deprivation. Mol Cell Biochem 2013 Feb;374(1–2):181–90.
[64] Zhu Y, Zhao L, Liu L, Gao P, Tian W, Wang X, et al. Beclin 1 cleavage by caspase-3 inactivates autophagy and promotes apoptosis. Protein Cell 2010 May;1(5):468–77.
[65] Pagliarini V, Wirawan E, Romagnoli A, Ciccosanti F, Lisi G, Lippens S, et al. Proteolysis of Ambra1 during apoptosis has a role in the inhibition of the autophagic pro-survival response. Cell Death Differ 2012 Sep;19(9):1495–504.
[66] Cho DH, Jo YK, Hwang JJ, Lee YM, Roh SA, Kim JC. Caspase-mediated cleavage of ATG6/Beclin-1 links apoptosis to autophagy in HeLa cells. Cancer Lett 2009 Feb 8;274(1):95–100.
[67] Lassen KG, Kuballa P, Conway KL, Patel KK, Becker CE, Peloquin JM, et al. Atg16L1 T300A variant decreases selective autophagy resulting in altered cytokine signaling and decreased antibacterial defense. Proc Natl Acad Sci U S A 2014 May 27;111(21):7741–6.

CHAPTER 6

Proteases are cut out to regulate acute and chronic inflammation

Luiz G.N. de Almeida[a,b] and Antoine Dufour[a,b,c]
[a]McCaig Institute for Bone and Joint Health, Calgary, AB, Canada, [b]Department of Biochemistry and Molecular Biology, University of Calgary, Calgary, AB, Canada, [c]Department of Physiology and Pharmacology, University of Calgary, Calgary, AB, Canada

Introduction to inflammation

Inflammation is a necessary physiological process initiated in response to harmful stimuli, such as infection, injury, or stress [1]. Upon initiation, a multitude of molecular pathways coordinate a response to return to homeostasis [2]. There are five classical signs of inflammation: redness, pain, swelling, heat, and loss of tissue function. When inflammation is initiated, the first stage is the acute phase, aiming to eliminate the immediate threats, followed by the resolution phase, to restore normal cell/tissue functions [3]. The acute phase requires a tight regulation since excessive inflammatory response can result in extensive tissue damage and acute inflammatory diseases, for example, sepsis [1]. Moreover, if the acute phase does not resolve properly, the persistence of inflammatory mediators leads to a chronic stage, where wide tissue damage can result in various pathologies, including inflammatory bowel disease, joint diseases, neurodegenerative diseases, and many others [3, 4].

Inflammation can be caused by an infection (pathogens), or it can be sterile (injury). The innate immune system adapts and responds according to the type of injury, either by recognizing pathogen-associated molecular patterns (PAMPs) derived from pathogenic microorganisms or via damage-associated molecular patterns (DAMPs), released by damaged, stressed, or dead cells [5]. Tissue-resident cells, such as dendritic cells, macrophages, and mast cells, release a cascade of signaling molecules, including cytokines, chemokines, and lipid mediators [5]. Proinflammatory molecules are released to promote vascular permeability, alterations in body temperature, and leukocyte recruitment [2]. Components of the adaptive immune response can contribute to the recruitment of leukocytes to the inflammatory site via the identification of antigens by resident memory T cells [6]. During the acute phase, neutrophils are usually the first cells to respond; they migrate to the inflammatory site and release proteases that modulate the microenvironment by

Proteolytic Signaling in Health and Disease. https://doi.org/10.1016/B978-0-323-85696-6.00003-2

regulating proinflammatory signals and by enhancing the recruitment of inflammatory monocytes [7]. For example, neutrophils secrete matrix metalloproteinase 8 (MMP8), which cleaves and inactivates the serine proteinase inhibitor α1-proteinase inhibitor (α1-PI) [8]. Subsequently, due to the diminished levels of α1-PI, neutrophil elastase can more easily process and activate C-X-C motif chemokine 5 (CXCL5), which is a potent chemoattractant for polymorphonuclear (PMN) leukocytes [8]. Importantly, at the site of inflammation, antigen-presenting cells (APCs), such as macrophages, phagocytize pathogens, produce reactive oxygen species (ROS), and have high-presenting antigen activity [9]. Besides the macrophage regulation of the inflammatory site, the processing and regulation of cytokines also contribute to transitioning the microenvironment to a resolution phase. For example, proinflammatory type 1 T-helper (Th1) cells and natural killer (NK) cells can secrete interferon gamma (IFN-γ), an activator of proinflammatory macrophages [10]. Interestingly, IFN-γ has a dichotomous and time-dependent effect regulated by proteolysis [11]. Within a proinflammatory microenvironment, macrophages secrete matrix metalloproteinase 12 (MMP12), resulting in IFN-γ truncation, between ^{135}Glu ↓ Leu^{136}, a site located at the C-terminus of IFN-γ where it interacts with the IFN receptor 1 (IFNGR1) [11]. Truncated IFN-γ fails to initiate the proinflammatory Janus kinase-signal transducer and activator of transcription proteins 1 (JAK-STAT1) signaling pathway, favoring the conversion to the antiinflammatory profile, followed by the positive feedback of IL-4 release and further expression of MMP12 (Fig. 1) [11]. Thus, the switch to the resolution phase is an active process, regulated by matrix metalloproteinases, neutrophil elastase, and many other proteases, and is critical for the transition to tissue homeostasis [12].

Cytokines

Cytokines are implicated in numerous biological processes by the immune system, such as host response to infection, inflammation, and the regulation of various signaling pathways [13]. The interleukin family comprises multiple cytokines that play a crucial role in the initiation and regulation of the inflammatory response, mediating the communication of innate and adaptive immunity, and is a hallmark of numerous inflammatory diseases [14]. The members of the interleukin 1 (IL-1) family have different activities, including agonist (IL-1α and IL-β, IL-18, IL-33, and IL-36α, β, and γ) and antagonist (IL-1Ra, IL-36Ra, and IL-38) ligands, receptors (IL-1R1, IL-1R2, IL-1R4, IL-1R5, and IL-1R6), accessory chains (IL-1R3 and IL-1R7), and antiinflammatory functions (IL-37) [15]. Interleukins are potent initiators of inflammation and must be tightly controlled to avoid inappropriate activation of proinflammatory signaling. For this reason, most of its members are synthesized as inactive precursors, or zymogens, with little to no activity. Proteolytic processing of inactive cytokines is a necessary step for their activation. For example, pro-IL-1β is converted into its mature form by caspase-1 [16]. The cleavage of pro-IL-1β generates two fragments, a 17-kDa fragment, which is the mature and functional form of IL-1β, and a 26-kDa

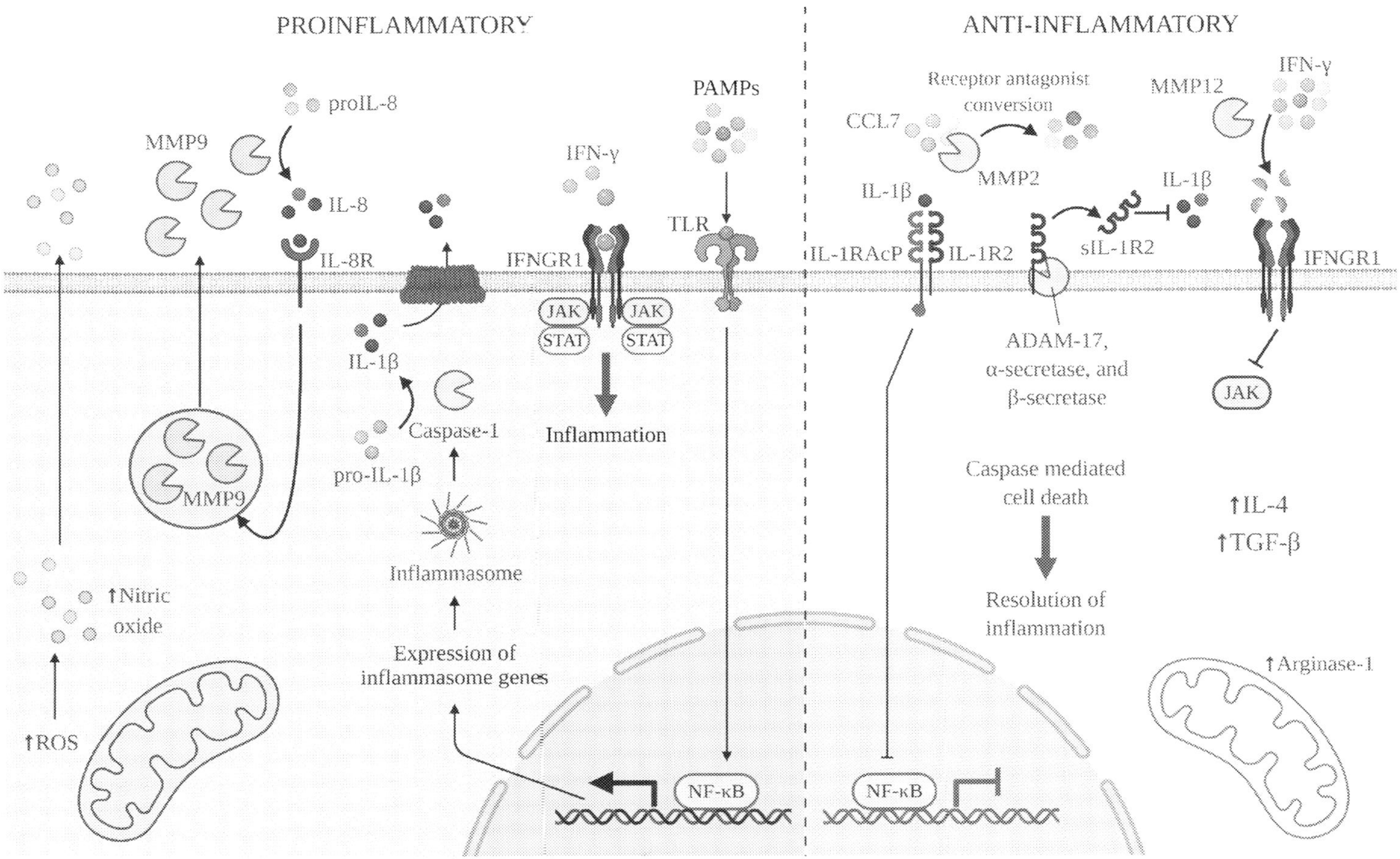

Fig. 1

Various proteases and their roles in proinflammatory and antiinflammatory conditions. During proinflammation (left panel), toll-like receptor (TLR) recognition of pathogen-associated molecular patterns (PAMPs) promotes pro-IL-1β expression and processing by caspase-1. As a result, mature IL-1β secretion nduces additional proinflammatory signals, such as IFN-γ and the production of reactive oxygen species (ROS). Proteolysis of cytokines and chemokines can initiate a positive feedback loop, resulting in acute inflammation. MMP9 processing of IL-8 promotes neutrophil activation and degranulation, increasing the protease release in a feed-forward mechanism. Conversely, proteases involved in the antiinflammatory environment (right panel) can negatively regulate cytokine activity, such as IFN-γ processing by MMP12. The interleukin receptor 2 (IL-1R2) lacks the toll/interleukin 1 receptor (TIR) domain, disrupting the signaling pathway induced by IL-1β. However, the proteolysis of IL-1R2 can result in the shedding of the ectodomain, which can inhibit IL-1β signaling functions. Proteases can also convert chemokines to a receptor antagonist, such as CCL7 processing by MMP2. Image was created with BioRender.

product that still has an undefined function (Fig. 1) [17, 18]. Another level of regulation is required as caspase-1 itself necessitates auto-proteolytic processing to be activated via the assembly of inflammasomes, a multiprotein intracellular complex comprised of a sensor molecule, an adaptor, and a pro-caspase-1 [19]. The inflammasome senses PAMPs and activates proinflammatory signaling, including the primary cleavage of caspase-1 and the noncanonical processing of caspase-4 and caspase-5 [19]. NACHR, LRR, and PYD domain-containing protein 3 (NLRP3) is one the main catalysts for the release of mature IL-1β, and it is primarily located in macrophages [20, 21]. Therefore, inflammasome activation is tightly linked with proteolysis to regulate the conversion of pro-IL-1β into its mature form [21]. Once the mature form of IL-1β is secreted, it generates a high-affinity complex by binding to the interleukin receptor type 1 (IL-1R1) and recruiting the IL-1 receptor accessory protein (IL-1RAcP) [22, 23]. The signaling cascade initiates after dimerization of IL-1R1 and IL-1RAcP, which promotes the recruitment of accessory proteins, including myeloid differentiation primary response 88 (MyD88) factor and IL-1R-associated protein kinase (IRAK), via the toll-IL-1R (TIR) domain [24–26]. Next, phosphorylated interleukin-1 receptor-associated kinase 1 (IRAK1) leaves the complex to cooperate with TNF receptor-associated factor 6 (TRAF6) and transforming growth factor-beta-activated kinase 1 (TAK1), also called mitogen-activated protein kinase kinase kinase 7 (MAP3K7), activating key factors, such as nuclear factor NF-kappa-B p105 subunit (NF-κB), that will ultimately produce an inflammatory response [24, 27, 28]. Similar to IL-1R1, the IL-1R2 receptor can bind with great affinity to IL-1β; however, it lacks the intracellular TIR domain, precluding the downstream signaling and activation of proinflammatory signals (Fig. 1) [29]. Thus, IL-1R2 acts as a regulator of the IL-1 response by competing with IL-1R1 for IL-1β and IL-1RAcP [30, 31]. Interestingly, this decoy receptor is also present as a soluble protein (sIL-1R2), binding to IL-1β in a nearly irreversible manner [32]. One of the mechanisms responsible for releasing the soluble protein is dependent on regulated intramembrane proteolysis (RIP) [33]. This cleavage occurs within the ectodomain, at a site in close proximity to the transmembrane region, shedding most of the protein's extracellular portion [34]. RIP also generates a membrane-embedded fragment that can undergo secondary cleavage by intramembrane proteolysis [35]. In the case of IL-1R2, various proteases are responsible for proteolysis, including a disintegrin and metalloproteinase 17 (ADAM17) and α- or β-secretase (BACE-1 and BACE-2), resulting in the ectodomain shedding and release of the soluble decoy receptor (Fig. 1) [33, 36–38]. Interestingly, sIL-1R2 is elevated in the serum of multiple inflammatory diseases, such as sepsis [39–41]. The treatment of septic patients with glucocorticoid hormones increases the concentration of sIL-1R2, indicating the upregulation of pathways that counterbalance the inflammatory effects of IL-1 [42]. Therefore, protease inhibition in inflammatory disease must be carefully evaluated, as proteases are also responsible for important protective mechanisms and not only detrimental ones.

Chemokines

Chemokines are essential for the regulation of inflammation by influencing the migration and homing patterns of various leukocytes [43]. Chemokines are produced in the tissue, and when released and activated, they can guide circulating immune cells to the inflammatory loci [44]. Inflammatory chemokines target mostly monocytes, macrophages, and T cells, being the mediators of many pathologies, including cancer, chronic, allergic, and autoimmune diseases [43, 45, 46]. Importantly, chemokine functions must be initiated but also terminated, as the recruitment of neutrophils, monocytes, NK cells, basophils, and eosinophils can result in the release of multiple molecules, including ROS, that are effective against pathogens but ultimately can cause tissue damage and failure to transition the immune response from an inflammatory to a resolving state [47–49]. Proteases regulate chemokines' functions in various ways: (1) they augment their activity, (2) diminish their activity, or (3) inactivate chemokines and turn them into receptor antagonists [49].

For example, monocytes and macrophages are the primary producers of the chemokine interleukin-8 (IL-8)/CXCL8 [50]. In leukocyte-conditioned media, IL-8 is present mainly with a truncated amino-terminal region, indicating an outcome of proteolytic processing [49]. A key function of IL-8 is the activation of neutrophils, which results in their degranulation [51]. This process creates a positive feedback loop as MMP9, stored in neutrophil granules in a tissue inhibitor of metalloprotease (TIMP)-free form, was shown to activate IL-8 (Fig. 1) [52]. MMP9 cleaves the full-length form of IL-8 by removing six amino acids from its N-terminal region. Importantly, the truncated form of IL-8 (7–77) product has a 10- to 27-fold higher potency in activating neutrophils [52]. Similarly, most if not all MMPs can cleave the CCL/monocyte chemoattractant protein (MCP) family, which comprises four members: MCP-1, MCP-2, MCP-3, and MCP4 (also known as CCL-2, CCL-8, CCL-7, and CCL-13, respectively). For example, MMP2 cleavage of MCP-3 converts the chemokine into a receptor antagonist (Fig. 1) [53]. Of note, multiple MMPs cleave various MCPs at the same site (amino acid 4 ↓ 5), further supporting a key regulation of proteases in chemokine activation and inactivation.

The roles of proteases in sepsis

Sepsis is a multifactorial and life-threatening syndrome associated with organ failure resulting from an uncontrolled host immune response to infection. Patients may develop a more severe form of the syndrome, termed septic shock, where the mortality risk is largely increased due to extensive deficiencies in cellular, metabolic, and circulatory processes [54]. To increase the survival rate, immediate diagnosis and treatment are necessary. Previously, sepsis was defined as a systemic inflammatory response syndrome (SIRS) to microbial infection, requiring at least two of the following clinical features: hypothermia or pyrexia (fever), leukocytosis, leukopenia or neutrophilia, tachypnea (accelerate breathing), and tachycardia (accelerated heartbeat) [55]. However, some of these symptoms were proven to be too broad; thus, the recent definition uses the sequential organ failure assessment (SOFA) score and states that an infection combined with a critical change in SOFA (score of 2 points or more) must be present for a sepsis diagnosis [56, 57]. Moreover, determining the incidence of sepsis can be challenging, as different countries may report different data depending on the disorder classification and infecting microorganism. Importantly, sepsis was demonstrated to be the leading cause of death worldwide, surpassing cancer and coronary disease, which was believed to be responsible for the most deaths throughout the world [58].

An uncontrolled immune response is the key feature of sepsis, where the pathogen infection is indirectly responsible for the disease symptoms. Sepsis can be initiated by an infection from any microorganism, including bacteria, viruses, and fungi, resulting in the release of PAMPs and DAMPs that are captured by specialized receptors from epithelial, endothelial, and innate immune cells responsible for identifying potential threats [59]. The pattern-recognition receptors (PRR), such as the toll-like receptors (TLR), C-type lectin receptors (CLRs), retinoic acid-inducible gene (RIG)-I-like receptors (RLRs), and NOD-like receptors (NLRs), promote multiple changes in inflammatory mediators that have complementary and redundant functions [59, 60]. For example, the molecular modifications include phosphorylation of JAK-STATs, mitogen-activated protein kinases (MAPKs), and NF-κB [61]. The alterations of inflammatory pathways result in the expression of cytokines, promotion of neutrophil-endothelial cell adhesion, activation of the complement, and promotion of the clotting cascades [57]. Importantly, sepsis leads to a hypercoagulable state where microvascular thrombi and fibrin deposition are commonly present, resulting in disseminated intravascular coagulation (DIC) and aggravation due to multiple organ failure [61]. Under healthy conditions, the endothelium has antithrombotic characteristics to avoid improper activation of the coagulation cascade [62]. For example, the endothelial cell layer expresses thrombomodulin (TM) and endothelial PC receptor (EPCR), promoting the generation of activated protein C (APC), which is a serine protease that cleaves activated clotting factors, such as Factor Va (FVa) and Factor VIIIa (FVIIIa) (Fig. 2) [63]. Additionally, endothelial cells express tissue factor pathway inhibitor (TFPI), a serine protease inhibitor with an anticoagulant function that hinders FX conversion to FXa (Fig. 2) [64]. TFPI seems to be protective in sepsis, as administration on septic primates attenuated the coagulopathy [65]. FXa is also inhibited by antithrombin (AT), an anticoagulant with antiinflammatory properties [64]. For example, AT reduces leukocyte recruitment by impairing their ability to roll on the endothelium [66]. Moreover, endothelial cells also secrete tissue plasminogen activator (tPA), another serine protease with anticoagulant properties and responsible for converting plasminogen to plasmin, which results in the breakdown of blood clots (Fig. 2) [67]. However, during sepsis, the proinflammatory signals override this protection mechanism and establish a systemic prothrombotic environment on the activated endothelium [64].

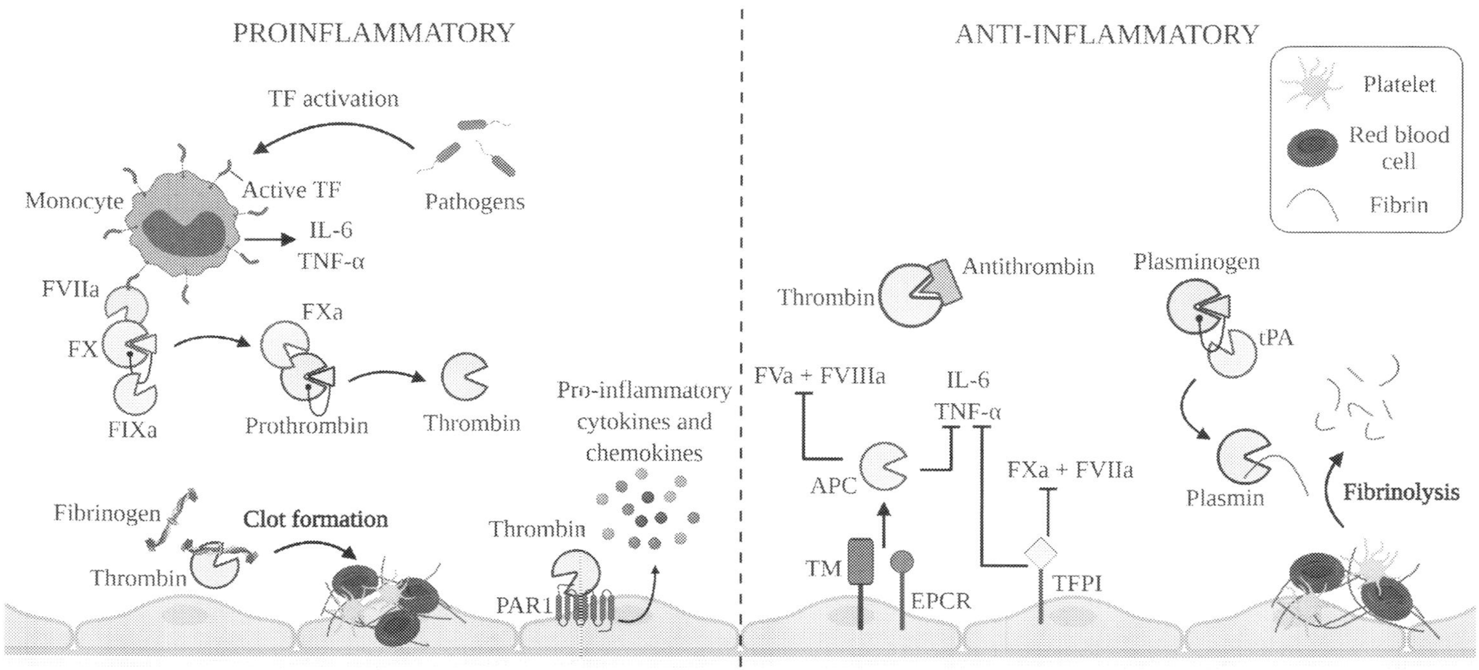

Fig. 2

Proteases of the coagulation cascade and their effect on inflammation. Pathogens can initiate an acute proinflammatory response (left panel), leading to cytokine release, endothelium dysfunction, and loss of barrier properties. Active tissue factor (TF)/Factor III, expressed on monocytes, binds to active Factor VII (FVII) and, in combination with Factor IXa (FIXa), can activate Factor X (FX). This process leads to thrombin generation, the release of proinflammatory signals, and clot formation. In contrast, to prevent nonspecific clotting, the endothelium can release anticoagulant and antiinflammatory molecules (right panel). Thrombomodulin (TM) and endothelial protein C receptor (EPCR) generate activated protein C (APC), which can inhibit multiple coagulant factors and proinflammatory cytokines. Tissue factor pathway inhibitor (TFPI) has similar functions but targets different coagulation factors. Additionally, tissue plasminogen activator (tPA) plays key roles in fibrinolysis, generating plasmin, resulting in the digestion of fibrin from blood clots. Image was created with BioRender.

Inflammation induces the release of cytokines, C-reactive protein (CRP), and various other factors that strongly upregulate tissue factor (TF) expression in the circulation, mostly on monocytes (Fig. 2) [68]. *Staphylococcus aureus* can upregulate TF expression and plasma activity in human blood via C5a, emphasizing the importance of the complement system in inflammation-induced coagulation [69]. TF, also known as Factor III, is a transmembrane protein in which the extracellular domain interacts with a serine protease, creating a complex that has proteolytic activities and initiates a coagulation cascade [70]. However, TF is not present in the circulatory system under normal conditions; rather, it is substantially expressed in blood tissue barriers, rapidly responding and initiating the coagulation process upon any injuries [71]. Once TF expression increases in the blood, it generates a complex with Factor VIIa (FVIIa) (Fig. 2). This newly assembled complex triggers FIX and FX's conversion from zymogens to active proteases. Cleavage of prothrombin by active FX (FXa) generates thrombin, which converts fibrinogen to fibrin, and results in fibrin cross-linking and clot formation (Fig. 2). In nonhuman primates, blocking of TF function was protective after intravenous injection of *Escherichia coli*, showing the importance of coagulation mechanisms during bacteremia [72]. For humans, the drug drotrecogin alfa (activated protein C), a recombinant form of APC, showed promising initial results, but further randomized controlled trials (RCT) failed to confirm its benefits, and the drug was withdrawn from the market [73]. Currently, the clinical efficacy of thrombomodulin, another drug that modulates the clotting cascade, was assessed on critically ill patients with sepsis-associated coagulopathy [74]. Phase III RCT showed that the drug did not significantly reduce the 28-day all-cause mortality. However, the study had several limitations as stated by the authors, including a sample size that might not have been sufficient for the statistical analysis, a post hoc test that was not planned beforehand and was advised to be interpreted with caution, and the effect on some variables, such as long-term mortality, was not fully assessed as the follow-up was not completed by the time the study was published. Therefore, studies with a more robust statistical approach should be considered for further analysis of thrombomodulin efficacy.

The relationship between inflammation and coagulation is multidirectional, as the clotting cascade can also regulate inflammatory signaling that may result in tissue damage [75]. For example, TM has antiinflammatory activity, besides being one of the protein C (PC) activators and an inhibitor of coagulation (Fig. 2). TM dampens the proinflammatory function of thrombin by binding to this serine protease; additionally, the TM-thrombin complex activates the thrombin-activatable fibrinolysis inhibitor (TAFI), which suppresses complement factor C3a, C5a, and bradykinin activity [76]. PC also has antiinflammatory properties as demonstrated in transgenic mice with low levels of this zymogen, where intraperitoneal injection with endotoxin induces higher neutrophil influx, severe DIC, and organ damage [77]. Furthermore, in vitro studies show that APC, the active protease, inhibits inflammation by downregulating multiple factors in monocytes and macrophages, such as cytokine signaling, NF-κB translocation, and the expression of TF and tumor necrosis factor (TNF)-α [78, 79].

Protease-activated receptors (PARs) are another pathway in which the coagulation cascade can regulate inflammatory signals. PAR belongs to the G-protein-coupled receptors family, and there are four variations of these transmembrane proteins encoded in the human genome, PAR1, PAR2, PAR3, and PAR4, localized on various cell types, including fibroblast, smooth muscle cells, endothelial cells, platelets, and others [80]. The mechanism for PAR activation depends on proteolytic processing at the N-terminus of the receptor's ectodomain. Proteolytic cleavage exposes a neo-N-termini that becomes available to fold back and activate the receptor, also called a tethered ligand mechanism [81]. This process was first described in 1991 by demonstrating a thrombin cleavage of PAR1 [82], followed by additional studies that demonstrated it could also cleave PAR3 and PAR4 [83, 84]. Thrombin cleavage of PARs is concentration-dependent since low amounts are sufficient for PAR1 processing while higher concentrations are necessary for PAR4 activation [80]. In PAR1-expressing cells, a concentration of thrombin as low as 10 pM was sufficient for an increase in cytoplasmic calcium, while PAR4-expressing cells required 1 nM of thrombin [85]. Additional members of the coagulation cascade can activate PAR signaling, such as PAR2 processing by the complexes of TF-FVIIa and TF-FVIIa-FXa [86]. Importantly, sepsis patients showed a 9-fold increase in plasma concentration of active MMP1, a protease that can activate PAR1 [87]. Even though thrombin and MMP1 cleave PAR1 at different sites ($^{39}D \downarrow P^{40}$ and $^{41}R \downarrow S^{42}$, respectively), both can activate the receptor and promote barrier dysfunction, disseminated intravascular coagulation, and worsening of the disease [87, 88]. Similarly, the activation of PAR1 and PAR2 results in detectable levels of interleukin 6 (IL-6) on the supernatant of cultured endothelial cells, and both receptors are capable of augmenting the cytokine levels upon lipopolysaccharide (LPS) and TNF-α stimulation (Fig. 2) [89]. Moreover, deficiency in TF expression and inhibition of the TF-FVIIa complex are associated with lowering IL-6 expression in endotoxemia and sepsis [90, 91]. One explanation for the low cytokine expression is via PAR signaling, as shown in a mice model of endotoxemia in which a combination of PAR1 deficiency and hirudin (inhibitor of PAR1 and PAR4) treatment resulted in deficient LPS-induced IL-6 inflammation [81]. Of note, neither $PAR2^{-/-}$ nor $PAR1^{-/-}$ mice affected IL-6 expression upon LPS treatment, showing a likely redundant control of PAR activation in inflammation [81, 90]. Furthermore, the APC processing of coagulation factors results in less thrombin generation, hence, less PAR processing and indirect downregulation of inflammation. However, APC also cleaves PAR1 and induces numerous transcriptional changes that result in downregulation of inflammatory signaling pathways and apoptosis inhibition, in a process dependent on EPCR [78, 92]. There are some conflicting reports on PAR1 processing by APC in vivo during sepsis onset; therefore, its importance remains to be elucidated [93, 94]. Altogether, multiple coagulation cascade proteases cleave PARs, which may promote inflammation during sepsis and endotoxemia.

In addition to the coagulation cascade proteases, gastrointestinal tract proteases have also been demonstrated to regulate immune responses in sepsis. For example, enteral administration of the protease inhibitor tranexamic acid (TXA) in a rat model of shock improved hemodynamics and vascular responsiveness [95]. A similar study treated three rat models of shock, namely hemorrhagic (bleeding), septic (infection), and toxic (endotoxin), with three enzymatic inhibitors, including TXA, resulting in diminished microhemorrhages and increased survival compared with untreated animals [96]. Even though the exact proteases and cleavage targets are uncertain, there is multiple evidence showing increased overall proteolysis in sepsis and septic shock [97]. Notably, there is probable association between the mortality of septic shock patients and their peptidomics patterns. The analysis of the peptide sequences and cleavage sites revealed a predominant role of serine proteases and metalloproteases in the proteolytic degradation present in the septic shock patient's plasma as compared to nonhospitalized healthy controls [98]. A multiomics serum analysis of patients with *Staphylococcus aureus* bacteremia showed a potential predictive power of mortality risk [99]. The potential clinical biomarkers were two protease inhibitors, fetuin-B and serpin D1, further demonstrating the importance of characterizing what proteases and protease inhibitors are dysregulated in sepsis.

The roles of proteases in Crohn's disease

Crohn's disease (CD) and ulcerative colitis (UC) comprise the two major inflammatory bowel diseases (IBD), which differ in symptoms, disease course, complications, affected site of the gastrointestinal (GI) tract, and treatment. Although some studies reported a stable incidence in western countries, the IBD rates are increasing in newly developed countries from regions such as Africa, Asia, and South America [100]. For example, the annual percentage change of CD in Brazil and Taiwan has increased by 11% from 1988 to 2012 (95% confidence interval 4.8–17.8) and by 4% from 1988 to 2018 (95% CI 1.0–7.1), respectively. Of note, the western lifestyle, including diet, industrialization, and urbanization, seems to play an important role in the disease incidence [101]. However, population-based studies are needed to evaluate the shared risk factors between developed and developing countries, which will aid in a better understanding of the disease [102].

To date, the etiology of CD is still unknown, and it is believed that environmental factors, genetic susceptibility, and intestinal microbiota are likely responsible for the disease pathogenesis [102]. CD is a damaging and progressive condition in which almost half of the patients develop extraintestinal manifestations, resulting in a higher risk of hospitalization, complications, and surgery [103]. The disease is characterized by a chronic, relapsing, and transmural inflammation that affects segments of the GI tract, resulting in diarrhea, persistent abdominal pain, bowel obstruction, and perianal injury [104]. Symptoms of CD can be manifested anywhere in the GI, but the terminal ileum and proximal colon are the most commonly affected areas [104]. At the time of diagnosis, approximately one-third of CD patients already present signs of bowel damage, with half of them requiring a surgical procedure in the first two decades within the diagnosis [105]. Complications such as stricture and fistulae affect > 50% of all patients in the first 10 years [106]. To optimize the disease diagnosis, multiple biomarkers have been proposed; however, they are susceptible to genetic and serological variability, present variable levels despite disease activity, or there is no consensus on the clinical cutoff [107–109]. Interestingly, protease activity has been proposed as a key biomarker for inflammatory conditions such as ankylosing spondylitis [110]; however, increased protease activity could result in biomarker degradation and hamper biomarker's diagnostic potential [111].

CD is a complex disorder that is challenging to diagnose and difficult to treat with currently available medications since the affected site, severity, and disease behavior are heterogeneous among patients. The current treatments aim to block the disease progression, avoiding complications and bowel damage. For mild to moderate disease, corticosteroids and immunosuppressants are typically the first-line therapy [104]. For moderate to severe CD, biologics and antiadhesion molecules are most often used. Biologic therapies, especially anti-TNF, have revolutionized CD treatment [112]. However, there is a large percentage of unresponsive patients, as the remission rates have been reported to be as low as 15% for vedolizumab [113], and as high as 55% for infliximab [104, 114]. Failure of biologics treatment efficacy may be attributed to the immunogenicity that the monoclonal antibodies can induce or due to interactions with dysregulated proteins, such as proteases [115]. For example, the effect of MMP3 and MMP12 on the integrity of three anti-TNF drugs, infliximab, adalimumab, and etanercept, has been investigated [115]. Both MMP3 and MMP12 are elevated in IBD patient's mucosa, and they have a broad range of substrates, including the Pro-Glu scissile bond at the lower hinge of IgG_1 [116]. After incubation of recombinant MMP3, MMP12, and tissue homogenate from IBD patients with each of the monoclonal antibodies, cleaved IgG and reduced activity of infliximab, adalimumab, and etanercept was demonstrated [115]. Moreover, a higher proportion of cleaved IgG was found in the serum of patients that are nonresponders to anti-TNF therapy when compared to responders. Interestingly, MMP3 and MMP12 are upregulated by IL-17A, which is significantly increased in CD patients with strictures, suggesting that this phenotype is less likely to benefit from anti-TNF therapies (Fig. 3) [115].

Proteases are central to CD pathogenesis, and the characterization of their roles could impact our understanding of the disease mechanisms, potentially resulting in new drug targets and biomarkers for disease diagnosis [111]. Interestingly, the GI tract is the organ with the highest exposure to proteases [117]. As expected, the lumen of the upper GI tract is rich in proteases to aid the digestive function; however, multiple studies have reported an association of pathogenic conditions to an increased protease activity in the mucosal tissue throughout the entire GI tract [117]. For example, neutrophils and lymphocytes in the inflamed mucosa of UC and CD patients have increased expression of MMP3 and MMP8, respectively. MMP9, which is expressed by neutrophils and macrophages, has been linked to leucocyte extravasation and serum protein permeation into the surrounding tissue. MMP3, MMP8, and MMP9 were also expressed in actively inflamed areas in the ulcer base, indicating that MMPs participate in tissue remodeling and leukocyte extravasation in inflamed areas of CD and UC patients' intestine [118]. Of note, MMP9 is linked to fistula in acute CD patients and correlates with disease activity [119]. Furthermore, the identification of proteases overly expressed in the GI tract does not necessarily translate to their active status; therefore, an increasing number of studies are focusing on the assessment of protease activity in IBD [120, 121]. Using chemical probes, Anderson et al. demonstrated increased activity of neutrophil elastase in inflamed biopsies of UC patients [122]. Similarly, the activity of the serine protease cathepsin G was found to be elevated in supernatants from IBD patients' tissues [121]. Cathepsin G can participate in the inflammatory response by cleaving and inactivating chemokines, cytokines, and growth factors, or it can process and activate such mediators, as is the case for the activation and alarmin conversion of IL-33 [121, 123]. PARs have also been implicated to play a role in the proinflammatory response of CD, and cathepsin G, secreted from macrophages, is one of the proteases regulating PAR activation (Fig. 3) [124]. The cleavage site of PAR2 by cathepsin G ($^{56}E \downarrow T^{57}$) is a different site from the one cleaved by trypsin or tryptase [124]. Interestingly, neutrophil elastase also cleaves PAR2 yet at a different site ($^{67}S \downarrow V^{68}$), near the transmembrane domain, resulting in a probable conformational change that activates the receptor [124]. The biased cleavage of PAR2 by cathepsin G and neutrophil elastase does not lead to the canonical receptor internalization but rather promotes inflammation, pain, and gastrointestinal leakiness [111, 124]. Thrombin concentration was found to be 100-fold higher in the colon of CD patients when compared to healthy controls, and it has been proposed to be detrimental for the disease progression [125]. Even though the authors used a mouse model of UC to evaluate the impact of protease activity in CD, they showed that PAR1 plays an important role in thrombin-mediated proinflammatory signaling and bowel damage, suggesting the protease as a possible target for future therapies [125]. These findings show that PAR activation is likely involved in the disease pathogenesis; however, little is currently known about its mechanism in CD (Fig. 3).

Conversely, proteases have also been linked to protective mechanisms in CD. A variety of short peptides generated by neutrophil elastase after proteolytic processing of the cellular adherens junction protein, E-cadherin, resulted in protective functions [126]. The short peptides were also observed in tissue samples from IBD patients and showed wound healing properties in an in vitro model, indicating that proinflammatory proteases may also promote mucosal healing (Fig. 3) [120, 126]. Interestingly, a mouse model of mucosal inflammation showed a higher susceptibility for disease development and severe inflammatory response when MMP2 was knocked out, indicating a potential protective role for this protease (Fig. 3) [127]. Importantly, additional studies are required to characterize which proteases are protective, their optimal concentration, and the ideal conditions that provide their protective response.

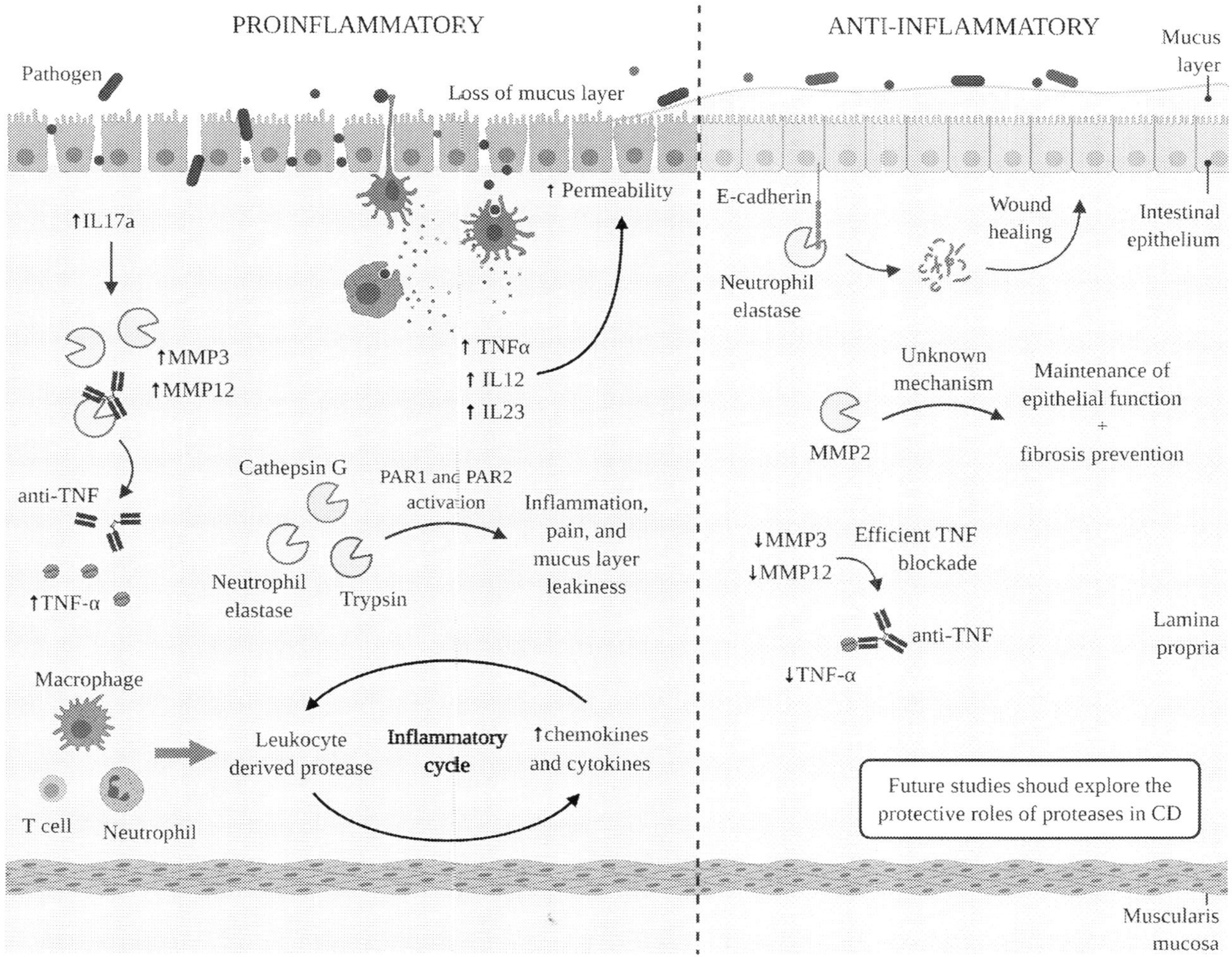

Fig. 3

The roles of various proteases in Crohn's disease and inflammatory bowel disease. During active Crohn's disease (left panel), IL-17a can upregulate the expression of multiple matrix metalloproteinases (MMPs) capable of inactivating monoclonal antibodies targeting tumor necrosis factor (TNF)-α. Proteases can also process and activate protease-activated receptors (PARs), leading to inflammation, pain, and increased intestinal permeability. In a feed-forward fashion, it can generate additional cytokine and chemokine release, promoting protease expression and secretion. The regulation of antiinflammatory mechanisms (right panel) also implicated various proteases. E-cadherin processing by neutrophil elastase may generate small peptides with wound healing properties, aiding in the reconstitution of the intestinal epithelium. The absence of MMP2 has been linked to the disease's worsening, but the exact mechanism remains elusive. The lack of proinflammatory signaling results in diminished MMP3 and MMP12 levels, potentially allowing efficient blockade of TNF-α by monoclonal antibodies. Image was created with BioRender.

Conclusions

Proteases have long been considered simple degradative enzymes only associated with protein breakdown [11, 128, 129]. We now know that proteases precisely and selectively regulate multiple biological processes, including the initiation and termination of various steps in inflammation. They are essential to support the intricate signaling network promoted by the immune system during homeostasis or responses to stress, infections, and harmful stimuli. The clinical use of protease inhibitors has been proven effective in various diseases; however, due to the broad impact of proteolysis on inflammation, the precise roles of each protease need to be carefully evaluated. One area that is still understudied is the role of proteases in bacterial, viral, and fungal infections. The generation of chemical and biochemical tools to investigate proteases in complex samples, such as clinical biopsies or animal models, has and will continue to provide key biological information about the multiple roles of proteases in inflammatory diseases.

References

[1] Medzhitov R. Inflammation 2010: new adventures of an old flame. Cell 2010;140(6):771–6. https://doi.org/10.1016/j.cell.2010.03.006.

[2] Kotas ME, Medzhitov R. Homeostasis, inflammation, and disease susceptibility. Cell 2015;160(5):816–27. https://doi.org/10.1016/j.cell.2015.02.010.

[3] Serhan CN, Gupta SK, Perretti M, et al. The Atlas of Inflammation Resolution (AIR). Mol Asp Med 2020;74:100894. https://doi.org/10.1016/j.mam.2020.100894.

[4] Chen L, Deng H, Cui H, et al. Inflammatory responses and inflammation-associated diseases in organs. Oncotarget 2018;9(6):7204–18. https://doi.org/10.18632/oncotarget.23208.

[5] Medzhitov R. Origin and physiological roles of inflammation. Nature 2008;454(7203):428–35. https://doi.org/10.1038/nature07201.

[6] Nourshargh S, Alon R. Leukocyte migration into inflamed tissues. Immunity 2014;41(5):694–707. https://doi.org/10.1016/j.immuni.2014.10.008.

[7] Borregaard N. Neutrophils, from marrow to microbes. Immunity 2010;33(5):657–70. https://doi.org/10.1016/j.immuni.2010.11.011.

[8] Fortelny N, Cox JH, Kappelhoff R, et al. Network analyses reveal pervasive functional regulation between proteases in the human protease web. Levchenko A, ed. PLoS Biol 2014;12(5). https://doi.org/10.1371/journal.pbio.1001869, e1001869.

[9] Muraille E, Leo O, Moser M. Th1/Th2 paradigm extended: macrophage polarization as an unappreciated pathogen-driven escape mechanism? Front Immunol 2014;5. https://doi.org/10.3389/fimmu.2014.00603.

[10] Schroder K, Hertzog PJ, Ravasi T, Hume DA. Interferon-γ: an overview of signals, mechanisms and functions. J Leukoc Biol 2004;75(2):163–89. https://doi.org/10.1189/jlb.0603252.

[11] Dufour A, Bellac CL, Eckhard U, et al. C-terminal truncation of IFN-γ inhibits proinflammatory macrophage responses and is deficient in autoimmune disease. Nat Commun 2018;9(1):2416. https://doi.org/10.1038/s41467-018-04717-4.

[12] Ortega-Gómez A, Perretti M, Soehnlein O. Resolution of inflammation: an integrated view. EMBO Mol Med 2013;5(5):661–74. https://doi.org/10.1002/emmm.201202382.

[13] Dinarello CA. Proinflammatory cytokines. Chest 2000;118(2):503–8. https://doi.org/10.1378/chest.118.2.503.

[14] Garlanda C, Dinarello CA, Mantovani A. The interleukin-1 family: back to the future. Immunity 2013;39(6):1003–18. https://doi.org/10.1016/j.immuni.2013.11.010.

[15] Dinarello CA. Overview of the IL-1 family in innate inflammation and acquired immunity. Immunol Rev 2018;281(1):8–27. https://doi.org/10.1111/imr.12621.

[16] Thornberry NA, Bull HG, Calaycay JR, et al. A novel heterodimeric cysteine protease is required for interleukin-1β processing in monocytes. Nature 1992;356(6372):768–74. https://doi.org/10.1038/356768a0.

[17] Mosley B, Urdal DL, Prickett KS, et al. The interleukin-1 receptor binds the human interleukin-1 alpha precursor but not the interleukin-1 beta precursor. J Biol Chem 1987;262(7):2941–4.

[18] Howard AD, Chartrain N, Ding GF, et al. Probing the role of interleukin-1 beta convertase in interleukin-1 beta secretion. Agents Actions Suppl 1991;35:77–83.

[19] Broz P, Dixit VM. Inflammasomes: mechanism of assembly, regulation and signalling. Nat Rev Immunol 2016;16(7):407–20. https://doi.org/10.1038/nri.2016.58.

[20] Mangan MSJ, Olhava EJ, Roush WR, Seidel HM, Glick GD, Latz E. Targeting the NLRP3 inflammasome in inflammatory diseases. Nat Rev Drug Discov 2018;17(8):588–606. https://doi.org/10.1038/nrd.2018.97.

[21] Afonina IS, Müller C, Martin SJ, Beyaert R. Proteolytic processing of interleukin-1 family cytokines: variations on a common theme. Immunity 2015;42(6):991–1004. https://doi.org/10.1016/j.immuni.2015.06.003.

[22] Greenfeder SA, Nunes P, Kwee L, Labow M, Chizzonite RA, Ju G. Molecular cloning and characterization of a second subunit of the interleukin 1 receptor complex. J Biol Chem 1995;270(23):13757–65. https://doi.org/10.1074/jbc.270.23.13757.

[23] Sims J, March C, Cosman D, et al. cDNA expression cloning of the IL-1 receptor, a member of the immunoglobulin superfamily. Science 1988;241(4865):585–9. https://doi.org/10.1126/science.2969618.

[24] Dunne A, O'Neill LAJ. The interleukin-1 receptor/Toll-like receptor superfamily: signal transduction during inflammation and host defense. Sci Signal 2003;2003(171):re3. https://doi.org/10.1126/stke.2003.171.re3.

[25] Huang J, Gao X, Li S, Cao Z. Recruitment of IRAK to the interleukin 1 receptor complex requires interleukin 1 receptor accessory protein. Proc Natl Acad Sci 1997;94(24):12829–32. https://doi.org/10.1073/pnas.94.24.12829.

[26] Wesche H, Henzel WJ, Shillinglaw W, Li S, Cao Z. MyD88: an adapter that recruits IRAK to the IL-1 receptor complex. Immunity 1997;7(6):837–47. https://doi.org/10.1016/S1074-7613(00)80402-1.

[27] Cao Z, Xiong J, Takeuchi M, Kurama T, Goeddel DV. TRAF6 is a signal transducer for interleukin-1. Nature 1996;383(6599):443–6. https://doi.org/10.1038/383443a0.

[28] Takaesu G, Kishida S, Hiyama A, et al. TAB2, a novel adaptor protein, mediates activation of TAK1 MAPKKK by linking TAK1 to TRAF6 in the IL-1 signal transduction pathway. Mol Cell 2000;5(4):649–58. https://doi.org/10.1016/S1097-2765(00)80244-0.

[29] McMahan CJ, Slack JL, Mosley B, et al. A novel IL-1 receptor, cloned from B cells by mammalian expression, is expressed in many cell types. EMBO J 1991;10(10):2821–32.

[30] Colotta F, Re F, Muzio M, et al. Interleukin-1 type II receptor: a decoy target for IL-1 that is regulated by IL-4. Science 1993;261(5120):472–5. https://doi.org/10.1126/science.8332913.

[31] Neumann D, Kollewe C, Martin MU, Boraschi D. The membrane form of the type II IL-1 receptor accounts for inhibitory function. J Immunol 2000;165(6):3350–7. https://doi.org/10.4049/jimmunol.165.6.3350.

[32] Arend WP, Malyak M, Smith MF, et al. Binding of IL-1 alpha, IL-1 beta, and IL-1 receptor antagonist by soluble IL-1 receptors and levels of soluble IL-1 receptors in synovial fluids. J Immunol 1994;153(10):4766–74.

[33] Kuhn P-H, Marjaux E, Imhof A, De Strooper B, Haass C, Lichtenthaler SF. Regulated intramembrane proteolysis of the interleukin-1 receptor II by α-, β-, and γ-secretase. J Biol Chem 2007;282(16):11982–95. https://doi.org/10.1074/jbc.M700356200.

[34] Brown MS, Ye J, Rawson RB, Goldstein JL. Regulated intramembrane proteolysis. Cell 2000;100(4):391–8. https://doi.org/10.1016/S0092-8674(00)80675-3.

[35] Weihofen A, Martoglio B. Intramembrane-cleaving proteases: controlled liberation of proteins and bioactive peptides. Trends Cell Biol 2003;13(2):71–8. https://doi.org/10.1016/S0962-8924(02)00041-7.

[36] Reddy P, Slack JL, Davis R, et al. Functional analysis of the domain structure of tumor necrosis factor-α converting enzyme. J Biol Chem 2000;275(19):14608–14. https://doi.org/10.1074/jbc.275.19.14608.

[37] Uchikawa S, Yoda M, Tohmonda T, et al. ADAM17 regulates IL-1 signaling by selectively releasing IL-1 receptor type 2 from the cell surface. Cytokine 2015;71(2):238–45. https://doi.org/10.1016/j.cyto.2014.10.032.

[38] Cui X, Rouhani FN, Hawari F, Levine SJ. Shedding of the type II IL-1 decoy receptor requires a multifunctional aminopeptidase, aminopeptidase regulator of TNF receptor type 1 shedding. J Immunol 2003;171(12):6814–9. https://doi.org/10.4049/jimmunol.171.12.6814.

[39] Giri JG, Wells J, Dower SK, et al. Elevated levels of shed type II IL-1 receptor in sepsis. Potential role for type II receptor in regulation of IL-1 responses. J Immunol 1994;153(12):5802–9.

[40] Jouvenne P, Vannier E, Dinarello CA, Miossec P. Elevated levels of soluble interleukin-1 receptor type II and interleukin-1 receptor antagonist in patients with chronic arthritis: correlations with markers of inflammation and joint destruction. Arthritis Rheum 1998;41(6):1083–9. https://doi.org/10.1002/1529-0131(199806)41:6<1083::AID-ART15>3.0.CO;2-9.

[41] Pietruczuk A, Świerzbińska R, Pancewicz S, Pietruczuk M, Hermanowska-Szpakowicz T. Serum levels of interleukin-18 (IL-18), interleukin-1β (IL-1β), its soluble receptor sIL-1RII and C-reactive protein (CRP) in patients with lyme arthritis. Infection 2006;34(3):158–62. https://doi.org/10.1007/s15010-006-5013-z.

[42] Müller B, Peri G, Doni A, et al. High circulating levels of the IL-1 type II decoy receptor in critically ill patients with sepsis: association of high decoy receptor levels with glucocorticoid administration. J Leukoc Biol 2002;72(4):643–9.

[43] Gerard C, Rollins BJ. Chemokines and disease. Nat Immunol 2001;2(2):108–15. https://doi.org/10.1038/84209.

[44] Foxman EF, Campbell JJ, Butcher EC. Multistep navigation and the combinatorial control of leukocyte chemotaxis. J Cell Biol 1997;139(5):1349–60. https://doi.org/10.1083/jcb.139.5.1349.

[45] Baggiolini M, Dahinden CA. CC chemokines in allergic inflammation. Immunol Today 1994;15(3):127–33. https://doi.org/10.1016/0167-5699(94)90156-2.

[46] Wang JM, Deng X, Gong W, Su S. Chemokines and their role in tumor growth and metastasis. J Immunol Methods 1998;220(1–2):1–17. https://doi.org/10.1016/S0022-1759(98)00128-8.

[47] Schluger NW, Rom WN. Early responses to infection: chemokines as mediators of inflammation. Curr Opin Immunol 1997;9(4):504–8. https://doi.org/10.1016/s0952-7915(97)80102-1.

[48] Baggiolini M. Chemokines and leukocyte traffic. Nature 1998;392(6676):565–8. https://doi.org/10.1038/33340.

[49] Struyf S, Proost P, Van Damme J. Regulation of the immune response by the interaction of chemokines and proteases. In: Advances in immunology, vol. 81. Elsevier; 2003. p. 1–44. https://doi.org/10.1016/S0065-2776(03)81001-5.

[50] Remick DG. Interleukin-8. Crit Care Med 2005;33(Suppl):S466–7. https://doi.org/10.1097/01.CCM.0000186783.34908.18.

[51] Baggiolini M, Walz A, Kunkel SL. Neutrophil-activating peptide-1/interleukin 8, a novel cytokine that activates neutrophils. J Clin Invest 1989;84(4):1045–9. https://doi.org/10.1172/JCI114265.

[52] Van den Steen PE, Proost P, Wuyts A, Van Damme J, Opdenakker G. Neutrophil gelatinase B potentiates interleukin-8 tenfold by aminoterminal processing, whereas it degrades CTAP-III, PF-4, and GRO-α and leaves RANTES and MCP-2 intact. Blood 2000;96(8):2673–81. https://doi.org/10.1182/blood.V96.8.2673.

[53] McQuibban GA, Gong JH, Tam EM, McCulloch CA, Clark-Lewis I, Overall CM. Inflammation dampened by gelatinase A cleavage of monocyte chemoattractant protein-3. Science 2000;289(5482):1202–6. https://doi.org/10.1126/science.289.5482.1202.

[54] Singer M, Deutschman CS, Seymour CW, et al. The third international consensus definitions for sepsis and septic shock (sepsis-3). JAMA 2016;315(8):801. https://doi.org/10.1001/jama.2016.0287.

[55] Bone RC, Sibbald WJ, Sprung CL. The ACCP-SCCM consensus conference on sepsis and organ failure. Chest 1992;101(6):1481–3. https://doi.org/10.1378/chest.101.6.1481.

[56] Vincent J-L, Moreno R, Takala J, et al. The SOFA (Sepsis-related Organ Failure Assessment) score to describe organ dysfunction/failure: on behalf of the Working Group on Sepsis-Related Problems of the European Society of Intensive Care Medicine. Intensive Care Med 1996;22(7):707–10. https://doi.org/10.1007/BF01709751.

[57] Cecconi M, Evans L, Levy M, Rhodes A. Sepsis and septic shock. Lancet 2018;392(10141):75–87. https://doi.org/10.1016/S0140-6736(18)30696-2.
[58] Rudd KE, Johnson SC, Agesa KM, et al. Global, regional, and national sepsis incidence and mortality, 1990–2017: analysis for the Global Burden of Disease Study. Lancet 2020;395(10219):200–11. https://doi.org/10.1016/S0140-6736(19)32989-7.
[59] Takeuchi O, Akira S. Pattern recognition receptors and inflammation. Cell 2010;140(6):805–20. https://doi.org/10.1016/j.cell.2010.01.022.
[60] Tang D, Kang R, Coyne CB, Zeh HJ, Lotze MT. PAMPs and DAMPs: signal 0s that spur autophagy and immunity. Immunol Rev 2012;249(1):158–75. https://doi.org/10.1111/j.1600-065X.2012.01146.x.
[61] Hotchkiss RS, Moldawer LL, Opal SM, Reinhart K, Turnbull IR, Vincent J-L. Sepsis and septic shock. Nat Rev Dis Primers 2016;2(1):16045. https://doi.org/10.1038/nrdp.2016.45.
[62] Bombeli T, Mueller M, Haeberli A. Anticoagulant properties of the vascular endothelium. Thromb Haemost 1997;77(3):408–23.
[63] Walker FJ. Regulation of activated protein C by a new protein. A possible function for bovine protein S. J Biol Chem 1980;255(12):5521–4.
[64] Schouten M, Wiersinga WJ, Levi M, van der Poll T. Inflammation, endothelium, and coagulation in sepsis. J Leukoc Biol 2008;83(3):536–45. https://doi.org/10.1189/jlb.0607373.
[65] Creasey AA, Chang AC, Feigen L, Wün TC, Taylor FB, Hinshaw LB. Tissue factor pathway inhibitor reduces mortality from Escherichia coli septic shock. J Clin Invest 1993;91(6):2850–60. https://doi.org/10.1172/JCI116529.
[66] Ostrovsky L, Woodman RC, Payne D, Teoh D, Kubes P. Antithrombin III prevents and rapidly reverses leukocyte recruitment in ischemia/reperfusion. Circulation 1997;96(7):2302–10. https://doi.org/10.1161/01.CIR.96.7.2302.
[67] Cesarman-Maus G, Hajjar KA. Molecular mechanisms of fibrinolysis. Br J Haematol 2005;129(3):307–21. https://doi.org/10.1111/j.1365-2141.2005.05444.x.
[68] Witkowski M, Landmesser U, Rauch U. Tissue factor as a link between inflammation and coagulation. Trends Cardiovas Med 2016;26(4):297–303. https://doi.org/10.1016/j.tcm.2015.12.001.
[69] Skjeflo EW, Christiansen D, Fure H, et al. *Staphylococcus aureus*- induced complement activation promotes tissue factor-mediated coagulation. J Thromb Haemost 2018;16(5):905–18. https://doi.org/10.1111/jth.13979.
[70] Rauch U, Nemerson Y. Circulating tissue factor and thrombosis. Curr Opin Hematol 2000;7(5):273–7. https://doi.org/10.1097/00062752-200009000-00003.
[71] Camerer E, Kolstø A-B, Prydz H. Cell biology of tissue factor, the principal initiator of blood coagulation. Thromb Res 1996;81(1):1–41. https://doi.org/10.1016/0049-3848(95)00209-X.
[72] Taylor FB, Chang A, Ruf W, et al. Lethal E. coli septic shock is prevented by blocking tissue factor with monoclonal antibody. Circ Shock 1991;33(3):127–34.
[73] Ranieri VM, Thompson BT, Barie PS, et al. Drotrecogin alfa (activated) in adults with septic shock. N Engl J Med 2012;366(22):2055–64. https://doi.org/10.1056/NEJMoa1202290.
[74] Vincent J-L, Francois B, Zabolotskikh I, et al. Effect of a recombinant human soluble thrombomodulin on mortality in patients with sepsis-associated coagulopathy: the SCARLET Randomized Clinical Trial. JAMA 2019;321(20):1993. https://doi.org/10.1001/jama.2019.5358.
[75] Esmon CT. The interactions between inflammation and coagulation. Br J Haematol 2005;131(4):417–30. https://doi.org/10.1111/j.1365-2141.2005.05753.x.
[76] Myles T, Nishimura T, Yun TH, et al. Thrombin activatable fibrinolysis inhibitor, a potential regulator of vascular inflammation. J Biol Chem 2003;278(51):51059–67. https://doi.org/10.1074/jbc.M306977200.
[77] Lay AJ, Donahue D, Tsai M-J, Castellino FJ. Acute inflammation is exacerbated in mice genetically predisposed to a severe protein C deficiency. Blood 2007;109(5):1984–91. https://doi.org/10.1182/blood-2006-07-037945.
[78] Joyce DE, Gelbert L, Ciaccia A, DeHoff B, Grinnell BW. Gene expression profile of antithrombotic protein C defines new mechanisms modulating inflammation and apoptosis. J Biol Chem 2001;276(14):11199–203. https://doi.org/10.1074/jbc.C100017200.

[79] Grey ST, Tsuchida A, Hau H, Orthner CL, Salem HH, Hancock WW. Selective inhibitory effects of the anticoagulant activated protein C on the responses of human mononuclear phagocytes to LPS, IFN-gamma, or phorbol ester. J Immunol 1994;153(8):3664–72.
[80] Coughlin SR. Thrombin signalling and protease-activated receptors. Nature 2000;407(6801):258–64. https://doi.org/10.1038/35025229.
[81] Pawlinski R, Mackman N. Tissue factor, coagulation proteases, and protease-activated receptors in endotoxemia and sepsis. Critical Care Med 2004;32(Suppl.):S293–7. https://doi.org/10.1097/01.CCM.0000128445.95144.B8.
[82] Vu T-KH, Hung DT, Wheaton VI, Coughlin SR. Molecular cloning of a functional thrombin receptor reveals a novel proteolytic mechanism of receptor activation. Cell 1991;64(6):1057–68. https://doi.org/10.1016/0092-8674(91)90261-V.
[83] Ishihara H, Connolly AJ, Zeng D, et al. Protease-activated receptor 3 is a second thrombin receptor in humans. Nature 1997;386(6624):502–6. https://doi.org/10.1038/386502a0.
[84] Xu W-f, Andersen H, Whitmore TE, et al. Cloning and characterization of human protease-activated receptor 4. Proc Natl Acad Sci 1998;95(12):6642–6. https://doi.org/10.1073/pnas.95.12.6642.
[85] Kahn ML, Nakanishi-Matsui M, Shapiro MJ, Ishihara H, Coughlin SR. Protease-activated receptors 1 and 4 mediate activation of human platelets by thrombin. J Clin Invest 1999;103(6):879–87. https://doi.org/10.1172/JCI6042.
[86] Camerer E, Huang W, Coughlin SR. Tissue factor- and factor X-dependent activation of protease-activated receptor 2 by factor VIIa. Proc Natl Acad Sci 2000;97(10):5255–60. https://doi.org/10.1073/pnas.97.10.5255.
[87] Tressel SL, Kaneider NC, Kasuda S, et al. A matrix metalloprotease-PAR1 system regulates vascular integrity, systemic inflammation and death in sepsis. EMBO Mol Med 2011;3(7):370–84. https://doi.org/10.1002/emmm.201100145.
[88] Austin KM, Covic L, Kuliopulos A. Matrix metalloproteases and PAR1 activation. Blood 2013;121(3):431–9. https://doi.org/10.1182/blood-2012-09-355958.
[89] Chi L, Li Y, Stehno-Bittel L, et al. Interleukin-6 production by endothelial cells via stimulation of protease-activated receptors is amplified by endotoxin and tumor necrosis factor-α. J Interf Cytokine Res 2001;21(4):231–40. https://doi.org/10.1089/107999001750169871.
[90] Pawlinski R, Pedersen B, Schabbauer G, et al. Role of tissue factor and protease-activated receptors in a mouse model of endotoxemia. Blood 2004;103(4):1342–7. https://doi.org/10.1182/blood-2003-09-3051.
[91] Taylor FB, Chang AC, Peer G, Li A, Ezban M, Hedner U. Active site inhibited factor VIIa (DEGR VIIa) attenuates the coagulant and interleukin-6 and -8, but not tumor necrosis factor, responses of the baboon to LD100 Escherichia coli. Blood 1998;91(5):1609–15.
[92] Cheng T, Liu D, Griffin JH, et al. Activated protein C blocks p53-mediated apoptosis in ischemic human brain endothelium and is neuroprotective. Nat Med 2003;9(3):338–42. https://doi.org/10.1038/nm826.
[93] Esmon CT. Is APC activation of endothelial cell PAR1 important in severe sepsis?: no. J Thromb Haemost 2005;3(9):1910–1. https://doi.org/10.1111/j.1538-7836.2005.01573.x.
[94] Ruf W. Is APC activation of endothelial cell PAR1 important in severe sepsis?: yes. J Thromb Haemost 2005;3(9):1912–4. https://doi.org/10.1111/j.1538-7836.2005.01576.x.
[95] Santamaria MH, Aletti F, Li JB, et al. Enteral tranexamic acid attenuates vasopressor resistance and changes in α1-adrenergic receptor expression in hemorrhagic shock. J Trauma Acute Care Surg 2017;83(2):263–70. https://doi.org/10.1097/TA.0000000000001513.
[96] DeLano FA, Hoyt DB, Schmid-Schonbein GW. Pancreatic digestive enzyme blockade in the intestine increases survival after experimental shock. Sci Transl Med 2013;5(169):169ra11. https://doi.org/10.1126/scitranslmed.3005046.
[97] Dufour A. Degradomics of matrix metalloproteinases in inflammatory diseases. Front Biosci (Schol Ed) 2015;7:150–67.
[98] Bauzá-Martinez J, Aletti F, Pinto BB, et al. Proteolysis in septic shock patients: plasma peptidomic patterns are associated with mortality. Br J Anaesth 2018;121(5):1065–74. https://doi.org/10.1016/j.bja.2018.05.072.

[99] Wozniak JM, Mills RH, Olson J, et al. Mortality risk profiling of Staphylococcus aureus bacteremia by multi-omic serum analysis reveals early predictive and pathogenic signatures. Cell 2020;182(5):1311–27. e14 https://doi.org/10.1016/j.cell.2020.07.040.

[100] Ng SC, Shi HY, Hamidi N, et al. Worldwide incidence and prevalence of inflammatory bowel disease in the 21st century: a systematic review of population-based studies. Lancet 2017;390(10114):2769–78. https://doi.org/10.1016/S0140-6736(17)32448-0.

[101] Linares de la Cal JA, Cantón C, Hermida C, Pérez-Miranda M, Maté-Jiménez J. Estimated incidence of inflammatory bowel disease in Argentina and Panama (1987–1993). Rev Esp Enferm Dig 1999;91(4):277–86.

[102] Roda G, Chien Ng S, Kotze PG, et al. Crohn's disease. Nat Rev Dis Primers 2020;6(1):22. https://doi.org/10.1038/s41572-020-0156-2.

[103] Vavricka SR, Brun L, Ballabeni P, et al. Frequency and risk factors for extraintestinal manifestations in the Swiss inflammatory bowel disease cohort. Am J Gastroenterol 2011;106(1):110–9. https://doi.org/10.1038/ajg.2010.343.

[104] Torres J, Mehandru S, Colombel J-F, Peyrin-Biroulet L. Crohn's disease. Lancet 2017;389(10080):1741–55. https://doi.org/10.1016/S0140-6736(16)31711-1.

[105] Thia KT, Sandborn WJ, Harmsen WS, Zinsmeister AR, Loftus EV. Risk factors associated with progression to intestinal complications of Crohn's disease in a population-based cohort. Gastroenterology 2010;139(4):1147–55. https://doi.org/10.1053/j.gastro.2010.06.070.

[106] Louis E. Behaviour of Crohn's disease according to the Vienna classification: changing pattern over the course of the disease. Gut 2001;49(6):777–82. https://doi.org/10.1136/gut.49.6.777.

[107] Duarte-Silva M, Afonso PC, de Souza PR, Peghini BC, Rodrigues-Júnior V, de Barros Cardoso CR. Reappraisal of antibodies against *Saccharomyces cerevisiae* (ASCA) as persistent biomarkers in quiescent Crohn's disease. Autoimmunity 2019;52(1):37–47. https://doi.org/10.1080/08916934.2019.1588889.

[108] Vermeire S. Laboratory markers in IBD: useful, magic, or unnecessary toys? Gut 2006;55(3):426–31. https://doi.org/10.1136/gut.2005.069476.

[109] Mumolo MG, Bertani L, Ceccarelli L, et al. From bench to bedside: fecal calprotectin in inflammatory bowel diseases clinical setting. WJG 2018;24(33):3681–94. https://doi.org/10.3748/wjg.v24.i33.3681.

[110] Skjøt-Arkil H, Schett G, Zhang C, et al. Investigation of two novel biochemical markers of inflammation, matrix metalloproteinase and cathepsin generated fragments of C-reactive protein, in patients with ankylosing spondylitis. Clin Exp Rheumatol 2012;30(3):371–9.

[111] Mainoli B, Hirota S, Edgington-Mitchell LE, Lu C, Dufour A. Proteomics and imaging in Crohn's disease: TAILS of unlikely allies. Trends Pharmacol Sci 2020;41(2):74–84. https://doi.org/10.1016/j.tips.2019.11.008.

[112] Kuek A, Hazleman BL, Ostor AJK. Immune-mediated inflammatory diseases (IMIDs) and biologic therapy: a medical revolution. Postgrad Med J 2007;83(978):251–60. https://doi.org/10.1136/pgmj.2006.052688.

[113] Sandborn WJ, Feagan BG, Rutgeerts P, et al. Vedolizumab as induction and maintenance therapy for Crohn's disease. N Engl J Med 2013;369(8):711–21. https://doi.org/10.1056/NEJMoa1215739.

[114] Present DH, Rutgeerts P, Targan S, et al. Infliximab for the treatment of fistulas in patients with Crohn's disease. N Engl J Med 1999;340(18):1398–405. https://doi.org/10.1056/NEJM199905063401804.

[115] Biancheri P, Brezski RJ, Di Sabatino A, et al. Proteolytic cleavage and loss of function of biologic agents that neutralize tumor necrosis factor in the mucosa of patients with inflammatory bowel disease. Gastroenterology 2015;149(6):1564–74. e3 https://doi.org/10.1053/j.gastro.2015.07.002.

[116] Brezski RJ, Jordan RE. Cleavage of IgGs by proteases associated with invasive diseases: an evasion tactic against host immunity? mAbs 2010;2(3):212–20. https://doi.org/10.4161/mabs.2.3.11780.

[117] Vergnolle N. Protease inhibition as new therapeutic strategy for GI diseases. Gut 2016;65(7):1215–24. https://doi.org/10.1136/gutjnl-2015-309147.

[118] Arihiro S, Ohtani H, Hiwatashi N, Torii A, Sorsa T, Nagura H. Vascular smooth muscle cells and pericytes express MMP-1, MMP-9, TIMP-1 and type I procollagen in inflammatory bowel disease. Histopathology 2001;39(1):50–9. https://doi.org/10.1046/j.1365-2559.2001.01142.x.

[119] Baugh MD, Perry MJ, Hollander AP, et al. Matrix metalloproteinase levels are elevated in inflammatory bowel disease. Gastroenterology 1999;117(4):814–22. https://doi.org/10.1016/S0016-5085(99)70339-2.
[120] Gordon MH, Anowai A, Young D, et al. N-terminomics/TAILS profiling of proteases and their substrates in ulcerative colitis. ACS Chem Biol 2019;14(11):2471–83. https://doi.org/10.1021/acschembio.9b00608.
[121] Denadai-Souza A, Bonnart C, Tapias NS, et al. Functional proteomic profiling of secreted serine proteases in health and inflammatory bowel disease. Sci Rep 2018;8(1):7834. https://doi.org/10.1038/s41598-018-26282-y.
[122] Anderson BM, Poole DP, Aurelio L, et al. Application of a chemical probe to detect neutrophil elastase activation during inflammatory bowel disease. Sci Rep 2019;9(1):13295. https://doi.org/10.1038/s41598-019-49840-4.
[123] Lefrancais E, Roga S, Gautier V, et al. IL-33 is processed into mature bioactive forms by neutrophil elastase and cathepsin G. Proc Natl Acad Sci 2012;109(5):1673–8. https://doi.org/10.1073/pnas.1115884109.
[124] Canals M, Poole DP, Veldhuis NA, Schmidt BL, Bunnett NW. G-protein–coupled receptors are dynamic regulators of digestion and targets for digestive diseases. Gastroenterology 2019;156(6). https://doi.org/10.1053/j.gastro.2019.01.266.
[125] Motta J-P, Palese S, Giorgio C, et al. Increased mucosal thrombin is associated with Crohn's disease and causes inflammatory damage through protease-activated receptors activation. J Crohn's Colitis 2020. https://doi.org/10.1093/ecco-jcc/jjaa229. jjaa229.
[126] Gordon MH, Chauvin A, Boisvert F-M, MacNaughton WK. Proteolytic processing of the epithelial adherens junction molecule E-cadherin by neutrophil elastase generates short peptides with novel wound-healing bioactivity. Cell Mol Gastroenterol Hepatol 2019;7(2):483–6. e8 https://doi.org/10.1016/j.jcmgh.2018.10.012.
[127] Garg P, Rojas M, Ravi A, et al. Selective ablation of matrix metalloproteinase-2 exacerbates experimental colitis: contrasting role of gelatinases in the pathogenesis of colitis. J Immunol 2006;177(6):4103–12. https://doi.org/10.4049/jimmunol.177.6.4103.
[128] Young D, Das N, Anowai A, Dufour A. Matrix metalloproteases as influencers of the cells' social media. IJMS 2019;20(16):3847. https://doi.org/10.3390/ijms20163847.
[129] Dufour A, Overall CM. Missing the target: matrix metalloproteinase antitargets in inflammation and cancer. Trends Pharmacol Sci 2013;34(4):233–42. https://doi.org/10.1016/j.tips.2013.02.004.

CHAPTER 7

Proteolytic processing of laminin and the role of cryptides in tumoral biology

Adriane Sousa de Siqueira[a,b], **Vanessa Morais Freitas**[a], **and Ruy Gastaldoni Jaeger**[a]

[a]*Department of Cell and Developmental Biology, Institute of Biomedical Sciences, University of Sao Paulo, São Paulo, SP, Brazil,* [b]*School of Health Sciences, Positivo University, Curitiba, PR, Brazil*

The tumor microenvironment represents a complex niche where neoplastic and non-neoplastic cells, soluble factors, signaling molecules, and components of the extracellular matrix (ECM) cooperatively interact and modulate cellular events related to tumor progression [1, 2]. During the invasion process, neoplastic cells come into close contact with the extracellular matrix, a three-dimensional network of macromolecules formed by well-organized fibrous polymers, represented by various isoforms of collagen and elastin. These polymers are embedded in an amorphous and highly hydrated mixture of proteoglycans and glycosaminoglycans, called fundamental substance [3, 4]. Fibrous and nonfibrous components proportion, fiber diameter, and organization of nonfibrous elements dictate the biomechanical properties of the extracellular matrix of a given tissue [5, 6].

ECM does not function merely as a structural support for cells but can influence cellular behavior affecting growth, differentiation, motility, and viability [3, 4, 6, 7]. ECM proteins are large and multifunctional and present domains that give several functions to the same macromolecule. Also, ECM can act as a reservoir of substances secreted by cells, including growth factors, cytokines, and matrix metalloproteinases (MMPs). Availability of these substances to cells can be regulated through matrix rearrangements, such as those that occur, for example, during healing and invasion of malignant tumors [8]. In addition, there are well-developed communication routes between cell surface and ECM, which allow cells to perceive the environment in which they are inserted [4]. In this process, cells depend mainly on integrins, a group of receptors that modulate their adhesion to substrates. Integrins are part of a family of transmembrane receptors involved in cell-cell and cell-matrix interactions, both in physiological and pathological processes [9, 10]. They are heterodimeric glycoproteins formed by two subunits, alpha and beta, which depend on bivalent cations (such as Ca^{2+} and Mg^{2+}) to establish connections with other molecules [11, 12].

Proteolytic Signaling in Health and Disease. https://doi.org/10.1016/B978-0-323-85696-6.00009-3

Structurally, ECM can be divided into two compartments: the basement membrane and the interstitial matrix. The interstitial matrix is synthesized mainly by fibroblasts and is rich in fibrillar collagen, proteoglycans, and glycoproteins (such as tenascin C and fibronectin). For this reason, it is highly hydrated and loaded, giving tissues the property of tensile strength. On the other hand, basement membrane represents a thin, flexible, and highly specialized layer of ECM, with a distinct macromolecular composition, being more compact and less porous than interstitial matrix [3, 4, 6, 13, 14]. The basement membrane is about 100–300 nm thick and acts as a kind of adhesive framework, contributing to anchoring cell layers and maintaining tissue architecture [13, 15, 16]. In addition to being located at the interface between epithelial cells and connective tissues, basement membrane can also be observed underlying endothelial and mesothelial cells and enmeshing muscle cells, Schwann cells, and adipocytes [13, 14, 17–19].

The basement membrane is related to the functions of support, cell anchoring, and tissue compartmentalization, as well as determines cell polarity. In the renal glomeruli and pulmonary alveoli, this structure functions as a kind of highly selective filter for salts and small molecules. It also takes part in cell adhesion, migration, growth, and differentiation, being essential for the formation and maintenance of tissues both in the early stages of development and in the adult phase [14, 15, 17, 20, 21]. Furthermore, the basement membrane represents an extension of the plasma membrane, working as an interactive interface between cells and the microenvironment in which they are inserted [13].

The basement membrane can contain more than 50 different components, being formed predominantly by: (1) type IV collagen, a nonfibrillar polymer; (2) laminin, an adhesive glycoprotein; (3) one or two types of nidogen [entactin]; and (4) perlecan and/or agrin, heparan sulfate proteoglycans. Small amounts of other proteoglycans and glycoproteins can also be found [14, 15, 17, 20, 21]. There are no significant architectural differences between basement membranes of various organs and tissues. However, its composition may vary since different isoforms of collagen IV and laminin are synthesized in different tissues [17, 21, 22].

Due to the fact that the basement membrane has pores of approximately 50 nm, only small molecules are able to passively diffuse through this structure [18]. Cells are typically on the order of 10 μm in size, too large to squeeze through the nanometer-scale pores of basement membrane [23]. Therefore, this structure serves as a barrier for large solutes and is impenetrable for most cells in normal adult tissue, except during inflammatory processes [4]. For this reason, it represents the first obstacle to be crossed by neoplastic cells during tumor invasion [2, 18, 24]. In order to overcome the basement membrane and migrate to the surrounding tissue, neoplastic cells undergo changes in their morphology, behavior, and relationship with the ECM [4, 8]. In this sense, they are capable of not only promoting basement membrane rupture, opening the way for its migration, but also of modifying the matrix composition, changing expression levels, or promoting degradation of different molecules. These events are largely mediated by proteolytic enzymes such as matrix metalloproteinases (MMPs) [2, 25–27].

MMPs are part of a family that contains more than 25 zinc-dependent endopeptidases, capable of degrading macromolecules present not only in ECM but also on the cell surface. Regarding their general structure, MMPs have three domains: the propeptide, the catalytic domain, and the C-terminal hemopexin domain [28, 29]. MMPs are initially expressed in an enzymatically inactive state due to the interaction of a cysteine residue from the prodomain with zinc ion from the catalytic domain. These enzymes become active when proteolytic removal of the prodomain or chemical modification of the cysteine residue occurs [25, 30–32]. Simply put, MMPs can be divided into two subclasses: soluble MMPs and membrane-anchored MMPs (MT-MMP, from membrane-type matrix metalloproteases) [2, 6, 30–33].

MMPs are involved in several physiological and pathological processes, including tissue remodeling and embryonic development and regulation of inflammatory events and cancer [29, 32, 34–36]. Since they are related to proteolysis, these enzymes can create spaces for migration, regulate cell architecture (through effects on the matrix and intercellular junctions), as well as activate, deactivate, or modify signaling molecules [32, 37].

Overexpression of several MMPs seems to be strongly related to tumor invasiveness [29, 34]. Initial attempts to identify which MMPs would be responsible for mediating the invasion of neoplastic cells revealed that secreted enzymes, such as MMPs-2 and -9, were the main ones involved in this event [18, 25]. However, this concept has changed in recent years since MMPs-2 and -9 do not appear to give tumor cells the direct ability to degrade or migrate through basement membrane [18, 38].

Currently, it is believed that MT-MMPs (especially MT1-MMP) assume an important role in basement membrane transmigration and tumor cell traffic through surrounding tissues [35, 39–49]. A key aspect of MT-MMP-mediated invasion is the ability to concentrate proteolytic activity on the cell surface region. It has been shown that soluble forms (not anchored to the membrane) of MT-MMPs, even when activated, are not related to basement membrane transmigration and invasion of interstitial matrix [18, 38, 50]. Unlike the constitutive secretion pathways that control molecule export, secretory vesicles containing active MT1-MMP are taken to the plasma membrane in cell-ECM contact points, especially in places where invadopodia (actin-rich membrane protrusions with proteolytic activity) are formed [35, 39, 41, 43, 44, 47, 51].

In the basement membrane, MMPs' performance would generate discontinuities or structural loss, supporting the formation of "channels" for cell migration, thus facilitating the invasion process [37, 52, 53]. However, MMPs have a much more complex role than simply opening paths in the matrix. These enzymes regulate access to growth factors and stimulate angiogenesis and cytoskeleton reorganization. MMPs can also promote profound modifications of ECM molecules, exposing cryptic epitopes and generating bioactive degradation products, which act as soluble ligands capable of modifying cellular genotype and phenotype [7, 24, 28, 30–33, 54–57]. These modifications, even if restricted to only a few molecules, can result in complex changes in cellular regulatory pathways [53].

Regulatory mechanisms ensure that ECM dynamics, assessed from the production, degradation, contraction, and remodeling of its components, remains normal during the development and function of organic tissues [32, 50]. Disruption of these control mechanisms promotes ECM deregulation and disorganization, which can lead to abnormal cellular behavior and finally to loss of homeostasis and function of a given organ [6, 8, 50]. Since they control growth factor availability and regulate, together with integrins and proteolytic enzymes, several cell signaling pathways, ECM and basement membrane constituents may play important roles in malignant tumor progression and metastasis formation [8, 26, 30–33].

For many years, our group has been studying the role of laminin, a basement membrane component, in the biological behavior of tumors [58–69]. For this reason, some relevant aspects about the structure and functions of this molecule will be addressed next.

Laminin

Laminin is the main adhesive glycoprotein that composes the basement membrane, and it shows both structural and biological activities [22, 70, 71]. This molecule was identified in 1979, after being isolated and purified from a mouse neoplasm with a large amount of basement membrane, the Engelbreth-Holm-Swarm (EHS) tumor [72]. Laminin is extremely important for the organization of the basement membrane throughout the body. It has been shown that mice that do not express this glycoprotein die soon after implantation of the embryo, whereas in the absence of molecules such as perlecan, collagen IV, or nidogen, this stage normally occurs [71].

Laminin is a heterotrimeric protein typically made up of intertwined alpha, beta, and gamma chains. Different types of laminin tertiary structures are formed, depending on trimer composition. The three most commonly described are cross-shaped, Y-shaped, or rod-shaped [73–75]. The trimer may present 2 or 3 short arms and a long arm. The long arm is formed by interlacing of portions of the three chains, generating a helix structure. The short arms are responsible for basement membrane assembly through interaction with other ECM molecules [22, 74–78]. Electron micrographs revealed that laminin presents rod-like and globular domains distributed by its structure [70, 74, 79].

Laminin chains have specific tissue distribution, which are susceptible to changes mediated by developmental and pathological processes [22, 74]. To date, five distinct genes have been identified for the alpha chain, four genes for the beta chain, and three genes for the gamma chain [17, 21, 22, 79]. The union of these different chains could result in 45 possible heterotrimeric combinations, without considering splicing variants. However, the actual number of combinations is much lower, due to restrictions of coupling between some chains [21, 80].

Currently, the laminin family consists of 16 isoforms in vertebrates. Each of them shows differences in their organization and interaction with receptors, allowing variations in the final structure, signaling, and basement membrane stability [13]. These isoforms are classified using a composite number that represents the chains by which they are formed (alpha, beta, gamma). Thus, laminin-111 (former laminin-1), for example, is so-called because it is composed of alpha1, beta1, and gamma1 chains. Laminin-332 (former laminin-5), in turn, gets its name from alpha3, beta3, and gamma2 chains. Other laminin isoforms are as follows: 121, 211, 213, 221, 311, 312, 321, 411, 421, 422, 423, 511, 521, and 523 [75, 80].

Laminin isoforms are synthesized by virtually all epithelial cells, in addition to smooth muscle cells, bone tissue, heart muscle, nerve cells, endothelial cells, and bone marrow. After synthesis, laminins are deposited mainly, but not exclusively, in the basement membrane [17]. Laminins have a distinct tissue distribution. Isoforms 511 and 521, for instance, are ubiquitously expressed in various tissues of the body. Isoforms 421 and 411, in turn, are found in the basement membrane that overlay endothelial cells [22, 81]. Due to its specific tissue location and structural diversity, laminin is capable of modulating specific physiological functions in each of the several basement membranes found in the body [74, 82].

Laminin is recognized as a cell adhesion molecule and binds to several components of ECM. Moreover, this glycoprotein presents a dual role in scaffolding and in signaling. Their N-terminal domain connects with distinct basement membrane biomacromolecules, determining its architecture. Their C-terminal domain, in turn, attaches to receptors anchored in the plasma membrane, like $\alpha3\beta1$, $\alpha6\beta1$, $\alpha6\beta7$, and $\alpha7\beta1$ integrins, generating a link between cells and biochemical signaling [13, 22, 74, 83]. Biological activities promoted by laminin include cell adhesion, migration, proliferation, neurite growth, and protease secretion [13, 17, 21, 74].

Due to its pronounced expression in the basement membrane, laminin possibly influences tumor biological behavior [84, 85]. This glycoprotein is considered an important autocrine factor produced by neoplastic cells to promote tumorigenesis [37]. Tumor cells injected with laminin-111 in mice were related to greater tumor growth rates compared to isolated neoplastic cells or cells injected with collagen I [86]. Laminin also has a chemotactic effect when in solution, which may explain its ability to stimulate cell movements related to invasion and metastasis processes [70, 87]. This molecule is also capable of inducing a malignant phenotype. It has been observed that cells cultured on laminin generate more tumors in vivo than nonadherent cells or cells adhering to fibronectin [88]. In addition, tumor cells co-injected with laminin in athymic rats promote an increase in the formation of lung metastatic colonies in these animals [89].

Laminin-111 cryptides

Laminin, like other molecules in the extracellular matrix, is able to regulate biological activities and function as a constant source of instructions for cells. These instructions are not static and appear according to the needs of a given tissue [90–92]. Biochemical dissection revealed that some of the laminin functions are related to specific domains, demonstrating that different parts of this molecule can have different effects on cells [93]. Additionally, studies have shown that small amino acid sequences derived from laminin are capable of stimulating biological functions. These sequences are known as bioactive peptides and some of the functions they perform are not the same as laminin in its intact form [90, 91].

Thus, the effects of laminin cannot be related only to the entire protein, since it has several active sites in its structure. These sites can be found in various components of the ECM and are generally not exposed on the surface of the molecule, remaining hidden within its intact structure. For this reason, they are called matricriptic sites or, more recently, cryptides [26, 92, 94–99]. Mechanisms that control cryptide exposure represent a pivotal step in the regulation of biological events by ECM and involve molecules' structural or conformational modifications. Among these mechanisms, we can highlight: (1) heterotypic attachment to other molecules, leading to conformational changes; (2) multimerization [organization of molecules in arrangement]; (3) cell-mediated mechanical forces; (4) denaturation; and (5) enzymatic degradation by proteolytic action [26, 92, 94–99].

This last mechanism is closely related to tumor progression, and laminin molecules can be mainly processed by MMPs. It has already been shown that MT1-MMP unmasks cryptic EGF-like fragments from the laminin gamma2 chain that activates EGFR [91, 100]. This protease also cleaves laminin-511, promoting prostate cancer cell migration and releasing putative biomarkers [101]. Additionally, laminin gamma2 chain proteolysis by MT1-MMP is related to cell migration and architectural organization in ovarian clear cell carcinoma [102]. Other proteases may also play a role in laminin processing, such as hepsin [103], matriptase [104], BMP-1 [105], plasmin [106], neutrophil elastase, and cathepsin G [107].

The most-accepted hypothesis for the existence of matricriptic sites is that they are part of an evolutionary strategy to control different cellular activities. In this way, certain instructions can remain hidden in the intact molecule until required or until a specific proteolytic enzyme is activated. Therefore, it is not necessary to produce inhibitors or block an activity that has not been requested at a given time [91, 99].

Different biological activities stimulated by laminin, both in normal and neoplastic cells, aroused interest in studying its different active sites. Therefore, proteolytic fragments and synthetic peptides corresponding to several structural domains of laminin have been used to locate and analyze such activities in order to demonstrate that laminin is a multifunctional protein with several biologically relevant domains [24, 108].

In general, to search for active domains in laminin molecules, systematic sequential scans were performed, with subsequent synthesis of peptides that had between 6 and 12 amino acids and that were superimposed on neighboring sequences by about 4 amino acids. Through this approach, approximately 673 synthetic peptides were produced only for laminin-111. Taking into account that cell adhesion is the main function related to laminins, the peptides, after being synthesized, were tested for their ability to mediate adhesion [109–113]. Later, through additional experiments, laminin peptides were characterized as modulators of the progression of malignant tumors [24]. From now on, we will discuss some relevant laminin-111-derived cryptides and their role in tumor behavior.

YIGSR (YIGSR, short arm of beta1 chain)

YIGSR, a sequence of five amino acids (tyrosine-isoleucine-glycine-serine-arginine), derived from beta1 chain, was one of the first peptides described for laminin. It binds to 32/67 kDa cell surface receptor and has several activities related to inhibition of tumor progression [114, 115]. Iwamoto and collaborators, in 1987, first investigated the relation between YIGSR and tumor progression, demonstrating that this sequence decreases lung colony formation in mice injected with melanoma cells [89]. In vivo, this peptide inhibits the formation of bone metastases [116], as well as inhibits angiogenesis in several assays [117]. Tumors grown in vivo with YIGSR present less blood vessels, which may explain their smaller size [24].

YIGSR conjugated to liposomes has also been used for directional administration of chemotherapy drugs. YIGSR-PEG liposomes carrying Adriamycin exhibited in vitro cytotoxicity for HT080 cells [118], while peptide-anchored liposomes containing 5-fluorouracil stimulated tumor regression in mice bearing B16F10 melanoma cells [119]. More recently, tumor growth was decreased in mice treated with YIGSR-displaying bacteriophages [120].

On the other hand, YIGSR has been used as a radiopharmaceutical for the location of neoplasms, appearing to be an exceptional radiotracer with high sensitivity and specificity in mice presenting Ehrlich ascites tumors [121]. Furthermore, studies involving YIGSR nanoparticles demonstrated that they present a twofold increase in uptake compared to nanoparticles containing scrambled peptides in neoplastic cells in vitro, while both peptides were not uptaken by normal lung cells. Those results indicate that YIGSR conjugated with a specific ligand can promote a targeted and effective delivery of this peptide for tumor regression [24, 122].

Findings regarding the mechanisms by which YIGSR acts are preliminary and require further investigations. But these mechanisms may be related to increased phosphorylation [123] and increased apoptosis [124]. In prostate cancer-derived cells (PC3), YIGSR restrained growth, migration, and ATP synthesis [125]. Our group reported that adenoid cystic carcinoma cells

(CAC2) demonstrated a fibroblast-like morphology in YIGSR presence, while untreated cells exhibited epithelioid characteristics. Thus, we may infer that the peptide interfered with epithelial-mesenchymal transition [64].

IKVAV (IKVAV, long arm of alpha1 chain)

The IKVAV peptide is located in alpha1 chain, just above the globular domain [126]. This peptide stimulates cell activities such as adhesion and migration and also may influence tumor behavior by prompting angiogenesis, MMP secretion, tumor cell growth, and metastasis [126–130]. IKVAV has already been related to elevated sprouting and high vessel number in in vitro assays [128] and promoted tumor growth and vessel formation in colon cancer cells co-injected with Matrigel in nude mice [131]. Furthermore, IKVAV can trigger invasive phenotype of melanoma K-1735 clones cultured in Matrigel, probably by stimulating MMP-2 activity [132]. B16F10 melanoma cells treated with IKVAV exhibited increased activity and production of tissue plasminogen activator (t-PA), a metastasis-associated protease, but this peptide presented no effect on the expression of this enzyme in nonmetastatic B16F1 cells [133]. When interacting with colon cancer cells, IKVAV stimulated metastasis formation in the liver [131].

Studies by our group demonstrated that IKVAV induces the secretion of MMPs in cells derived from adenoid cystic carcinoma [61, 62]. In monocytes and macrophages, IIKVAV induced the secretion of prostaglandin E_2 and MMPs, but the intact laminin molecule failed to influence these monocyte responses [134]. Additionally, IKVAV incorporation on polymer-modified adenovirus enabled virus entry into PC3 cells through a6β1 integrin, suggesting that it acts through this receptor to localize in neoplasms [135]. This peptide was also related to amyloid fibril formation, associated with Alzheimer's and Parkinson's diseases [136].

In addition to the analysis of its effects on tumor development, IKVAV has been employed in studies involving biomaterial engineering for tissue regeneration in the central nervous system and in skin wound healing [137–139].

AG73 (RKRLQVQLSIRT, LG4 domain of alpha1 chain)

AG73 bioactive sequence is possibly available in the tumor microenvironment since the LG4 domain of alpha1 chain (which contains the peptide) was found in the basement membrane extract (Matrigel) and in laminin-111 extracted from EHS tumors [24]. Thus, it is not surprising that functions related to this peptide have been studied in a wide variety of tumor cell lines. AG73 was first identified due to its ability to promote cell adhesion of tumor cell lines HT1080 (fibrosarcoma), B16F10 (melanoma), and SW480 (colon adenocarcinoma) [109]. In vitro, AG73 stimulates cell adhesion, migration, invasion, and MMP secretion in melanoma cells [140], promotes angiogenesis in different assays [141], and alters filopodium formation in breast cancer cells [142].

In vivo, this sequence promotes tumor growth and metastasis in the lung and liver [143, 144]. Breast cancer and melanoma cell metastasis to the bone is stimulated by AG73 [140], likewise ovarian cancer growth and spread [145]. AG73 may also prompt tumor development because of its capacity of promoting angiogenesis [141]. Studies by our group have shown that AG73 stimulates biological activities related to tumor progression, such as invasion, protease secretion, and invadopodia formation in cell lines derived from salivary gland neoplasms [63, 65] and oral squamous cell carcinoma [146].

AG73 sequence plays an important role in binding the laminin LG4 domain to syndecans, transmembrane proteoglycans that may cooperate with chemokine receptors, integrins, and growth factors [147, 148]. AG73 is capable of increasing invadopodia of adenoid cystic carcinoma cells, and this event is restrained by β1 integrin silencing and RAC1 and ERK inhibition [65]. Moreover, this peptide induces the colocalization of syndecan-1 and β1 integrin in oral squamous carcinoma cells [146] and malignant and benign salivary gland tumors [63], suggesting that syndecans and integrins synergically cooperate to the capacity of AG73 to stimulate tumor cell invasion.

AG73-coated liposomes have shown promising results in gene delivery to cancer cells [149], enhanced antitumor drug delivery [150], and as a contrast in ultrasound imaging of tumor vasculature [151]. Also, AG73 can be conjugated to chitosan and alginate or mixed with agarose gel and employed in cell and tissue engineering [152–154].

C16 (KAFDITYVRLKF, gamma1 chain)

When it was identified as an active sequence, the C16 peptide, found in the first globular domain of the gamma1 chain, showed strong adhesive activity [110]. Integrins αvβ3 and α5β1 have already been identified as receptors for this peptide, by means of affinity chromatography and adhesion assays with blocking antibodies [155]. Because it does not have an RGD sequence (the usual ligand of the integrins mentioned above), it is believed that C16 may stimulate distinct signaling pathways from those activated by fibronectin and other molecules containing this domain [24].

It has been shown that C16 stimulates neurite growth, hepatocyte differentiation, decreased acini formation and differentiation, MMP secretion, and formation of lung metastases from melanoma cells [110, 155–158]. Furthermore, C16 can prevent leukocyte infiltration and inflammation in allergic encephalomyelitis models [159], inducing extravascular migration of melanoma cells [160], as well as stimulating membrane hyperpolarization in a hybrid cell line of neuroblastoma and glioblastoma [161]. Its remarkable proangiogenic activity [155, 158, 162] stimulated the conjugation of this peptide to polymeric implants used in tissue regeneration, with satisfactory results [163]. Peptide C16 increases invadopodia activity of adenoid cystic carcinoma cells [65] and squamous carcinoma and fibrosarcoma cells probably by interacting with β1 integrin and activating Src and ERK 1/2 signaling pathways [68]. In prostate cancer cells, the C16 sequence increased Tks5 and reactive oxygen species, further explaining invadopodia activity mechanisms induced by laminin peptides [58]. In breast cancer cells, in turn, C16 induced GPNMB gene overexpression, which may be associated with malignant phenotype [69].

Recently, C16 was related to increased vascularization and reduced inflammation in neuromyelitis optica (an autoimmune condition that causes disturbance in the spinal cord and optic nerve) [164], and has shown an important therapeutic potential for psoriasis [165] (Fig. 1).

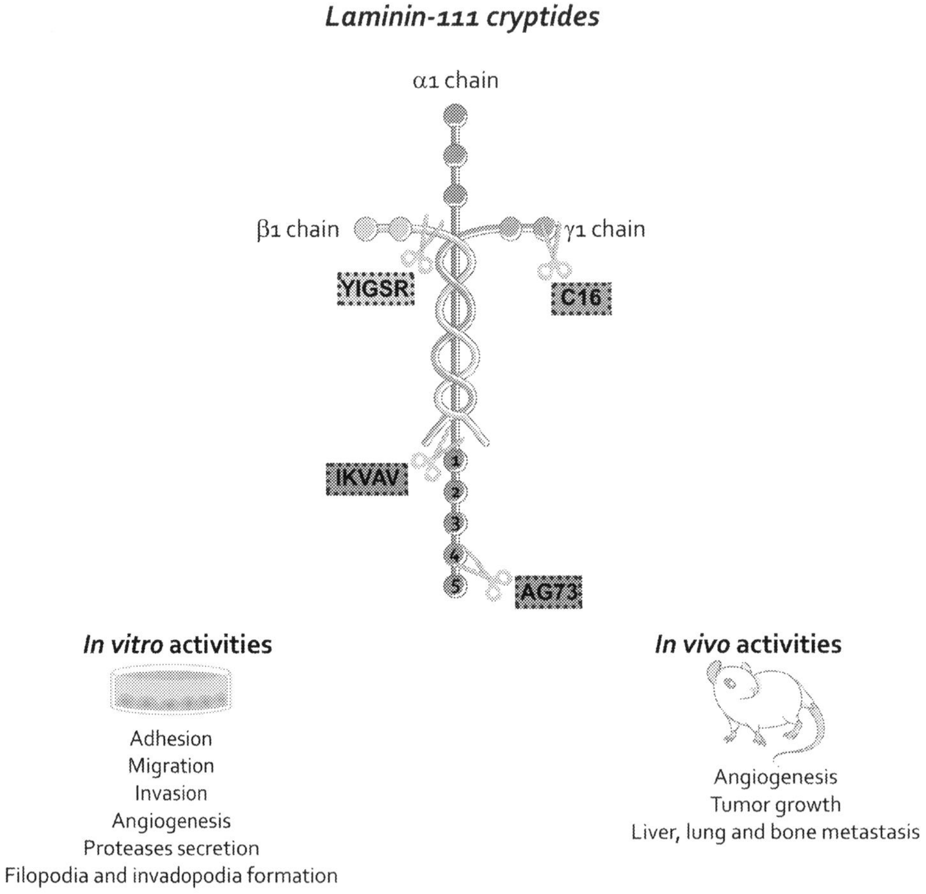

Fig. 1
Laminin-111 schematic model exhibiting the location of cryptides with relevant roles in tumor behavior and main cryptide activities detected in in vivo and in vitro studies. Scissors represent protease-mediated laminin cleavage.

Conclusion

It is known that laminins can undergo controlled proteolysis during basal lamina degradation induced by tumor cells, releasing peptides endowed with biological activities [24, 26, 93]. Laminin fragments have already been detected in serological exams of patients with squamous cell carcinoma of the head and neck [166], which corroborates the hypothesis that enzymatic degradation mediated by neoplastic cells can promote laminin fragment release, making them available to the tumor. The role of laminin-derived peptides in tumor behavior is not fully elucidated, but we may infer that they may be released from the intact molecule and modulate cellular functions, sometimes in a greater way than the whole molecule itself. These effects are often associated with a more aggressive and invasive cancer phenotype, such as invasion, angiogenesis, and metastasis. We also believe that laminin peptides are generated by protease action and can induce a positive feedback loop by increasing MMP production, for example, contributing to cancer progression.

References

[1] Catalano V, Turdo A, Di Franco S, Dieli F, Todaro M, Stassi G. Tumor and its microenvironment: a synergistic interplay. Semin Cancer Biol 2013;23(6 Pt B):522–32.

[2] Rowe RG, Weiss SJ. Navigating ECM barriers at the invasive front: the cancer cell-stroma interface. Annu Rev Cell Dev Biol 2009;25:567–95.

[3] Frantz C, Stewart KM, Weaver VM. The extracellular matrix at a glance. J Cell Sci 2010;123(Pt 24):4195–200.

[4] Tanzer ML. Current concepts of extracellular matrix. J Orthop Sci 2006;11(3):326–31.

[5] Ozbek S, Balasubramanian PG, Chiquet-Ehrismann R, Tucker RP, Adams JC. The evolution of extracellular matrix. Mol Biol Cell 2010;21(24):4300–5.

[6] Lu P, Weaver VM, Werb Z. The extracellular matrix: a dynamic niche in cancer progression. J Cell Biol 2012;196(4):395–406.

[7] Hynes RO. The extracellular matrix: not just pretty fibrils. Science 2009;326(5957):1216–9.

[8] Miles FL, Sikes RA. Insidious changes in stromal matrix fuel cancer progression. Mol Cancer Res 2014.

[9] Wolfenson H, Lavelin I, Geiger B. Dynamic regulation of the structure and functions of integrin adhesions. Dev Cell 2013;24(5):447–58.

[10] Hynes RO. Integrins: versatility, modulation, and signaling in cell adhesion. Cell 1992;69(1):11–25.

[11] Harburger DS, Calderwood DA. Integrin signalling at a glance. J Cell Sci 2009;122(Pt 2):159–63.

[12] Missan DS, DiPersio M. Integrin control of tumor invasion. Crit Rev Eukaryot Gene Expr 2013;22(4):309–24.

[13] Yurchenco PD. Basement membranes: cell scaffoldings and signaling platforms. Cold Spring Harb Perspect Biol 2011;3(2).

[14] Candiello J, Balasubramani M, Schreiber EM, Cole GJ, Mayer U, Halfter W, et al. Biomechanical properties of native basement membranes. FEBS J 2007;274(11):2897–908.

[15] Halfter W, Candiello J, Hu H, Zhang P, Schreiber E, Balasubramani M. Protein composition and biomechanical properties of in vivo-derived basement membranes. Cell Adh Migr 2013;7(1):64–71.

[16] Yurchenco PD, Amenta PS, Patton BL. Basement membrane assembly, stability and activities observed through a developmental lens. Matrix Biol 2004;22(7):521–38.

[17] Bosman FT, Stamenkovic I. Functional structure and composition of the extracellular matrix. J Pathol 2003;200(4):423–8.
[18] Rowe RG, Weiss SJ. Breaching the basement membrane: who, when and how? Trends Cell Biol 2008;18(11):560–74.
[19] Kalluri R. Basement membranes: structure, assembly and role in tumour angiogenesis. Nat Rev Cancer 2003;3(6):422–33.
[20] Suzuki N, Yokoyama F, Nomizu M. Functional sites in the laminin alpha chains. Connect Tissue Res 2005;46(3):142–52.
[21] Colognato H, Yurchenco PD. Form and function: the laminin family of heterotrimers. Dev Dyn 2000;218(2):213–34.
[22] Durbeej M. Laminins. Cell Tissue Res 2010;339(1):259–68.
[23] Chang J, Chaudhuri O. Beyond proteases: basement membrane mechanics and cancer invasion. J Cell Biol 2019;218(8):2456–69.
[24] Kikkawa Y, Hozumi K, Katagiri F, Nomizu M, Kleinman HK, Koblinski JE. Laminin-111-derived peptides and cancer. Cell Adh Migr 2013;7(1):150–256.
[25] Sternlicht MD, Werb Z. How matrix metalloproteinases regulate cell behavior. Annu Rev Cell Dev Biol 2001;17:463–516.
[26] Schenk S, Quaranta V. Tales from the crypt[ic] sites of the extracellular matrix. Trends Cell Biol 2003;13(7):366–75.
[27] Mott JD, Werb Z. Regulation of matrix biology by matrix metalloproteinases. Curr Opin Cell Biol 2004;16(5):558–64.
[28] Overall CM, Blobel CP. In search of partners: linking extracellular proteases to substrates. Nat Rev Mol Cell Biol 2007;8(3):245–57.
[29] Kessenbrock K, Plaks V, Werb Z. Matrix metalloproteinases: regulators of the tumor microenvironment. Cell 2010;141(1):52–67.
[30] Winkler J, Abisoye-Ogunniyan A, Metcalf KJ, Werb Z. Concepts of extracellular matrix remodelling in tumour progression and metastasis. Nat Commun 2020;11(1):5120.
[31] Lawson DA, Kessenbrock K, Davis RT, Pervolarakis N, Werb Z. Tumour heterogeneity and metastasis at single-cell resolution. Nat Cell Biol 2018;20(12):1349–60.
[32] Page-McCaw A, Ewald AJ, Werb Z. Matrix metalloproteinases and the regulation of tissue remodelling. Nat Rev Mol Cell Biol 2007;8(3):221–33.
[33] Nguyen-Ngoc KV, Cheung KJ, Brenot A, Shamir ER, Gray RS, Hines WC, et al. ECM microenvironment regulates collective migration and local dissemination in normal and malignant mammary epithelium. Proc Natl Acad Sci U S A 2012;109(39):E2595–604.
[34] Egeblad M, Werb Z. New functions for the matrix metalloproteinases in cancer progression. Nat Rev Cancer 2002;2(3):161–74.
[35] Gifford V, Itoh Y. MT1-MMP-dependent cell migration: proteolytic and non-proteolytic mechanisms. Biochem Soc Trans 2019;47(3):811–26.
[36] Niland S, Eble JA. Hold on or cut? Integrin- and MMP-mediated cell-matrix interactions in the tumor microenvironment. Int J Mol Sci 2020;22(1).
[37] Sharma M, Sah P, Sharma SS, Radhakrishnan R. Molecular changes in invasive front of oral cancer. J Oral Maxillofac Pathol 2013;17(2):240–7.
[38] Hotary K, Li XY, Allen E, Stevens SL, Weiss SJ. A cancer cell metalloprotease triad regulates the basement membrane transmigration program. Genes Dev 2006;20(19):2673–86.
[39] Watanabe A, Hoshino D, Koshikawa N, Seiki M, Suzuki T, Ichikawa K. Critical role of transient activity of MT1-MMP for ECM degradation in invadopodia. PLoS Comput Biol 2013;9(5), e1003086.
[40] Hoshino D, Nagano M, Saitoh A, Koshikawa N, Suzuki T, Seiki M. The phosphoinositide-binding protein ZF21 regulates ECM degradation by invadopodia. PLoS One 2013;8(1), e50825.
[41] Branch KM, Hoshino D, Weaver AM. Adhesion rings surround invadopodia and promote maturation. Biol Open 2012;1(8):711–22.

[42] Hoshino D, Koshikawa N, Suzuki T, Quaranta V, Weaver AM, Seiki M, et al. Establishment and validation of computational model for MT1-MMP dependent ECM degradation and intervention strategies. PLoS Comput Biol 2012;8(4), e1002479.
[43] Pinheiro JJ, Nascimento CF, Freitas VM, de Siqueira AS, Junior SM, Jaeger RG. Invadopodia proteins, cortactin and membrane type I matrix metalloproteinase (MT1-MMP) are expressed in ameloblastoma. Histopathology 2011;59(6):1266–9.
[44] Takino T, Tsuge H, Ozawa T, Sato H. MT1-MMP promotes cell growth and ERK activation through c-Src and paxillin in three-dimensional collagen matrix. Biochem Biophys Res Commun 2010;396(4):1042–7.
[45] Sabeh F, Li XY, Saunders TL, Rowe RG, Weiss SJ. Secreted versus membrane-anchored collagenases: relative roles in fibroblast-dependent collagenolysis and invasion. J Biol Chem 2009;284(34):23001–11.
[46] Barbolina MV, Stack MS. Membrane type 1-matrix metalloproteinase: substrate diversity in pericellular proteolysis. Semin Cell Dev Biol 2008;19(1):24–33.
[47] Bravo-Cordero JJ, Marrero-Diaz R, Megias D, Genis L, Garcia-Grande A, Garcia MA, et al. MT1-MMP proinvasive activity is regulated by a novel Rab8-dependent exocytic pathway. EMBO J 2007;26(6):1499–510.
[48] Artym VV, Zhang Y, Seillier-Moiseiwitsch F, Yamada KM, Mueller SC. Dynamic interactions of cortactin and membrane type 1 matrix metalloproteinase at invadopodia: defining the stages of invadopodia formation and function. Cancer Res 2006;66(6):3034–43.
[49] Itoh Y. MT1-MMP: a key regulator of cell migration in tissue. IUBMB Life 2006;58(10):589–96.
[50] Larsen M, Artym VV, Green JA, Yamada KM. The matrix reorganized: extracellular matrix remodeling and integrin signaling. Curr Opin Cell Biol 2006;18(5):463–71.
[51] Clark ES, Weaver AM. A new role for cortactin in invadopodia: regulation of protease secretion. Eur J Cell Biol 2008;87(8-9):581–90.
[52] Ziober AF, Falls EM, Ziober BL. The extracellular matrix in oral squamous cell carcinoma: friend or foe? Head Neck 2006;28(8):740–9.
[53] Alexandrova AY. Evolution of cell interactions with extracellular matrix during carcinogenesis. Biochemistry (Mosc) 2008;73(7):733–41.
[54] Wang X, Yu YY, Lieu S, Yang F, Lang J, Lu C, et al. MMP9 regulates the cellular response to inflammation after skeletal injury. Bone 2013;52(1):111–9.
[55] Nakasone ES, Askautrud HA, Kees T, Park JH, Plaks V, Ewald AJ, et al. Imaging tumor-stroma interactions during chemotherapy reveals contributions of the microenvironment to resistance. Cancer Cell 2012;21(4):488–503.
[56] Ewald AJ, Huebner RJ, Palsdottir H, Lee JK, Perez MJ, Jorgens DM, et al. Mammary collective cell migration involves transient loss of epithelial features and individual cell migration within the epithelium. J Cell Sci 2012;125(Pt 11):2638–54.
[57] Lu P, Takai K, Weaver VM, Werb Z. Extracellular matrix degradation and remodeling in development and disease. Cold Spring Harb Perspect Biol 2011;3(12).
[58] Caires-Dos-Santos L, da Silva SV, Smuczek B, de Siqueira AS, Cruz KSP, Barbuto JAM, et al. Laminin-derived peptide C16 regulates Tks expression and reactive oxygen species generation in human prostate cancer cells. J Cell Physiol 2020;235(1):587–98.
[59] Capuano AC, Jaeger RG. The effect of laminin and its peptide SIKVAV on a human salivary gland myoepithelioma cell line. Oral Oncol 2004;40(1):36–42.
[60] Freitas VM, Jaeger RG. The effect of laminin and its peptide SIKVAV on a human salivary gland adenoid cystic carcinoma cell line. Virchows Arch 2002;441(6):569–76.
[61] Freitas VM, Scheremeta B, Hoffman MP, Jaeger RG. Laminin-1 and SIKVAV a laminin-1-derived peptide, regulate the morphology and protease activity of a human salivary gland adenoid cystic carcinoma cell line. Oral Oncol 2004;40(5):483–9.
[62] Freitas VM, Vilas-Boas VF, Pimenta DC, Loureiro V, Juliano MA, Carvalho MR, et al. SIKVAV, a laminin alpha1-derived peptide, interacts with integrins and increases protease activity of a human

salivary gland adenoid cystic carcinoma cell line through the ERK 1/2 signaling pathway. Am J Pathol 2007;171(1):124–38.

[63] Gama-de-Souza LN, Cyreno-Oliveira E, Freitas VM, Melo ES, Vilas-Boas VF, Moriscot AS, et al. Adhesion and protease activity in cell lines from human salivary gland tumors are regulated by the laminin-derived peptide AG73, syndecan-1 and beta1 integrin. Matrix Biol 2008;27(5):402–19.

[64] Morais Freitas V, Nogueira da Gama de Souza L, Cyreno Oliveira E, Furuse C, Cavalcanti de Araujo V, Gastaldoni Jaeger R. Malignancy-related 67kDa laminin receptor in adenoid cystic carcinoma. Effect on migration and beta-catenin expression. Oral Oncol 2007;43(10):987–98.

[65] Nascimento CF, de Siqueira AS, Pinheiro JJ, Freitas VM, Jaeger RG. Laminin-111 derived peptides AG73 and C16 regulate invadopodia activity of a human adenoid cystic carcinoma cell line. Exp Cell Res 2011;317(18):2562–72.

[66] Nascimento CF, Gama-De-Souza LN, Freitas VM, Jaeger RG. Role of MMP9 on invadopodia formation in cells from adenoid cystic carcinoma. Study by laser scanning confocal microscopy. Microsc Res Tech 2010;73(2):99–108.

[67] Siqueira AS, Gama-de-Souza LN, Arnaud MV, Pinheiro JJ, Jaeger RG. Laminin-derived peptide AG73 regulates migration, invasion, and protease activity of human oral squamous cell carcinoma cells through syndecan-1 and beta1 integrin. Tumour Biol 2010;31(1):46–58.

[68] Siqueira AS, Pinto MP, Cruz MC, Smuczek B, Cruz KS, Barbuto JA, et al. Laminin-111 peptide C16 regulates invadopodia activity of malignant cells through beta1 integrin, Src and ERK 1/2. Oncotarget 2016;7(30):47904–17.

[69] Smuczek B, Santos ES, Siqueira AS, Pinheiro JJV, Freitas VM, Jaeger RG. The laminin-derived peptide C16 regulates GPNMB expression and function in breast cancer. Exp Cell Res 2017;358(2):323–34.

[70] Martin GR, Timpl R. Laminin and other basement membrane components. Annu Rev Cell Biol 1987;3:57–85.

[71] Sasaki T, Fassler R, Hohenester E. Laminin: the crux of basement membrane assembly. J Cell Biol 2004;164(7):959–63.

[72] Timpl R, Rohde H, Robey PG, Rennard SI, Foidart JM, Martin GR. Laminin—a glycoprotein from basement membranes. J Biol Chem 1979;254(19):9933–7.

[73] Rasmussen DGK, Karsdal MA. Chapter 29—Laminins. In: Karsdal MA, editor. Biochemistry of Collagens, Laminins and Elastin. 2nd ed. Academic Press; 2019. p. 209–63.

[74] Aumailley M. The laminin family. Cell Adh Migr 2013;7(1):48–55.

[75] Aumailley M, Bruckner-Tuderman L, Carter WG, Deutzmann R, Edgar D, Ekblom P, et al. A simplified laminin nomenclature. Matrix Biol 2005;24(5):326–32.

[76] Roig-Rosello E, Rousselle P. The human epidermal basement membrane: a shaped and cell instructive platform that aging slowly alters. Biomolecules 2020;10(12).

[77] Krieg T, Aumailley M. The extracellular matrix of the dermis: flexible structures with dynamic functions. Exp Dermatol 2011;20(8):689–95.

[78] Aumailley M, Rousselle P. Laminins of the dermo-epidermal junction. Matrix Biol 1999;18(1):19–28.

[79] Miner JH, Yurchenco PD. Laminin functions in tissue morphogenesis. Annu Rev Cell Dev Biol 2004;20:255–84.

[80] Macdonald PR, Lustig A, Steinmetz MO, Kammerer RA. Laminin chain assembly is regulated by specific coiled-coil interactions. J Struct Biol 2010;170(2):398–405.

[81] Yousif LF, Di Russo J, Sorokin L. Laminin isoforms in endothelial and perivascular basement membranes. Cell Adh Migr 2013;7(1):101–10.

[82] Aumailley M, Smyth N. The role of laminins in basement membrane function. J Anat 1998;193(Pt 1):1–21.

[83] Karamanos NK, Theocharis AD, Piperigkou Z, Manou D, Passi A, Skandalis SS, et al. A guide to the composition and functions of the extracellular matrix. FEBS J 2021.

[84] Givant-Horwitz V, Davidson B, Reich R. Laminin-induced signaling in tumor cells. Cancer Lett 2005;223(1):1–10.

[85] Kulasekara KK, Lukandu OM, Neppelberg E, Vintermyr OK, Johannessen AC, Costea DE. Cancer progression is associated with increased expression of basement membrane proteins in three-dimensional in vitro models of human oral cancer. Arch Oral Biol 2009;54(10):924–31.
[86] Terranova VP, Liotta LA, Russo RG, Martin GR. Role of laminin in the attachment and metastasis of murine tumor cells. Cancer Res 1982;42(6):2265–9.
[87] Malinda KM, Kleinman HK. The laminins. Int J Biochem Cell Biol 1996;28(9):957–9.
[88] Terranova VP, Williams JE, Liotta LA, Martin GR. Modulation of the metastatic activity of melanoma cells by laminin and fibronectin. Science 1984;226(4677):982–5.
[89] Iwamoto Y, Robey FA, Graf J, Sasaki M, Kleinman HK, Yamada Y, et al. YIGSR, a synthetic laminin pentapeptide, inhibits experimental metastasis formation. Science 1987;238(4830):1132–4.
[90] Davis GE, Bayless KJ, Davis MJ, Meininger GA. Regulation of tissue injury responses by the exposure of matricryptic sites within extracellular matrix molecules. Am J Pathol 2000;156(5):1489–98.
[91] Schenk S, Hintermann E, Bilban M, Koshikawa N, Hojilla C, Khokha R, et al. Binding to EGF receptor of a laminin-5 EGF-like fragment liberated during MMP-dependent mammary gland involution. J Cell Biol 2003;161(1):197–209.
[92] Tran KT, Lamb P, Deng JS. Matrikines and matricryptins: Implications for cutaneous cancers and skin repair. J Dermatol Sci 2005;40(1):11–20.
[93] Faisal Khan KM, Laurie GW, McCaffrey TA, Falcone DJ. Exposure of cryptic domains in the alpha 1-chain of laminin-1 by elastase stimulates macrophages urokinase and matrix metalloproteinase-9 expression. J Biol Chem 2002;277(16):13778–86.
[94] Iavarone F, Desiderio C, Vitali A, Messana I, Martelli C, Castagnola M, et al. Cryptides: latent peptides everywhere. Crit Rev Biochem Mol Biol 2018;53(3):246–63.
[95] Gaglione R, Cesaro A, Dell'Olmo E, Di Girolamo R, Tartaglione L, Pizzo E, et al. Cryptides identified in human apolipoprotein B as new weapons to fight antibiotic resistance in cystic fibrosis disease. Int J Mol Sci 2020;21(6).
[96] Hattori T, Mukai H. Cryptides: biologically active peptides hidden in protein structures. Nihon Yakurigaku Zasshi 2014;144(5):234–8.
[97] Pimenta DC, Lebrun I. Cryptides: buried secrets in proteins. Peptides 2007;28(12):2403–10.
[98] Ueki N, Someya K, Matsuo Y, Wakamatsu K, Mukai H. Cryptides: functional cryptic peptides hidden in protein structures. Biopolymers 2007;88(2):190–8.
[99] Maquart FX, Pasco S, Ramont L, Hornebeck W, Monboisse JC. An introduction to matrikines: extracellular matrix-derived peptides which regulate cell activity. Implication in tumor invasion. Crit Rev Oncol Hematol 2004;49(3):199–202.
[100] Ding BS, Nolan DJ, Guo P, Babazadeh AO, Cao Z, Rosenwaks Z, et al. Endothelial-derived angiocrine signals induce and sustain regenerative lung alveolarization. Cell 2011;147(3):539–53.
[101] Bair EL, Chen ML, McDaniel K, Sekiguchi K, Cress AE, Nagle RB, et al. Membrane type 1 matrix metalloprotease cleaves laminin-10 and promotes prostate cancer cell migration. Neoplasia 2005;7(4):380–9.
[102] Kato N, Motoyama T. Relation between laminin-5 gamma 2 chain and cell surface metalloproteinase MT1-MMP in clear cell carcinoma of the ovary. Int J Gynecol Pathol 2009;28(1):49–54.
[103] Tripathi M, Nandana S, Yamashita H, Ganesan R, Kirchhofer D, Quaranta V. Laminin-332 is a substrate for hepsin, a protease associated with prostate cancer progression. J Biol Chem 2008;283(45):30576–84.
[104] Tripathi M, Potdar AA, Yamashita H, Weidow B, Cummings PT, Kirchhofer D, et al. Laminin-332 cleavage by matriptase alters motility parameters of prostate cancer cells. Prostate 2011;71(2):184–96.
[105] Amano S, Scott IC, Takahara K, Koch M, Champliaud MF, Gerecke DR, et al. Bone morphogenetic protein 1 is an extracellular processing enzyme of the laminin 5 gamma 2 chain. J Biol Chem 2000;275(30):22728–35.
[106] Goldfinger LE, Stack MS, Jones JC. Processing of laminin-5 and its functional consequences: role of plasmin and tissue-type plasminogen activator. J Cell Biol 1998;141(1):255–65.

[107] Heck LW, Blackburn WD, Irwin MH, Abrahamson DR. Degradation of basement membrane laminin by human neutrophil elastase and cathepsin G. Am J Pathol 1990;136(6):1267–74.
[108] Yamada Y, Kleinman HK. Functional domains of cell adhesion molecules. Curr Opin Cell Biol 1992;4(5):819–23.
[109] Nomizu M, Kim WH, Yamamura K, Utani A, Song SY, Otaka A, et al. Identification of cell binding sites in the laminin alpha 1 chain carboxyl-terminal globular domain by systematic screening of synthetic peptides. J Biol Chem 1995;270(35):20583–90.
[110] Nomizu M, Kuratomi Y, Song SY, Ponce ML, Hoffman MP, Powell SK, et al. Identification of cell binding sequences in mouse laminin gamma1 chain by systematic peptide screening. J Biol Chem 1997;272(51):32198–205.
[111] Nomizu M, Kuratomi Y, Malinda KM, Song SY, Miyoshi K, Otaka A, et al. Cell binding sequences in mouse laminin alpha1 chain. J Biol Chem 1998;273(49):32491–9.
[112] Nomizu M, Kuratomi Y, Ponce ML, Song SY, Miyoshi K, Otaka A, et al. Cell adhesive sequences in mouse laminin beta1 chain. Arch Biochem Biophys 2000;378(2):311–20.
[113] Nomizu M, Yokoyama F, Suzuki N, Okazaki I, Nishi N, Ponce ML, et al. Identification of homologous biologically active sites on the N-terminal domain of laminin alpha chains. Biochemistry 2001;40(50):15310–7.
[114] Graf J, Ogle RC, Robey FA, Sasaki M, Martin GR, Yamada Y, et al. A pentapeptide from the laminin B1 chain mediates cell adhesion and binds the 67,000 laminin receptor. Biochemistry 1987;26(22):6896–900.
[115] Yamamura K, Kibbey MC, Jun SH, Kleinman HK. Effect of Matrigel and laminin peptide YIGSR on tumor growth and metastasis. Semin Cancer Biol 1993;4(4):259–65.
[116] Nakai M, Mundy GR, Williams PJ, Boyce B, Yoneda T. A synthetic antagonist to laminin inhibits the formation of osteolytic metastases by human melanoma cells in nude mice. Cancer Res 1992;52(19):5395–9.
[117] Grant DS, Tashiro K, Segui-Real B, Yamada Y, Martin GR, Kleinman HK. Two different laminin domains mediate the differentiation of human endothelial cells into capillary-like structures in vitro. Cell 1989;58(5):933–43.
[118] Lopez-Barcons LA, Polo D, Reig F, Fabra A. Pentapeptide YIGSR-mediated HT-1080 fibrosarcoma cells targeting of adriamycin encapsulated in sterically stabilized liposomes. J Biomed Mater Res A 2004;69(1):155–63.
[119] Dubey PK, Singodia D, Vyas SP. Liposomes modified with YIGSR peptide for tumor targeting. J Drug Target 2010;18(5):373–80.
[120] Dąbrowska K, Kaźmierczak Z, Majewska J, Miernikiewicz P, Piotrowicz A, Wietrzyk J, et al. Bacteriophages displaying anticancer peptides in combined antibacterial and anticancer treatment. Future Microbiol 2014;9(7):861–9.
[121] Hu J, Zhang YX, Lan XL, Qin GM, Zhang J, Hu ZH. An imaging study using laminin peptide 99mTc-YIGSR in mice bearing Ehrlich ascites tumour. Chin Med J (Engl) 2005;118(9):753–8.
[122] Sarfati G, Dvir T, Elkabets M, Apte RN, Cohen S. Targeting of polymeric nanoparticles to lung metastases by surface-attachment of YIGSR peptide from laminin. Biomaterials 2011;32(1):152–61.
[123] Bushkin-Harav I, Littauer UZ. Involvement of the YIGSR sequence of laminin in protein tyrosine phosphorylation. FEBS Lett 1998;424(3):243–7.
[124] Kim WH, Schnaper HW, Nomizu M, Yamada Y, Kleinman HK. Apoptosis in human fibrosarcoma cells is induced by a multimeric synthetic Tyr-Ile-Gly-Ser-Arg (YIGSR)-containing polypeptide from laminin. Cancer Res 1994;54(18):5005–10.
[125] Yu HN, Zhang LC, Yang JG, Das UN, Shen SR. Effect of laminin tyrosine-isoleucine-glycine-serine-arginine peptide on the growth of human prostate cancer (PC-3) cells in vitro. Eur J Pharmacol 2009;616(1–3):251–5.
[126] Tashiro K, Sephel GC, Weeks B, Sasaki M, Martin GR, Kleinman HK, et al. A synthetic peptide containing the IKVAV sequence from the A chain of laminin mediates cell attachment, migration, and neurite outgrowth. J Biol Chem 1989;264(27):16174–82.

[127] Kanemoto T, Reich R, Royce L, Greatorex D, Adler SH, Shiraishi N, et al. Identification of an amino acid sequence from the laminin A chain that stimulates metastasis and collagenase IV production. Proc Natl Acad Sci U S A 1990;87(6):2279–83.
[128] Grant DS, Kinsella JL, Fridman R, Auerbach R, Piasecki BA, Yamada Y, et al. Interaction of endothelial cells with a laminin A chain peptide (SIKVAV) in vitro and induction of angiogenic behavior in vivo. J Cell Physiol 1992;153(3):614–25.
[129] Sweeney TM, Kibbey MC, Zain M, Fridman R, Kleinman HK. Basement membrane and the SIKVAV laminin-derived peptide promote tumor growth and metastases. Cancer Metastasis Rev 1991;10(3):245–54.
[130] Balion Z, Sipailaite E, Stasyte G, Vailionyte A, Mazetyte-Godiene A, Seskeviciute I, et al. Investigation of cancer cell migration and proliferation on synthetic extracellular matrix peptide hydrogels. Front Bioeng Biotechnol 2020;8:773.
[131] Bresalier RS, Schwartz B, Kim YS, Duh QY, Kleinman HK, Sullam PM. The laminin alpha 1 chain Ile-Lys-Val-Ala-Val (IKVAV)-containing peptide promotes liver colonization by human colon cancer cells. Cancer Res 1995;55(11):2476–80.
[132] Royce LS, Martin GR, Kleinman HK. Induction of an invasive phenotype in benign tumor cells with a laminin A-chain synthetic peptide. Invasion Metastasis 1992;12(3-4):149–55.
[133] Stack MS, Gray RD, Pizzo SV. Modulation of murine B16F10 melanoma plasminogen activator production by a synthetic peptide derived from the laminin A chain. Cancer Res 1993;53(9):1998–2004.
[134] Corcoran ML, Kibbey MC, Kleinman HK, Wahl LM. Laminin SIKVAV peptide induction of monocyte/macrophage prostaglandin E2 and matrix metalloproteinases. J Biol Chem 1995;270(18):10365–8.
[135] Stevenson M, Hale AB, Hale SJ, Green NK, Black G, Fisher KD, et al. Incorporation of a laminin-derived peptide (SIKVAV) on polymer-modified adenovirus permits tumor-specific targeting via alpha6-integrins. Cancer Gene Ther 2007;14(4):335–45.
[136] Yamada M, Kadoya Y, Kasai S, Kato K, Mochizuki M, Nishi N, et al. Ile-Lys-Val-Ala-Val (IKVAV)-containing laminin alpha1 chain peptides form amyloid-like fibrils. FEBS Lett 2002;530(1–3):48–52.
[137] Itoh S, Yamaguchi I, Suzuki M, Ichinose S, Takakuda K, Kobayashi H, et al. Hydroxyapatite-coated tendon chitosan tubes with adsorbed laminin peptides facilitate nerve regeneration in vivo. Brain Res 2003;993(1–2):111–23.
[138] Chen S, Zhang M, Shao X, Wang X, Zhang L, Xu P, et al. A laminin mimetic peptide SIKVAV-conjugated chitosan hydrogel promoting wound healing by enhancing angiogenesis, re-epithelialization and collagen deposition. J Mater Chem B 2015;3(33):6798–804.
[139] Yin Y, Wang W, Shao Q, Li B, Yu D, Zhou X, et al. Pentapeptide IKVAV-engineered hydrogels for neural stem cell attachment. Biomater Sci 2021.
[140] Engbring JA, Hossain R, VanOsdol SJ, Kaplan-Singer B, Wu M, Hibino S, et al. The laminin alpha-1 chain derived peptide, AG73, increases fibronectin levels in breast and melanoma cancer cells. Clin Exp Metastasis 2008;25(3):241–52.
[141] Mochizuki M, Philp D, Hozumi K, Suzuki N, Yamada Y, Kleinman HK, et al. Angiogenic activity of syndecan-binding laminin peptide AG73 (RKRLQVQLSIRT). Arch Biochem Biophys 2007;459(2):249–55.
[142] Puchalapalli M, Mu L, Edwards C, Kaplan-Singer B, Eni P, Belani K, et al. The laminin. Anal Cell Pathol (Amst) 2019;2019, 9192516.
[143] Kim WH, Nomizu M, Song SY, Tanaka K, Kuratomi Y, Kleinman HK, et al. Laminin-alpha1-chain sequence Leu-Gln-Val-Gln-Leu-Ser-Ile-Arg (LQVQLSIR) enhances murine melanoma cell metastases. Int J Cancer 1998;77(4):632–9.
[144] Song SY, Nomizu M, Yamada Y, Kleinman HK. Liver metastasis formation by laminin-1 peptide (LQVQLSIR)-adhesion selected B16-F10 melanoma cells. Int J Cancer 1997;71(3):436–41.
[145] Yoshida Y, Hosokawa K, Dantes A, Kotsuji F, Kleinman HK, Amsterdam A. Role of laminin in ovarian cancer tumor growth and metastasis via regulation of Mdm2 and Bcl-2 expression. Int J Oncol 2001;18(5):913–21.
[146] Siqueira AS, Gama-de-Souza LN, Arnaud MVC, Pinheiro JJV, Jaeger RG. Laminin-derived peptide AG73 regulates migration, invasion, and protease activity of human oral squamous cell carcinoma cells through syndecan-1 and beta 1 integrin. Tumor Biol 2010;31(1):46–58.

[147] Hozumi K, Suzuki N, Nielsen PK, Nomizu M, Yamada Y. Laminin alpha1 chain LG4 module promotes cell attachment through syndecans and cell spreading through integrin alpha2beta1. J Biol Chem 2006;281(43):32929–40.
[148] Bass MD, Morgan MR, Humphries MJ. Syndecans shed their reputation as inert molecules. Sci Signal 2009;2(64), pe18.
[149] Iijima H, Negishi Y, Omata D, Nomizu M, Aramaki Y. Cancer cell specific gene delivery by laminin-derived peptide AG73-labeled liposomes. Bioorg Med Chem Lett 2010;20(15):4712–4.
[150] Hamano N, Negishi Y, Omata D, Takahashi Y, Manandhar M, Suzuki R, et al. Bubble liposomes and ultrasound enhance the antitumor effects of AG73 liposomes encapsulating antitumor agents. Mol Pharm 2013;10(2):774–9.
[151] Negishi Y, Hamano N, Tsunoda Y, Oda Y, Choijamts B, Endo-Takahashi Y, et al. AG73-modified Bubble liposomes for targeted ultrasound imaging of tumor neovasculature. Biomaterials 2013;34(2):501–7.
[152] Hozumi K, Otagiri D, Yamada Y, Sasaki A, Fujimori C, Wakai Y, et al. Cell surface receptor-specific scaffold requirements for adhesion to laminin-derived peptide-chitosan membranes. Biomaterials 2010;31(12):3237–43.
[153] Yamada Y, Hozumi K, Aso A, Hotta A, Toma K, Katagiri F, et al. Laminin active peptide/agarose matrices as multifunctional biomaterials for tissue engineering. Biomaterials 2012;33(16):4118–25.
[154] Mochizuki M, Kadoya Y, Wakabayashi Y, Kato K, Okazaki I, Yamada M, et al. Laminin-1 peptide-conjugated chitosan membranes as a novel approach for cell engineering. FASEB J 2003;17(8):875–7.
[155] Ponce ML, Nomizu M, Kleinman HK. An angiogenic laminin site and its antagonist bind through the alpha(v)beta3 and alpha5beta1 integrins. FASEB J 2001;15(8):1389–97.
[156] Kikkawa Y, Kataoka A, Matsuda Y, Takahashi N, Miwa T, Katagiri F, et al. Maintenance of hepatic differentiation by hepatocyte attachment peptides derived from laminin chains. J Biomed Mater Res A 2011;99(2):203–10.
[157] Kuratomi Y, Nomizu M, Tanaka K, Ponce ML, Komiyama S, Kleinman HK, et al. Laminin gamma 1 chain peptide, C-16 (KAFDITYVRLKF), promotes migration, MMP-9 secretion, and pulmonary metastasis of B16-F10 mouse melanoma cells. Br J Cancer 2002;86(7):1169–73.
[158] Ponce ML, Hibino S, Lebioda AM, Mochizuki M, Nomizu M, Kleinman HK. Identification of a potent peptide antagonist to an active laminin-1 sequence that blocks angiogenesis and tumor growth. Cancer Res 2003;63(16):5060–4.
[159] Fang M, Sun Y, Hu Z, Yang J, Davies H, Wang B, et al. C16 peptide shown to prevent leukocyte infiltration and alleviate detrimental inflammation in acute allergic encephalomyelitis model. Neuropharmacology 2013;70:83–99.
[160] Lugassy C, Kleinman HK, Vernon SE, Welch DR, Barnhill RL. C16 laminin peptide increases angiotropic extravascular migration of human melanoma cells in a shell-less chick chorioallantoic membrane assay. Br J Dermatol 2007;157(4):780–2.
[161] Kowtha VC, Bryant HJ, Pancrazio JJ, Stenger DA. Influence of extracellular matrix proteins on membrane potentials and excitability in NG108-15 cells. Neurosci Lett 1998;246(1):9–12.
[162] Ponce ML, Kleinman HK. Identification of redundant angiogenic sites in laminin alpha1 and gamma1 chains. Exp Cell Res 2003;285(2):189–95.
[163] Otagiri D, Yamada Y, Hozumi K, Katagiri F, Kikkawa Y, Nomizu M. Cell attachment and spreading activity of mixed laminin peptide-chitosan membranes. Biopolymers 2013;100(6):751–9.
[164] Chen H, Fu X, Jiang J, Han S. C16 peptide promotes vascular growth and reduces inflammation in a neuromyelitis optica model. Front Pharmacol 2019;10:1373.
[165] Ho TC, Yeh SI, Chen SL, Tsao YP. The psoriasis therapeutic potential of a novel short laminin peptide C16. Int J Mol Sci 2019;20(13).
[166] Kuratomi Y, Sato S, Monji M, Shimazu R, Tanaka G, Yokogawa K, et al. Serum concentrations of laminin gamma2 fragments in patients with head and neck squamous cell carcinoma. Head Neck 2008;30(8):1058–63.

CHAPTER 8

Proteolytic signaling in cutaneous wound healing

Konstantinos Kalogeropoulos, Louise Bundgaard, and Ulrich auf dem Keller
Technical University of Denmark, Department of Biotechnology and Biomedicine, Kongens Lyngby, Denmark

Introduction

Cutaneous wound healing is an extremely intricate process involving a plethora of molecular mechanisms and a wide range of cell lineages [1, 2]. Upon injury, cellular and molecular systems are activated in order to repair the skin structure. These systems cross talk and interplay within a complex web of physiological events that last for weeks or even months before the wound is healed and the skin morphology is adequately restored [3–5] (Fig. 1).

The wound healing process can be chronologically classified into four distinct but overlapping phases, namely hemostasis, inflammation, cell proliferation, and structural remodeling [1, 6]. Primary responses of the organism to tissue injury are platelet activation and blood coagulation through the coagulation cascade signaling pathway. The formation of a fibrin clot at the wound site serves as an important step in controlling bleeding, as well as provisionally restoring the skin barrier to prevent infection and additional damage. Coagulation is followed by the onset of the inflammatory phase, where inflammatory cells migrate to the site of injury, exerting signals for phagocytosis of necrotic cells and pathogenic agents. Subsequently, additional immune cells are recruited to the area, secreting proteins and hormones that promote repair mechanisms and cell growth. While the inflammatory stage is still ongoing, the next phase of cell proliferation is initiated, where proliferating cells, keratinocytes, and fibroblasts migrate to the wound edges, facilitating reepithelialization and granulation tissue formation. This phase is also characterized by neovascularization and angiogenesis [7, 8]. The last and most prolonged phase of wound healing is the remodeling of the wound matrix, where the provisional fibrin matrix is replaced with collagen and other extracellular matrix (ECM) components. Cross-linking and maturation of the tissue complete the wound healing process, resulting in wound closure and scar formation [3, 4].

Disruption of these tightly regulated and dynamic processes leads to healing impairments [9], which are characterized by substantially delayed or insufficient healing and closure of

Proteolytic Signaling in Health and Disease. https://doi.org/10.1016/B978-0-323-85696-6.00006-8

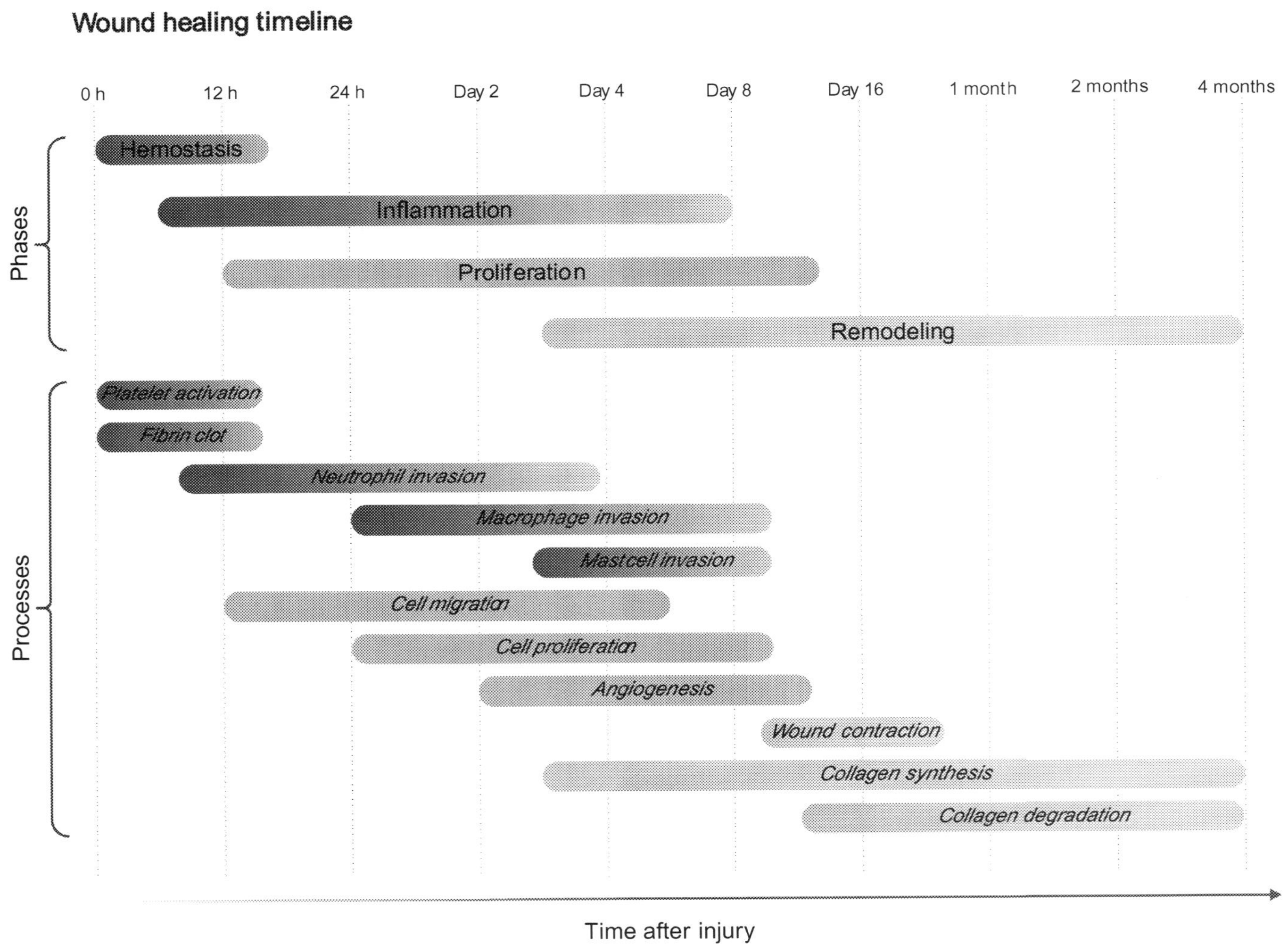

Fig. 1

Timeline of wound healing phases (*top*) and processes contributing to wound healing (*bottom*), in an exponential timescale. Processes are colored according to the phase they are mostly associated with. The scheme describes processes between injury and up to 4 months after; the matrix remodeling phase could potentially last longer.*Adapted from Clark RAF. The molecular and cellular biology of wound repair [Internet]. Boston, MA: Springer US; 1988 [cited 2021 Jan 26]. Available from: http://link.springer.com/10.1007/978-1-4899-0185-9. Created with BioRender.com.*

the skin defect [10]. Such wounds are termed chronic, and they pose a growing economic and social burden on Western societies [11]. Medical costs of chronic wounds are estimated as 1%–3% of healthcare system expenditure in developed countries, with increasing costs in recent years [12]. In the United States alone, it has been calculated that healthcare costs for chronic wounds exceed US$25 billion annually, with over 6.5 million affected patients [13]. In severe cases, amputations must be performed, contributing to the morbidity of the condition and considerably influencing the psychological and physical status of affected patients. Chronic wounds are prevalent in elder populations, although a significant portion of chronic wounds are ulcers attributed to or stemming from complications of lifestyle diseases, such as type 2 diabetes. With both demographics expected to increase in the future, treatment costs for chronic wounds are anticipated to increase in the upcoming years [14]. Although the etiologies of the healing deficiencies observed in chronic wounds are yet to be fully described, it is widely believed that the imbalance of biochemical factors and the inability of the wound to escape the inflammatory stage are the culprits [15–17].

Wound healing phases

Before discussing proteolytic signaling in detail, it is important to become familiar with the convoluted physiological and molecular events that occur during the process of wound healing. The healing phases overlap chronologically, and biological processes and signals might be present in multiple phases [1] (Fig. 2). Nevertheless, they are useful in monitoring the healing progression and studying the numerous molecular mechanisms involved.

Phase I: Hemostasis

The first response to tissue damage is vasoconstriction, which can temporarily prevent excessive bleeding. Platelets and other blood constituents such as von Willebrand factor leaking from damaged vessels come in contact with the exposed ECM and get activated through interaction with ECM components, mainly collagen. Stimulated platelets release coagulation factors and form aggregates, resulting in platelet clots [18]. Blood coagulation is essential for hemostasis and is a rigorously regulated pathway with several signaling checkpoints [19]. The majority of signaling steps in the coagulation process is mediated by proteases of the serine protease family, with several activation events that are catalyzed by proteolytic processing, discussed in detail in the next section. The central part of coagulation is the activity of thrombin or coagulation factor II, a well-known serine protease that cleaves and activates fibrinogen, the precursor of fibrin, which polymerizes and forms fibrin clots [20]. Fibrin clots also contain trapped platelets, fibronectin, vitronectin, and thrombospondin. The formed clot serves as a provisional matrix that facilitates cell adhesion and migration, assisting in the transition to the subsequent phases of wound healing [21].

Platelets also secrete a multitude of growth factors, promoting inflammation, cell migration, and proliferation. These include platelet-derived growth factor (PDGF), transforming growth factor (TGF), vascular endothelial growth factor (VEGF), and fibroblast growth factor (FGF) [22].

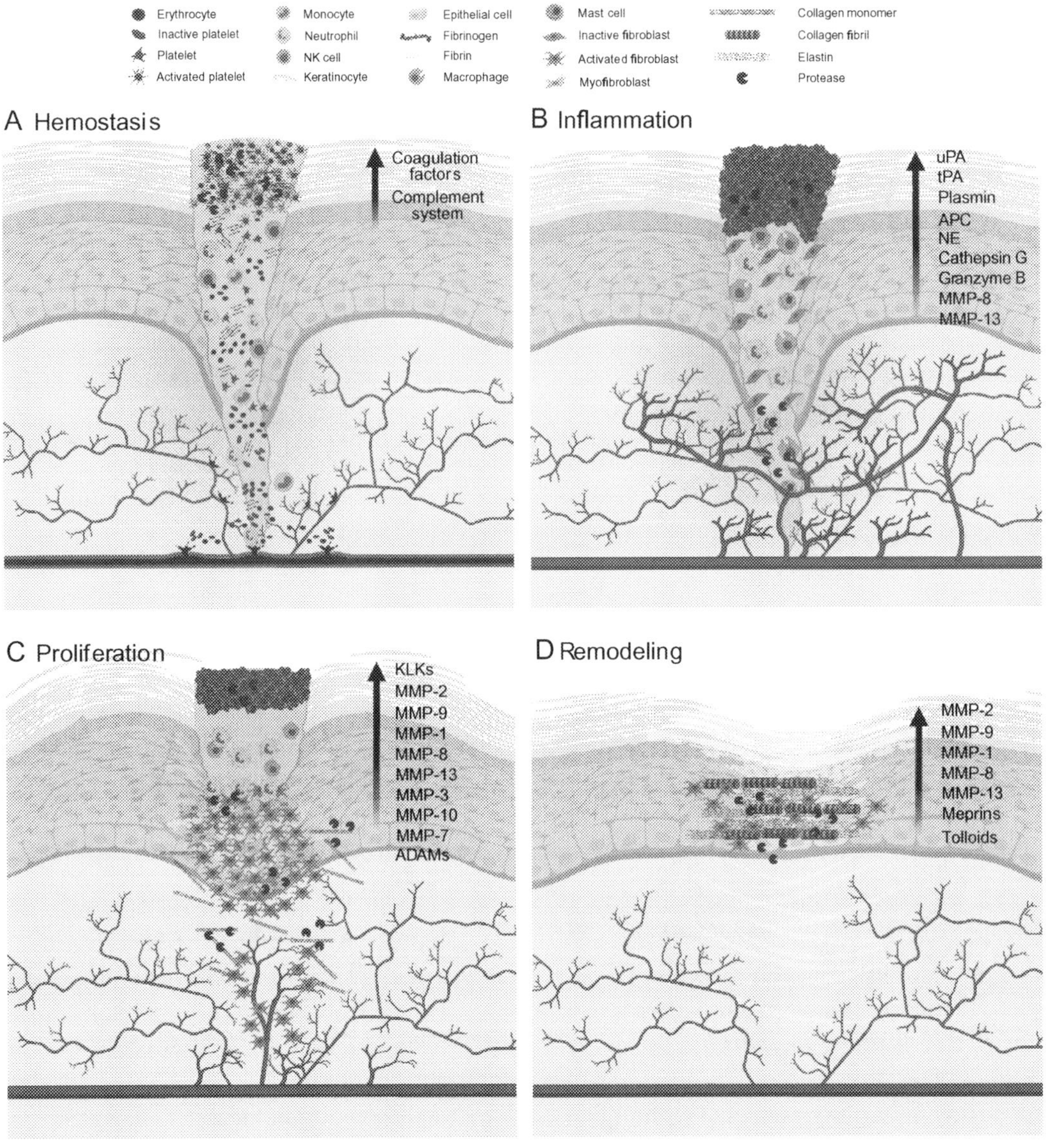

Fig. 2

Illustration of wound healing phases, with annotation of proteolytic enzymes or pathways that are involved in each phase. (A) The phase of hemostasis begins immediately after injury and results in the formation of a clot that prevents further bleeding and seals the wound. (B) Immune cells are recruited to the site and begin clearance of pathogens and phagocytosis. Granulocytes release cytokines and proteases that induce activation of proinflammatory pathways. Fibroblast migration is initiated in the early inflammation stage, along with neovascularization. (C) Fibroblasts and keratinocytes further migrate to the wound site and become activated, resulting in differentiation and proliferation. Inflammation and angiogenesis signals recede. Extracellular matrix components such as collagen and fibronectin are synthesized and secreted into the extracellular space. (D) Closing of the wound through contraction is followed by the formation of scar tissue. Collagen fibrils are cross-linked and oriented, and additional ECM components are organized.*Created with BioRender.com.*

Phase II: Inflammation

The inflammatory phase is subdivided into two stages; the early and the late inflammatory phase. After the initial vasoconstriction, a time span of vasodilation occurs, expediting the invasion of inflammatory cells to the area. Complement system activation, which involves multiple proteolytic processing events, stimulates the inflammatory response [8]. Neutrophil granulocytes are the first responders recruited to the wound site by the chemotactic signals of platelets, coagulation cascade products, and complement activation, approximately 24 h after the tissue injury [3]. Upon arrival, they adhere to endothelial cells and relocate to the extravascular space. Neutrophils have phagocytotic functions and contribute to the removal of infectious agents and damaged tissue from the wound site. Moreover, neutrophils produce various growth factors and cytokines that escalate inflammation by attracting additional inflammatory cells such as more neutrophils and monocytes to the region [23]. They also promote the recruitment of fibroblasts and epithelial cells by releasing TNF-alpha and IL-1-beta [24] and secrete numerous proteases, among which are neutrophil elastase, MMPs, and urokinase-type plasminogen activator (uPA) [25]. In particular, MMPs can regulate chemokine expression or alter their activity through proteolysis [26].

In the late inflammatory phase, monocytes infiltrate the wound site and become activated macrophages through acute signals mediated by pathogen-associated molecular patterns (PAMPs) and damage-associated molecular patterns (DAMPs) [27]. They contribute to phagocytosis, facilitating pathogen and cell debris cleanup, as well as the elimination of neutrophils and apoptotic cells. Macrophages also release proteases, cytokines, and growth factors that promote further inflammation and progression to the proliferative phase [28]. Notable macrophage products are TGF-beta, PDGF, beta-FGF, TNF-alpha, and interleukins [24]. Growth factors and chemoattractant agents stimulate angiogenesis and attract epithelial cells, contributing to the subsequent proliferative phase. In advanced inflammatory stages, activated mast cells and regulatory T-cells are also recruited to the wound site, attenuating inflammation and stimulating neovascularization [3].

Phase III: Proliferation

The proliferative phase is characterized by simultaneous advancement of multiple processes crucial for acute wound healing. Following the successful completion of hemostasis and inflammatory phases, the next stages in cutaneous tissue repair are reepithelization, granulation tissue formation, ECM component synthesis, cell adhesion, and migration [1]. The provisional fibrin-rich matrix that serves as a protective barrier during the initial wound healing phases must be degraded and replaced with newly synthesized ECM. Plasmin is the main protease that mediates degradation of the fibrin network and allows for disposal of the fibrin matrix as eschar at the wound edge [19]. Other key players of this phase are proteolytic enzymes of the MMP family, along with additional collagenolytic enzymes that degrade damaged collagen and ECM components [29].

A few days after the injury, growth factors act as chemoattractant agents, recruiting several cell types to the site. Fibroblast migration is supported by the fibrin network. The abundant proliferation of fibroblasts ensues, aiming at covering the entire surface of the wound. Primarily induced by TGF-beta, the proliferating fibroblasts synthesize various ECM proteins, mainly collagen, forming an ECM structural scaffold that facilitates further migration of other cell types [2]. Migration and proliferation of epithelial cells are stimulated by several growth factors including VEGF, PDGF and TGF-beta [5], and orchestrated with ECM synthesis and neovascularization within the newly formed tissue. The last cells to migrate to the area are keratinocytes that proliferate, assist in wound closure, and finally differentiate to form multiple layers of the stratified neo-epidermis. The newly formed vascular network is connected, the basement membrane is synthesized, and the provisional tissue structure is established. Differentiation of fibroblasts into myofibroblasts and myofibroblast contraction aid in wound closure and cell-cell interactions are formed, while redundant cells are removed by apoptosis and ongoing phagocytosis [8].

Phase IV: Remodeling

Tissue remodeling and scar formation are the last and most prolonged phase of wound healing, lasting for months or even years after injury. Transition of the acute wound healing process to the remodeling phase occurs 2–3 weeks after wounding [14]. Wound maturation and replacement of provisional components of the endothelial matrix are the prominent features in this phase, and these processes are characterized by a tightly regulated balance between degradation and synthesis of ECM components [30]. The temporary collagen III-rich ECM changes to the more resilient collagen I, and the collagen fibers are reorganized, cross-linked, and rearranged in bundles [31]. Furthermore, elastin fibers are deposited and the plentiful vasculature regresses to homeostatic levels. The tensile strength of the newly formed tissue increases, albeit to a lower level than prior to injury [22]. The formation of a scar in place marks the final step of wound repair, whereby the developing scar surface is decreased through wound contractions and fibroblast-ECM interaction [3].

Wound healing disruption and chronicity

Chronic wounds are wounds that are delayed in healing and show limited or no healing progression. These wounds are prevalent in patients with comorbidities and pre-existing conditions. In chronic wounds, one or more wound healing phases present defects in their physiological processes and come to a gridlock, believed to be caused by a disparity in expression or regulation of relevant proteins such as growth factors and proteases [10]. Fibroblast proliferation has been shown to be greatly reduced in chronic wounds, possibly due to downregulation of growth factors, e.g., FGF and EGF. The characteristic chronic wound phenotype exhibits an inability to escape the inflammatory stage of the repair process. The inflammatory environment may be potentiated by a multitude of factors. The excessive influx and prolonged maintenance of neutrophil and macrophage cells in the wound have been heavily accused [32]. Differentiation of macrophages to angiogenic and proliferating signaling cell lineages is essential in the late inflammatory phase, and disruption of this transformation might hinder healing progression [3]. In addition, uncontrolled release of bioactive molecules from neutrophils and macrophages and their disproportionate activity can contribute to wound healing impairments. Increased complement activity is also one of the candidates for the prolongment of the inflammatory phase. Elevated levels of pro-inflammatory cytokines such as IL1-beta and TNF-alpha and proteolytic enzymes, particularly of the MMP family, have been discovered in chronic wounds [33]. Increased activity of MMP-2 and MMP-9 has been found in impaired healing, while their inhibitors are present in decreased levels [34, 35]. The consequence of the increased protease/inhibitor ratio is an imbalance in degradation and synthesis of ECM components. The shifted ratio between MMPs and TIMPs is widely believed to be the main factor for the inability of the wound healing process to progress within a reasonable timeframe [17].

Proteases

Proteases, also often called proteinases or peptidases, are important actors in a wide spectrum of physiological and biological processes, such as differentiation, apoptosis, repair, and regeneration [8, 36, 37]. More than 2% of all genes have been attributed to encode for proteases in the human genome [38]. Proteolytic processing and signaling are vital for wound healing progression and homeostasis, with proteolytic enzymes playing pivotal roles in all four wound healing phases [21, 25, 39–42]. Proteases are enzymes that catalyze the hydrolysis of peptidyl bonds, thereby breaking a peptide chain into smaller fragments. They are termed endo- or exopeptidases based on their mode of action, cleaving proteins in the middle or terminal parts of their amino acid chains, respectively [43], and can have a limited processing or degrading function. Degradation of proteins is a natural process that assists in protein turnover, regulation, and renewal of the molecular machinery [44]. Proteolytic degradation is ubiquitous and is catalyzed by proteases that generally have a broad or generic substrate specificity profile with regard to target amino acid sequences. Limited proteolytic cleavages occur at conserved positions of the substrate molecule with distinct residue sequence profiles. Perhaps the most prominent proteolytic function is the removal of the signal peptide from newly synthesized proteins in the secretory pathway. A myriad of proteins requires proteolytic processing for conversion to their biologically active form [45]. Proteolysis activating a protein that assumes a specific and novel biological function post cleavage is a form of posttranslational modification [46, 47]. Due to its irreversible nature, the proteolytic activity must be under stringent regulation in biological systems, and consequently, aberrations in proteolysis levels are associated with pathological conditions in a high number of diseases [36, 48–51].

The overwhelming majority of extracellular protease species are themselves expressed and secreted as inactive proteoforms—a precursor form called zymogen—containing a propeptide whose removal is required for conversion to the active form of the enzyme [52]. Therefore, regulation of proteolytic activity is achieved at three distinct levels—transcriptional control of gene expression, activation of the inactive zymogen, and biochemical abrogation of enzymatic activity by physiological inhibitors [53]. Three major attributes are important for protease substrate specificity. First, the protease and its substrate have to be in physical proximity in order to initiate a biochemical contact and form a complex [54]. Therefore, the two components of the reaction must be colocalized spatially and temporally. Second, molecular interactions must be favorable for substrate binding [55]. However, this biochemical interaction needs to allow for subsequent release of the products after the enzymatic reaction. Hence, to avoid irreversible binding and inhibition of enzymatic activity [56], the binding energy of the complex should be within a restricted range of weak and strong molecular interactions. It can be envisioned that the substrate should fit into the enzyme's active site pocket similar to a key in a lock [57]. The conformation of the substrate-binding site in the enzyme active site pocket determines the structural features the substrate should possess [58]. Substrate sequence specificity is an intrinsic property of a protease and widely varies, with some proteases requiring highly specific residue sequences for cleavage, while others exhibit rather ubiquitous cleavage profiles. Third, the substrate-binding site must be accessible in terms of protein folding. Solvent accessibility is crucial for protease binding to a substrate and must always be considered when evaluating candidate substrates [59].

Endopeptidases are classified based on their mechanism of action and the catalytic residue in their active site [60]. The most widely characterized proteases are serine proteases, threonine proteases, metalloproteases, cysteine proteases, and aspartate proteases. Metalloproteases contain a metal ion, which coordinates the residues in the active site. All other groups are eponymously named after the catalytic residue that initiates the nucleophilic attack on the protease substrate. The only exception is aspartate proteases, in which the aspartate residue activates a water molecule that attacks the peptide bond on the substrate, similar to the activation mechanism that is coordinated by the metal ion in metalloproteases [43]. Relevant proteases in the process of wound healing are serine, cysteine, and metalloproteases; the biological role and mechanism of which will be the focus of the following sections.

Serine proteases

Serine proteases are a diverse and ubiquitous class of proteases, present in all forms of life. The most abundant group is the proteases of mixed nucleophile, superfamily A (PA clan), often described as trypsin fold containing proteases. Serine proteases are defined by a catalytic triad in the active site, consisting of a histidine, a serine, and an aspartate residue [43]. The hydroxyl group of the serine residue is the nucleophile that initiates the attack on the scissile bond of the substrate, assisted by the other two residues in the catalytic triad.

Serine proteases exert multiple functions, most notably digestion, coagulation, inflammation, and immunity. Serine proteases are further subcategorized as trypsin-like, thrombin-like, and chymotrypsin-like enzymes, among others [61]. As an example, trypsin-like specificity is defined by proteolytic cleavage after basic residues, while chymotrypsin-like enzymes cleave peptide bonds after aromatic residues. Serine proteases are indispensable components of the coagulation cascade that results in blood clotting, as well as all complement activation [62, 63]. Both pathways involve proteases that cleave bonds primarily C-terminal to arginine residues. Other notable serine proteases in wound healing include cathepsin G, granzyme B, and plasmin.

Coagulation cascade

Blood coagulation is an elaborate signaling pathway with multiple cleavage events and is commonly used as a paradigm of proteolytic signaling (Fig. 3). Following injury, the transmembrane tissue factor (TF) or coagulation factor III, a cell surface receptor constitutively expressed in epithelial cells, interacts with the inactive form of factor VII (FVII) present in the blood [64]. Injury enables the colocalization of the two proteins, which naturally are present in different environments. The complex formation markedly enhances the removal of the propeptide and the autocatalytic cleavage of the Arg^{192}-Ile^{193} peptide bond (human sequence), upon which the FVII is activated to FVIIa and assumes protease catalytic activity [65]. The two chains generated by the cleavage are kept together by a disulfide bond [64]. Cleavage of the precursor protein and association of the two products, along with a change in conformation, is a common pattern in coagulation factor enzymes. More recently, another serine protease aptly named factor VII activating protease (FSAP) able to activate FVII was discovered [66] that itself is activated proteolytically in an autocatalytic manner [67]. TF can also be generated by neutrophils, the primary responders from the immune system after tissue damage, whereby its activity is regulated by tissue factor pathway inhibitor (TFPI) protein [68]. Compared to FVIIa alone, the TF:FVIIa complex possesses orders of magnitude increased affinity for the circulating factors IX and X of the coagulation cascade and is able to activate them by converting their zymogen precursor to the active serine protease enzymes IXa and Xa [69]. TF:FVIIa complex activity requires phospholipids and calcium ions as cofactors and TFPI is able to inhibit both the FVIIa:TF complex and free Xa [68]. The factor X processing cleavage site lies between residues Arg^{194}-Ile^{195} [70, 71], releasing Xa, which in turn activates the precursor coagulation factor V, another plasma glycoprotein with three distinct cleavages, and forms the protease complex Xa:Va. Factor V can also be activated by thrombin upon cleavages of scissile bonds C-terminal to Arg^{709}, Arg^{1018}, and Arg^{1545} [72]. The Xa:Va or prothrombinase complex catalyzes the conversion of the zymogen prothrombin (coagulation factor II) to the serine protease thrombin in a calcium- and phospholipid-dependent manner [73]. Prothrombin is converted to thrombin by two consequent cleavages by the prothrombinase complex [74]. Any of the two peptide bonds located after residues Arg^{271} and Arg^{320} can be hydrolyzed first. Depending on the order

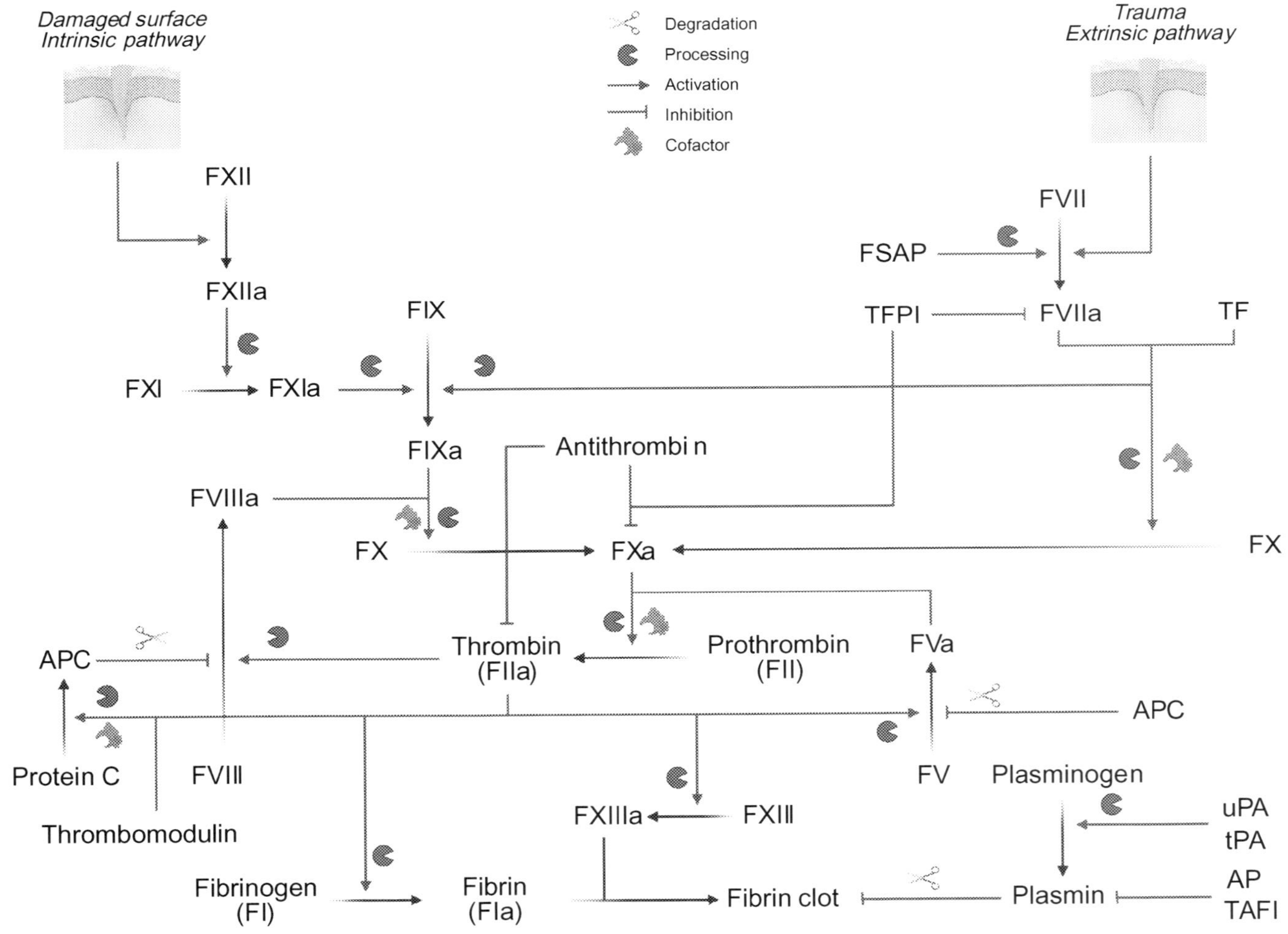

Fig. 3

Diagram of the coagulation cascade and fibrinolysis. The two coagulation-initiating pathways, the intrinsic and extrinsic, are to the left and right of the diagram, respectively. The two pathways converge in the activation of factor X, which forms a complex with FVa and activates thrombin, the main enzyme in the coagulation cascade whose activation culminates in the establishment of the fibrin clot. Resolution of the fibrin clot is mediated by plasmin. Proteolytic processing events, degrading cleavages, and cofactor roles are indicated with icons as shown in the figure (*middle top*). Activation events are indicated by *green point arrows* (gray in the print version), while inhibition events are represented by the *red block arrow* (dark gray in the print version). Products of the interaction are shown by *black point arrows*. (For interpretation of the references to color in this figure legend, the reader is referred to the web version of this article.)*Created with BioRender.com.*

of cleavage, the products are two distinct intermediates, prothrombin and meizothrombin, respectively. The prothrombin zymogen exists in different structural conformations, the open and the closed form, of which the latter is more abundant and promotes activation through the meizothrombin pathway [75]. However, it has been demonstrated that prothrombin activation on the platelet surface is proceeding through the prothrombin pathway [76, 77] and is greatly enhanced by membrane binding [78]. Thrombin activity is mainly regulated by antithrombin (AT), a potent thrombin inhibitor. AT belonging to serpins, a well-characterized family of serine protease inhibitors, can inhibit virtually all proteases of the coagulation cascade and its potency is significantly increased in conjunction with heparin [79]. Thrombin proteolytically activates factor XIII to its active form XIIIa with a single cleavage of the peptide bond after Arg^{37} [20]. In the last proteolytic step of the coagulation cascade, thrombin cleaves fibrinogen or coagulation factor I (FI) at conserved sites of the N-terminal fibrinopeptide sequence [80]. This event exposes functional sites of fibrin, which changes conformation and associates with other fibrin monomers to form fibrin clots [81, 82]. Finally, factor XIIIa cross-links fibrin polymers and stabilizes the newly formed clot [83, 84].

A second signaling cascade, the intrinsic activation of the coagulation cascade as opposed to the extrinsic activation described above, is triggered through contact with a damaged surface [85–87]. The serine protease factor XII (FXII) is activated to the active form XIIa by a conformational change upon contact with negatively charged surfaces and an autocatalytic cleavage after Arg^{372} [88]. Activation of FXII is enhanced by a positive feedback loop of activation with prekallikrein, another serine protease zymogen [89]. FXIIa catalyzes the activation of the serine protease factor XI (FXI) to XIa by proteolytic cleavage after Arg^{387}, which then is able to activate the coagulation factor IX to IXa with two cleavage events after Arg^{145} and Arg^{180} and removal of the internal peptide sequence [90, 91]. Activated factor IXa forms a complex with the glycoprotein factor VIIIa (FVIII) and forms the IXa:VIIIa complex. The IXa:VIIIa complex is also capable of activating factor X in a phospholipid- and calcium-ion-dependent manner [92]. FVIII is activated by multiple thrombin cleavages, after Arg^{372}, Arg^{740}, Arg^{1649}, and Arg^{1689}, producing a trimer from the cleavage products [93]. FVIIa is also able to activate FVIII [94], and the remarkably similar processing of FV and FVIII reflects their genomic and structural homology, as well as the specific, precise function of proteases in the coagulation cascade. With the proteolytic activation of factor X to Xa by FIXa, the intrinsic pathway converges with the extrinsic pathway in a common pathway that results in thrombin activation and fibrinogenolysis [95].

There is considerable evidence that the coagulation cascade signaling is in fact a complex web of activation events that induce coagulation, rather than a discreet, sequential proteolytic pathway [96]. As discussed, multiple coagulation factors can activate proteolytic or nonproteolytic coagulation components. For instance, factors IXa and Xa and thrombin are all known to induce activation of factor VIIa [65], and thrombin also activates factors V, VIII, XI, and XIII [63], while factor VIII can be activated by factor Xa as well [92]. This certain dispensability of several activation steps can protect against genetic defects and irregularities in coagulation factor levels during hemostasis. The coagulation cascade proteases exhibit a great preference toward scissile bonds between arginine and a hydrophobic amino acid. The conformational change inflicted upon cleavage activation and the formation of the enzyme active site pocket are the main determinants of this specificity. As mentioned, accessibility of the substrate peptide sequence containing the scissile bond allows for the cleavage event to occur. Another essential event during the coagulation cascade is the formation of complexes that contain calcium ions or coagulation factor IV (FIV) and phospholipids of a surface membrane [97]. Interaction of coagulation factors and calcium ions is mediated by a conserved domain in the N-terminal part of the proteases [98, 99]. The positively charged calcium ions serve as a facilitator, binding with the negatively charged phospholipids and the negatively charged N-terminal peptide sequences of the serine proteases of the coagulation cascade [100]. The requirement for a phospholipid surface ensures that the proteolytic activation events are spatially restricted and happen near the site of injury. This limitation ensures the activation of coagulation factors and other hemostatic molecules without unconstrained, excessive proteolytic activity in other sites that might produce deleterious effects on different biological processes.

Apart from their involvement in the coagulation cascade, the coagulation factors with serine protease activity possess additional biological functions and also cleave noncoagulation substrates. Thrombin has a central role in angiogenesis, whereby treatment of platelet-containing plasma with thrombin leads to elevated VEGF levels [101]. Moreover, thrombin is known to promote platelet activation and aggregation through activation of integrin and release of platelet activators [102]. In endothelial cells, thrombin stimulates platelet adhesion by inducing expression of P-selectin and release of von Willebrand factor [103]. Thrombin also promotes inflammation by protein C activation, upregulation of cytokines, and immune cell recruitment [102, 104]. This activity is at least partly regulated by the activation of protease-activated receptors (PARs) [105, 106], a family of transmembrane G-protein-coupled receptors, present in platelets and endothelial cells [102, 107]. PARs are activated by cleavage of their N-terminal sequence located in the extracellular space, revealing a neo-N-terminus, which acts as a tethered ligand to the receptor, thereby activating downstream signaling [108]. Ligand binding induces the exchange of bound GTP for GDP in the intracellular part of the receptor initiating a phosphorylation cascade, which results in activation of mitogen-activated protein kinase (MAPK) and phosphokinase C (PKC), among others, [109] subsequently catalyzing changes in thousands of phosphorylation sites [110]. With the exception of FIXa, all proteases of the coagulation cascade can activate PARs [107]. Additionally, other proteases such as activated protein C (APC) and neutrophil elastase are able to activate PARs in different sites and induce different signaling responses [111–114]. Aside from PARs activation, FXIIa and thrombin can stimulate inflammation by complement activation [115]. FXII-deficient mice demonstrated significantly faster rates of wound closure marked by reduced neutrophil recruitment and higher reepithelization rates [116].

Anticoagulant signaling and fibrinolysis

Once clotting has been achieved, amplification of the coagulation proteolytic signaling pathway must be attenuated. Since hydrolysis is irreversible, termination of proteolytic signaling poses particular challenges. Where inhibitors do not provide sufficient reduction in activity, a negative feedback mechanism must be in place for proteolysis to remain under control. One of the main components of the anticoagulant process is activated protein C (APC), which is expressed as a zymogen and gets activated by removal of its propeptide through cleavage C-terminal to Arg^{169} by the thrombin:thrombomodulin complex [20]. Thrombomodulin is a membrane protein that serves as a cofactor for thrombin and alters its substrate profile [117]. Therefore, thrombin clearly demonstrates coagulant activities, but upon binding to thrombomodulin, it is transformed into an anticoagulant mediator [19]. In addition, thrombin inhibition is markedly increased by antithrombin and protein C inhibitor, when the enzyme is bound to thrombomodulin [79, 118]. APC activation is significantly enhanced when inactive protein C is bound to the endothelial cell protein C receptor (EPCR) [119]. APC exerts its anticoagulant activity by proteolytically inactivating the two glycoprotein cofactors of the coagulation cascade, FVa and FVIIIa [120], catalyzing these reactions with the assistance of two main cofactors, inactive FV and protein S [121]. As mentioned, APC is also an activator of PARs, whereby the interaction with EPCR allows for PAR-1 signaling activation orders of magnitude higher compared to free APC activation [107]. APC-mediated PAR-1 activation induces cytoprotective and anti-inflammatory signals by downregulation of cytokines through nuclear factor kappa-light-chain-enhancer of activated B cells (NF-κB) pathway inhibition and expression of protective genes [122, 123]. APC has also been found to stimulate angiogenesis and reepithelization in animal wound models [124], under the tight control of the protein C inhibitor [125].

Resolution of the fibrin-rich hemostatic plug following fibrin degradation is a critical step for the initiation of cell migration, ECM synthesis, and remodeling in the later stages of wound healing. The fibrinolytic system is another example of a tightly regulated proteolytic cascade with a series of cleavage activation events. Fibrinolysis is mediated primarily by plasmin, another enzyme secreted as a serine protease zymogen. Plasmin is activated in the fibrinolytic pathway by two upstream serine proteases, uPA and tissue-type plasminogen activator (tPA), via a cleavage event C-terminal to Arg^{560}, upon which a conformational change and release of the N-terminal 77 residue long peptide sequence with an additional cleavage are promoted [126, 127]. Once activated, plasmin degrades fibrin by proteolysis in multiple sites, promoting the dissolution of the fibrin mesh. Cleavage events also expose C-terminal lysine residues which bind to plasminogen domains and propagate plasmin activation [128]. uPA and tPA are also circulating zymogens, which in wound healing are expressed by macrophages and endothelial cells, respectively. The two plasmin activators undergo activation through different mechanisms. Upon binding to its membrane receptor, uPAR, uPA is activated by adjacent proteases, including plasmin, by two subsequent cleavages at sites C-terminal to K^{158} and K^{135} [129]. Plasmin also activates tPA after cleavage of the peptide bond at Arg^{275}-Ile^{276}, producing two chains that associate. Therefore, a positive feedback loop is established by plasmin activity on its two main activators. tPA is unique in the sense that the single chain and the two chain forms of the enzyme have comparable activities [130]. tPA activity on plasminogen is increased by two orders of magnitude in the presence of fibrin [131]. Plasmin can also activate other proteases, cleave ECM components, and be involved in TGF-beta signaling [132]. Plasmin activity is regulated by its main inhibitor of the serpin family, alpha-2-plasmin inhibitor [128]. Additional inhibition of the fibrinolysis pathway is achieved by the carboxypeptidase thrombin activatable fibrinolysis inhibitor (TAFI) that removes terminal lysine residues in fibrin and prevents plasmin binding on the polymerized fibrin network [133, 134] and by the serpin family member plasminogen activator inhibitors 1 and 2, which abrogate activity of tPA and uPA [135, 136]. Importantly, plasmin can activate several other proteases involved in tissue remodeling, including MMPs [137].

Neutrophil proteases, tissue kallikreins, and complement system

Neutrophil elastase (NE) and cathepsin G are serine proteases released in circulation by neutrophil granulocytes. Neutrophils are migratory cells recruited to the wound site following inflammatory stimuli [138]. Neutrophil elastase has been named after its first substrate discovered, elastin, but it is a protease with numerous functions. NE is known to activate PARs, initiate the expression of inflammatory mediators [114], and directly activate cytokines via proteolytic processing [139]. NE is also a collagenase, able to degrade collagen substrates, participates in ECM remodeling [140] and together with other neutrophil proteases mediates chemotaxis and migration by cleavage of adhesion proteins in cell membranes [141]. NE activity is regulated by inhibitors of the serpin family, mainly alpha-1-antitrypsin [142], whereas elevated levels of NE have been frequently reported in patients with healing impairments, implicating NE in chronic wound pathology and a candidate biomarker of wound healing progression [17, 143, 144]. NE is expressed as an active protease and similar to cathepsin G does not require activation. Both NE and cathepsin G induce platelet activation and aggregation [145], but cathepsin G has also demonstrated the ability to convert angiotensin I to angiotensin II, stimulating smooth muscle contraction and vascular permeability [146]. Moreover, cathepsin G proteolytically activates chemotactic factors and inflammatory mediators and initiates upregulation of cytokine expression through PAR activation [147]. Another serine protease thought to be involved in wound healing is granzyme B that is secreted by granules from T-cells and natural killer cells but has also been found to be expressed by activated keratinocytes [148]. Granzyme B is thought to induce apoptosis through perforin-mediated pathway, but multiple studies have also indicated that granzyme B has the ability to cleave multiple ECM components and adhesion proteins, in addition to cytokine processing, playing an important role in inflammation and ECM remodeling [149, 150]. For instance, granzyme B has been shown to cleave fibronectin, vitronectin, and laminin, influencing cell detachment and signaling pathways by releasing growth factors trapped in the ECM network [151, 152].

Kallikreins (KLKs) are an important family of serine proteases with fifteen members (excluding plasma kallikrein, not considered anymore to be a member of the kallikrein family), possessing a regulating role in various biological processes [153–155]. Kallikreins are secreted as zymogens, with some being able to become activated through autocatalysis, while others are activated by other proteases. In fact, the kallikrein interaction and activation network is extremely complex, spanning multiple families of different classes of proteases [154, 156, 157]. In the healing skin wound, kallikreins are known to promote vasodilation and angiogenesis by interacting with coagulation factor components and by the generation of bioactive peptides, called kinins, from kininogen [158–161]. KLK5 and KLK7 are the main proteolytic enzymes in the upper part of the epidermis, the *stratum corneum*, and have been found to be responsible for keratinocyte detachment and migration [162, 163]. KLKs are also involved in fibrinolysis through the mediation of plasmin activation, as well as inflammation by activation of other zymogens and cytokine processing [164–166]. KLK activity in the skin is regulated by LEKTI inhibitor fragments [165].

Complement activation is one of the primary processes triggered after injury. Initiation of the complement cascade is pivotal for immune cell recruitment to the site of injury and subsequent clearance of the wound from pathogenic agents through phagocytosis. It also regulates the removal of necrotic tissue, acts as an inflammatory mediator, and plays a role in angiogenesis [41, 167]. In the initial stages of wound healing, the complement system interacts with the coagulation system and is activated through DAMP and PAMP signals [27, 115]. The complement is heavily involved in the inflammatory phase of wound healing, and aberrant activity has been shown to be part of the impaired healing pathogenesis due to prolonged inflammation [168]. Complement serine proteases consist of C1r, C1s, C2, factor B, factor D, and mannose-binding lectin-associated serine proteases (MASP) 1, 2, and 3 that are all expressed as zymogens. Complement activation involves three activation pathways—the classical, lectin, and alternative pathways. In the classical pathway, the C1q component is activated through the binding of antibody-antigen complexes. C1q then interacts with two molecules of C1s and C1r, which are proteolytically activated to form the C1 complex ($C1qr_2s_2$). Upon binding of C1q to immune complexes, e.g., IgG or IgM molecules bound to their antigen, autocatalytic activation of C1r is induced, which in turn cleaves and activates the zymogen form of C1s. The C1s-activated subunit of the C1 complex cleaves the propeptide of the zymogen C2 with a single cleavage after Arg^{222}, thereby producing its enzymatically active form C2a and C2b. The C1 complex also activates C4, producing four distinct products. The products C2a and C4b associate and form the C3 convertase complex (C4b2a), which is able to convert C3 to C3a and C3b. The activated C3b can subsequently bind to the C3 convertase complex to form the C5 convertase complex (C4b2a3b). The C5 convertase catalyzes the cleavage of C5 to C5b, which initiates the assembly of the membrane attack complex comprising C5b, and C6-C9. The membrane attack complex forms pores in target membranes, causing cell lysis. Fragments of the complement activation cascade that are not proteolytically active or are involved in complex formation, such as C3a, C4a, and C5a, act as anaphylatoxins and induce inflammatory responses [169].

The lectin activation pathway is initiated by binding of mannan-binding lectin (MBL) to carbohydrate structures in pathogens or antibodies, causing MBL to form a complex with MASPs, which acts similarly to the C1 complex, converging to the cleavage of C2 and C4. MASP2 is the protease responsible for these activation events and is therefore the equivalent of C1s in the MBL-MASP complex [170]. Another protein family called ficolins, structurally similar to C1q and MBL, can form complexes with MASPs, which can also activate C2 and C4 [171].

In the alternative pathway, the activated C3 associates with factor B and forms the C3iB complex. Factor D cleaves the zymogen factor B subunit of the complex and activates it. After activation of factor B to the enzymatically active Bb, it catalyzes the cleavage of C3 to C3a and C3b. C3b binds to surfaces and can recruit additional molecules of factor B that are activated by factor D, subsequently forming the active C3b:factor Bbb (C3bBb) complex. The C3bBb complex serves as an analog to the C3 convertase complex, propagating the conversion of more C3 into its active products. Proteolytic activity of C1r, C1s, and MASPs 1 and 2 is abrogated by interaction with their serpin C1-inhibitor [169].

Metalloproteases

Extracellular metalloproteases are almost entirely members of the metzincin superfamily. They are characterized by a conserved binding motif [HEXXHXX (G/N)XX (H/D)] in their active site and a loop Met-turn in close proximity [172]. The three histidines in this motif act as ligands to the zinc ion along with a labile water molecule that mediates the nucleophilic attack on the peptide carbonyl group, assisted by the glutamate residue in the second position of the motif which acts as a proton acceptor. Interaction of the zinc ion with a water molecule and with the substrate in the active site is critical for proteolysis, as metal chelators inhibit catalytic activity. Substrate profiles are generally not highly specific, but there is a preference toward hydrophobic residues C-terminal to the scissile bond [173]. Prominent metalloprotease families include astacins, adamalysins, and matrixins. Astacins contain protease families such as meprins and tolloid proteases, both of which are believed to be involved in wound healing since their main substrates are ECM proteins. Adamalysins is the family comprising A disintegrin and metalloprotease (ADAM) proteases and A disintegrin and metalloproteinase with thrombospondin motifs (ADAMTSs). ADAMs, often described as sheddases, are a family of transmembrane and secreted proteases implicated in processes like organism development, inflammation, signaling, and cell adhesion [37]. Members of the matrixins or matrix metalloproteinase (MMP) family are the predominant proteases involved in both acute and chronic wound healing. MMPs are an intensively probed protease family throughout the scientific literature, mainly implicated in inflammation, immune response, and matrix remodeling [26, 174]. Proteases of the MMP family are regulated by the tissue inhibitors of metalloproteinase (TIMP) family that have also been shown to inhibit ADAM activity.

Matrixins

Matrix metalloproteases (or matrix metalloproteinases) play pivotal roles in numerous physiological processes, and differential expression or imbalanced regulation of MMPs has been implicated in a multitude of pathologies and malignancies [26, 175–177]. The MMP family encompasses a large number of proteolytic enzymes, many of which are involved in wound healing. These are the MMP subgroups of gelatinases (MMP2 and MMP9), collagenases (MMP-1, MMP-8, and MMP-13), stromelysins (MMP-3, MMP-10), matrilysins (MMP-7 and MMP-26), and macrophage metalloelastase (MMP-12) [35]. MMP activity during tissue repair is one of the hallmarks of inflammation, angiogenesis, reepithelization, and matrix remodeling processes. Extracellular MMPs are secreted as zymogens from a multitude of cell lineages involved in wound healing, predominantly leukocytes, platelets, and fibroblasts [178, 179]. All aforementioned MMPs, with the exception of matrylisins, contain an additional hemopexin-like domain believed to assist in substrate recognition and convertase binding [180]. Their fascinating activation mechanism entails a cleavage in a propeptide bait region and a subsequent trans- or autocatalytic cleavage event, resulting in the release of the propeptide from the active site of the enzyme [181, 182]. The initial cleavage only removes part of the propeptide, while the secondary cleavage is essential for the release of a conserved sequence of the propeptide, containing a cysteine residue that binds to the zinc ion in the catalytic site. Exposal of the zinc ion through this "cysteine-switch" mechanism allows it to interact with a water molecule, a process necessary for enzymatic activity [183]. MMPs can be activated by various proteases, among which are plasmin, furin, KLKs, and other MMPs [184]. The MMP activation network is still not fully elucidated and remains an active field of research. The main activity of MMPs is exerted upon ECM components, but MMPs may also act on growth factors, chemokines, and cytokines [184, 185]. MMP substrates predominantly contain hydrophobic residues following the scissile bond and prefer a proline in P3, i.e., three positions N-terminal to the cleavage site. The enzyme active site pocket of MMPs varies with respect to structural features, explaining the wide substrate and residue specificity.

The main substrates of collagenases are regarded to be interstitial collagens, namely fibrillar types I, II, and III. Their ability to recognize, bind, and unwind the triple helical structure of fibrillar collagens is thought to be mediated by the hemopexin domain. The hydrolysis of collagen by collagenases is defined by a single cleavage, producing products that span ¾ and ¼ of the full precursor length, respectively. Collagenases can also act upon other ECM molecules such as laminin and fibronectin. MMP-1 is highly expressed by keratinocytes and facilitates migration through collagen binding and cleavage in an integrin-dependent manner. Membrane integrin receptors are associated with intact collagen in basal keratinocytes. MMP-1 cleaves the collagen molecular assembly, causing the release of integrin cell contacts due to weakened interactions with the unwound helix. This action facilitates the forward motion of keratinocytes, where repeated cycles of integrin binding, cleavage by MMP-1, and collagen dissociation allow them to migrate into the wound edge [39]. Increased expression of MMP-1 has been shown in impaired healing, potentially inhibiting wound repair by hindering reepithelization [186–188]. MMP-8 is considered to be a prevalent protease in healing wounds, assisting in wound debridement [189]. MMP-8 was found to be expressed in fibroblasts, neutrophils, and macrophages among others [15] and can cleave additional substrates beyond collagen, including aggrecan and angiotensin [40]. Furthermore, MMP-8 was found in elevated levels in wound ulcers [189, 190], while MMP-8-deficient mice exhibited a significantly delayed wound healing phenotype [191]. These results indicate that MMP-8 is a quintessential protease for wound healing, although the absence of MMP-8 can presumably be ameliorated, at least partially, by MMP-13, as indicated by increased levels of MMP-8 in an MMP-13-deficient murine model [192]. MMP-13 has also been found to be overexpressed in chronic wounds [187, 188, 193], and an MMP-9/MMP-13 double knockout murine model indicated that MMP-13 plays a key role in angiogenesis, cell migration, and wound contraction [194]. In malignancies, MMP-13 is implicated in tumor growth, mediating keratinocyte transformation [195].

The gelatinases MMP-2 and MMP-9 were named after their ability to readily digest gelatin, a form of denatured collagen. MMP-2 expression is mainly attributed to platelets and fibroblasts [196, 197] and is mostly constitutive, while MMP-9 is an inflammatory protease, produced by neutrophils and macrophages and highly increased in abundance during immune responses [15]. With regard to ECM components, gelatinases have the ability to process fibrillar collagens, as well as nonfibrillar collagen-like collagen type IV and also other substrates including laminin, fibronectin, aggrecan, and elastin [40]. A unique structural feature of gelatinases is the presence of three repeats of fibronectin type II domains in the metalloprotease domain, which allow binding to collagen, gelatin, and laminin [198]. In wound healing, gelatinases have the important role of degrading cleaved and damaged collagen, facilitating ECM remodeling and directly inducing cell migration [199–201]. The ability of gelatinases to cleave collagen IV has been associated with cell movement and reepithelization. Although MMP-2 and MMP-9 do not degrade type IV collagen with high catalytic efficiency, their activity toward collagen IV and other substrates indicates a key role in cell migration [39]. Decreased MMP-2 expression from fibroblasts has been associated with defective ECM assembly [202]. Increased levels of both MMP-2 and MMP-9 have been extensively reported in chronic wounds and their exudates, and a negative correlation of gelatinase levels with successful wound healing outcomes has been discovered [203–206]. It has been hypothesized that excessive MMP-9 activity contributes to the proliferation of the inflammatory environment in impaired healing that renders the wound incapable of progressing from the inflammatory to the proliferative phase [15]. The addition of exogenous MMP-9 in murine wounds delays wound healing, exhibiting significantly slower progression compared to control wounds [207]. However, at the same time, gelatinase activity is essential for wound healing progression [208], as indicated by delayed reepithelization of cutaneous wounds in MMP-9-deficient mice [194]. Therefore, it can be assumed that although uncontrolled gelatinase proteolytic activity is at least partly responsible for wound chronicity, it is indispensable in the wound healing process. Excessive degradation of key components essential for healing progression due to this elevated activity may be one of the causalities of impaired healing. As previously stated, reduced regulation of MMPs from their natural inhibitors, TIMPs, is commonly inculpated when inferring chronic wound etiologies. This is particularly important for MMP-9, as MMP-9/TIMP ratios have been found to be elevated in impaired healing and proposed as a biomarker of clinical outcome besides increased MMP levels [209, 210].

The group of stromelysins comprises highly homologs MMP-3 and MMP-10, as well as MMP-11, which is referred to as stromelysin-3 but is frequently grouped with other MMPs due to its significant deviation from the other two stromelysins in sequence and substrate specificity [211]. MMP-3 and MMP-10 are both considered to be involved in wound healing, expressed by proliferating keratinocytes and keratinocytes at the wound edge, respectively [212, 213]. Although MMP-10 activity is not well studied, different expression and localization of two stromelysins indicate distinct roles for the two proteases in wound healing. Stromelysins demonstrate a more generic profile with respect to substrate preference, as they are able to cleave a wide variety of ECM components and nonfibrillar-type collagens, as well as denatured fibrillar collagens [174, 214]. The most well-characterized member of the group is MMP-3 or stromelysin-1. MMP-3 can activate several other MMP zymogens, and its substrate specificity is comparable to MMP-10 [211, 215]. MMP-3 has been found to be upregulated in chronic wounds compared to acute cutaneous wounds [216], and a correlation of increased MMP-3 levels with nonhealing wounds has been suggested [208]. Mice deficient in MMP-3 have shown an impaired wound contraction phenotype [217], and the same effect has been observed with MMP-3 deficient fibroblasts *in vitro* [218]. MMP-3 and MMP-7 have also exhibited activity against immunoglobulins G (IgGs), indicating a role in antibody clearance and prevention of complement activation [219]. MMP-10 in wound healing has been associated with cleavage of cell junction proteins during cell migration [220], but mice deficient in MMP-10 did not show significantly abnormal healing [212]. However, a recent study also using deficient mice demonstrated that macrophage-derived MMP-10 plays a critical role in wound repair by regulating collagenolytic activity and expression of the collagenases MMP-8 and MMP-13 [221].

Matrilysins (MMP-7 and MMP-26) are also implicated in wound healing, and similar to stromelysins, matrilysin substrates include many ECM components and adhesion molecules, such as laminin, elastin, and cadherin. MMP-7 is not expressed in epidermal wounds, but is found in injured epithelia and is known to possess high potency for the cleavage of basement membrane components [25]. MMP-7 appears as an essential protease for reepithelization since it can act on cell adhesion and surface molecules among other substrates [222, 223] and several murine models have demonstrated that deficiency of MMP-7 hinders the healing process dramatically [223, 224]. MMP-26 expression during skin repair seems to be restricted to migrating keratinocytes, but its expression has also been found to be upregulated in chronic wounds, warranting more extensive studies [190]. The macrophage metalloelastase, or MMP-12, is expressed exclusively by macrophages infiltrating the site of injury [225, 226]. MMP-12 contributes to inflammation and might be involved in angiogenesis since it can generate angiostatin by proteolysic processing of type XVIII collagen [227].

MMP activity is tightly regulated by TIMPs, a family of endogenous protein inhibitors with four members, TIMP-1 to -4. TIMPs bind to the active site cleft of the zymogen or the activated proteoform and are able to inhibit enzymatic activity with 1:1 stoichiometry [211]. TIMP-1 and TIMP-2 are the most potent MMP inhibitors [25], and they are present in the wound bed regulating proteolytic activity [35]. Controlled ratios of MMP-1/TIMP-1 have been shown to be critical for tissue repair [228]. Multiple studies have reported decreased activity of TIMPs in chronic wounds and ulcers [28, 187, 203, 229], and the MMP/TIMP ratio has been considered as a biomarker for accessing wound healing progression [230]. However, all TIMPs appear to inhibit all MMPs, showcasing the remarkable homology of MMPs and the considerable difficulty of developing monospecific inhibitors for the MMP family. Another regulating mechanism of MMPs is the induction of expression by inflammatory mediators. In particular, TNF-alpha and interleukins are known to stimulate MMP expression and downregulation of TIMPs [15, 17]. MMPs can also activate PARs like many other proteases and induce expression of proinflammatory molecules, among other processes [231, 232].

Adamalysins and astacins

The majority of ADAMs are expressed as membrane proteins, and their primary role is considered to be the shedding of extracellular domains of transmembrane proteins. Apart from the metalloprotease domain, ADAMs contain a disintegrin and a cysteine-rich domain which participate in substrate recognition. A few members of this diverse protease family contain additional domains, such as the EGF-like domain [37]. The disintegrin domain was shown to bind to integrins, indicating a role in cell adhesion and movement [233], and the cysteine-rich domain can interact with ECM proteins and proteoglycans. ADAMTSs are secreted molecules and contain thrombospondin motif repeats and both ADAMs and ADAMTSs are activated intracellularly via propeptide removal by proprotein convertases in the Golgi system [132]. Their main involvement in wound healing is in the potentiation of the inflammatory environment, neovascularization, and cell migration. The ADAMs substrate repertoire is diverse both in substrate and biological function. Human leukocytes express multiple ADAMs, including ADAM-8, ADAM-15, and ADAM-17, of which ADAM-8 can influence neutrophil migration, presumably through degradation of basement membrane components or the release of bioactive molecules from the cell surface [234]. ADAM-15 can regulate cell adhesion by binding to integrins with the disintegrin domain and was found to be expressed in intestinal epithelial cells, where it could inhibit wound healing in this environment [235]. ADAM-17, also known as TNF-alpha-converting enzyme based on its ability to shed and thereby activate TNF-alpha, has a central role in inflammatory responses [48]. Other notable substrates of ADAM-17 are interleukin-6 and endothelial growth factor receptor ligands [233, 236]. The proteolytic activity of most ADAMs is also modulated by TIMPs. ADAMTSs in wound healing have more specific functions and are involved in antiangiogenic signaling and collagen synthesis through the release of antiangiogenic factors and procollagenases [132].

Astacins are thought to be involved in wound healing, but their role in the different processes of tissue repair and regeneration has not been fully elucidated. Meprins are synthesized as transmembrane proteins, can be shed by proteolysis, and are localized in the skin, where their expression increases in inflammatory conditions. Their substrates include ECM components, such as collagen type IV and laminin, and cell adhesion molecules such as cadherin [132]. Meprin activation is inhibited by mannan-binding lectin [237].

Tolloids are a family of soluble proteases that find their main role in tissue remodeling and morphogenesis [238]. Tolloid proteases are associated with maturation of fibrillar collagens, where they cleave C-terminal propeptides and prime collagen for fibril assembly in the ECM. Tolloids can also cleave other ECM components such as laminin [239], promote growth factor signaling through proteolytic processing [240], and play a role in the regulation of angiogenesis by activating antiangiogenic factors [241].

Cysteine proteases

Cysteine proteases are another large class of proteolytic enzymes. The most noteworthy cysteine proteases are encompassed in the superfamily of papain-like proteases [60]. A cysteine residue in the catalytic site of the enzyme acts as the nucleophile that attacks the substrate scissile bond, after deprotonation by a basic residue. Cysteine proteases are mainly involved in programmed cell death but also in immune responses with caspases as critical parts of inflammasomes and activators of inflammatory cytokines IL-1beta and IL-18 [36]. They are heavily implicated in disease states, with altered levels in autoimmune diseases and cancer [242]. The cathepsin family, which mostly consists of cysteine proteases, is implicated in protein turnover through lysosomal degradation as well as matrix remodeling in wound healing [243].

Cathepsins

Cysteine proteases of the cathepsin family comprise eleven members, among which are cathepsins B, K, L, S, and V. Their role in wound healing is still under investigation, but initially, they were thought to only participate in proteolysis in the endolysosomal system, where they are crucial for immune complex activation and protein breakdown during immune responses, as well as protein turnover in cell and tissue homeostasis. However, in recent decades, cysteine cathepsins have arisen also as significant players in the extracellular space [243]. In wound healing, cathepsins possibly contribute to the inflammatory phase and to ECM remodeling. Cathepsin activity is regulated by pH, and an acidic environment is required for proteolytic activity. The neutral pH of ECM can be overcome by pericellular acidification of the area, in which cysteine cathepsins can act on numerous substrates [244]. Inflammatory mediators and proinflammatory signaling pathways can induce upregulation of cysteine cathepsins in the extracellular space [243]. These proteases can degrade collagens either extracellularly or through endocytosis [245]. For instance, cathepsin K was found to possess collagenolytic activity and generate unique collagen fragmentation patterns compared to gelatinases [246]. Cathepsin V has potent activity against elastin and is expressed in activated macrophages [247]. Cysteine cathepsins have also been identified as convertases of chemokines and interleukins [248, 249].

Exogenous proteases

As a result of the rupture and disruption of normal skin structure following an injury, the organism becomes prone to infections. Until the hemostatic plug is formed and the skin barrier is reinstated, pathogenic agents can enter the organism through the wound site, accessing deeper dermal layers and possibly entering circulation. Pathogens can also release proteases that act on host substrates, either accommodating infection or activating host proteases that act on ECM components. Bacterial proteases can promote inflammation, reduce the activity of phagocytotic mechanisms, and inhibit complement system activation [250]. A number of bacterial proteases of the thermolysin family have been found to activate MMP-1, MMP-8, and MMP-9 by propeptide removal [251]. Recently, a correlation between levels of pathogens and MMP activity in chronic ulcers has also been reported [252]. Other bacterial proteases can hinder immune responses by degradation of complement components C3 and C5 or by degradation of cytokines, such as interferon-gamma and TNF-alpha [250].

Conclusion

Proteolytic enzymes are ubiquitous in numerous processes and all living forms, indicating their indispensable role in signaling pathways and the paramount importance of proteolytic processing as a posttranslational modification. Protease activation, their enzymatic mechanism of action, and cofactor requirements are considerably complex mechanisms, resulting from evolutionary pressure in the need for controlling proteolytic activity, especially when considering the irreversibility of peptide bond hydrolysis. Proteases are involved in all wound healing phases and are crucial for tissue repair and regeneration. Regulation of protease activity is achieved at multiple levels, and disruption of the checks and balances in proteolytic activity can be detrimental to wound healing progression. Proteases can activate or degrade their targets via proteolytic processing. Notably, a growing number of proteases are known to activate other proteases, highlighting the incredibly convoluted network of interactions, potentiation of activity, and feedback mechanisms between these enzymes. Protease levels or their activity on substrates can serve as valuable prognostic biomarkers during the assessment of wound healing trajectories [253]. Although proteases act on a quite diverse repertoire of protein targets that participate in various physiological processes, the most important protease substrates during wound healing are coagulation cascade proteins, inflammatory mediators, and ECM components. Proteases are central players in the induction and modulation of immune responses. They also play a pivotal role in cell migration, either by degrading ECM proteins and paving the road for proliferating cells, or by facilitating cell movement through cell adhesion molecule cleavages. The degradation and synthesis of ECM molecules are directly controlled by proteolytic enzymes, and the balance of those two processes is crucial for tissue remodeling and regeneration. Due to their role in virtually all wound healing mechanisms, proteases have frequently been considered as therapeutic targets. The field of protease substrate discovery has been a field attracting great interest in recent decades, particularly after the advancement of investigating techniques, such as mass spectrometry-based proteomics and degradomics [42, 254, 255]. Targeting protease activity in chronic wounds and ulcers presents immense potential for prognosis and treatment. Successful development of pharmaceutical agents that can modulate protease activity has the direct promise of impacting positively innumerable lives suffering from impaired wound healing.

References

[1] RAF C, editor. The Molecular and Cellular Biology of Wound Repair [Internet]. Boston, MA: Springer US; 1988. [cited 2021 Jan 26]. Available from: http://link.springer.com/10.1007/978-1-4899-0185-9.
[2] Gurtner GC, Werner S, Barrandon Y, Longaker MT. Wound repair and regeneration. Nature 2008 May;453(7193):314–21.
[3] Sorg H, Tilkorn DJ, Hager S, Hauser J, Mirastschijski U. Skin wound healing: an update on the current knowledge and concepts. Eur Surg Res 2017;58(1–2):81–94.
[4] Martin P, Nunan R. Cellular and molecular mechanisms of repair in acute and chronic wound healing. Br J Dermatol 2015 Aug;173(2):370–8.
[5] Reinke JM, Sorg H. Wound repair and regeneration. Eur Surg Res 2012;49(1):35–43.
[6] Singer AJ. Cutaneous wound healing. N Engl J Med 1999;9.
[7] Tonnesen MG, Feng X, Clark RAF. Angiogenesis in wound healing. J Investig Dermatol Symp Proc 2000 Dec;5(1):40–6.
[8] Toriseva M, Kähäri V-M. Proteinases in cutaneous wound healing. Cell Mol Life Sci 2009 Jan;66(2):203–24.
[9] Velnar T, Bailey T, Smrkolj V. The wound healing process: an overview of the cellular and molecular mechanisms. J Int Med Res 2009 Oct;37(5):1528–42.

[10] Goldberg SR, Diegelmann RF. What makes wounds chronic. Surg Clin North Am 2020 Aug;100(4):681–93.
[11] Eming SA, Martin P, Tomic-Canic M. Wound repair and regeneration: Mechanisms, signaling, and translation. Sci Transl Med 2014 Dec 3;6(265):265sr6.
[12] Brem H, Stojadinovic O, Diegelmann RF, Entero H, Lee B, Pastar I, et al. Molecular markers in patients with chronic wounds to guide surgical debridement. Mol Med 2007 Jan 1;13(1):30–9.
[13] Sen CK, Gordillo GM, Roy S, Kirsner R, Lambert L, Hunt TK, et al. Human skin wounds: a major and snowballing threat to public health and the economy. Wound Repair Regen 2009 Nov;17(6):763–71.
[14] Han G, Ceilley R. Chronic wound healing: a review of current management and treatments. Adv Ther 2017 Mar;34(3):599–610.
[15] Rayment EA, Upton Z. Review: finding the culprit: a review of the influences of proteases on the chronic wound environment. Int J Low Extrem Wounds 2009 Mar;8(1):19–27.
[16] Snyder RJ, Cullen B, Nisbet LT. An audit to assess the perspectives of US wound care specialists regarding the importance of proteases in wound healing and wound assessment. Int Wound J 2013 Dec;10(6):653–60.
[17] Trengove NJ, Stacey MC, Macauley S, Bennett N, Gibson J, Burslem F, et al. Analysis of the acute and chronic wound environments: the role of proteases and their inhibitors. Wound Repair Regen 1999 Nov;7(6):442–52.
[18] Gremmel T, Frelinger A, Michelson A. Platelet physiology. Semin Thromb Hemost 2016 Feb 29;42(03):191–204.
[19] Chapin JC, Hajjar KA. Fibrinolysis and the control of blood coagulation. Blood Rev 2015 Jan;29(1):17–24.
[20] Lane DA, Philippou H, Huntington JA. Directing thrombin. Blood 2005 Oct 15;106(8):2605–12.
[21] Opneja A, Kapoor S, Stavrou EX. Contribution of platelets, the coagulation and fibrinolytic systems to cutaneous wound healing. Thromb Res 2019 Jul;179:56–63.
[22] Shah JMY, Omar E, Pai DR, Sood S. Cellular events and biomarkers of wound healing. Indian J Plast Surg Off Publ Assoc Plast Surg India 2012 May;45(2):220–8.
[23] Strbo N, Yin N, Stojadinovic O. Innate and adaptive immune responses in wound epithelialization. Adv Wound Care 2014 Jul 1;3(7):492–501.
[24] Patel S, Maheshwari A, Chandra A. Biomarkers for wound healing and their evaluation. J Wound Care 2016 Jan 2;25(1):46–55.
[25] Barrick B, Campbell EJ, Owen CA. Leukocyte proteinases in wound healing: roles in physiologic and pathologic processes. Wound Repair Regen 1999 Nov;7(6):410–22.
[26] Nissinen L, Kähäri V-M. Matrix metalloproteinases in inflammation. Biochim Biophys Acta BBA Gen Subj 2014 Aug;1840(8):2571–80.
[27] Huber-Lang M, Kovtun A, Ignatius A. The role of complement in trauma and fracture healing. Semin Immunol 2013 Feb;25(1):73–8.
[28] Baker EA, Leaper DJ. Proteinases, their inhibitors, and cytokine profiles in acute wound fluid. Wound Repair Regen Off Publ Wound Heal Soc Eur Tissue Repair Soc 2000 Oct;8(5):392–8.
[29] Gill SE, Parks WC. Metalloproteinases and their inhibitors: regulators of wound healing. Int J Biochem Cell Biol 2008;40(6–7):1334–47.
[30] Page-McCaw A, Ewald AJ, Werb Z. Matrix metalloproteinases and the regulation of tissue remodelling. Nat Rev Mol Cell Biol 2007 Mar;8(3):221–33.
[31] Wang P-H, Huang B-S, Horng H-C, Yeh C-C, Chen Y-J. Wound healing. J Chin Med Assoc 2018 Feb;81(2):94–101.
[32] Yager DR, Nwomeh BC. The proteolytic environment of chronic wounds. Wound Repair Regen Off Publ Wound Heal Soc Eur Tissue Repair Soc 1999 Dec;7(6):433–41.
[33] Menke NB, Ward KR, Witten TM, Bonchev DG, Diegelmann RF. Impaired wound healing. Clin Dermatol 2007 Feb;25(1):19–25.
[34] Greener B, Hughes AA, Bannister NP, Douglass J. Proteases and pH in chronic wounds. J Wound Care 2005 Feb;14(2):59–61.

[35] Lazaro JL, Izzo V, Meaume S, Davies AH, Lobmann R, Uccioli L. Elevated levels of matrix metalloproteinases and chronic wound healing: an updated review of clinical evidence. J Wound Care 2016 May 2;25(5):277–87.
[36] Van Opdenbosch N, Lamkanfi M. Caspases in cell death, inflammation, and disease. Immunity 2019 Jun;50(6):1352–64.
[37] Edwards D, Handsley M, Pennington C. The ADAM metalloproteinases. Mol Aspects Med 2008 Oct;29(5):258–89.
[38] Rawlings ND, Morton FR, Kok CY, Kong J, Barrett AJ. MEROPS: the peptidase database. Nucleic Acids Res 2008 Jan;36(Database issue):D320–5.
[39] Rohani MG, Parks WC. Matrix remodeling by MMPs during wound repair. Matrix Biol 2015 May;44–46:113–21.
[40] Steffensen B, Häkkinen L, Larjava H. Proteolytic events of wound-healing—coordinated interactions among matrix metalloproteinases (MMPs), integrins, and extracellular matrix molecules. Crit Rev Oral Biol Med 2001 Sep;12(5):373–98.
[41] Cazander G, Jukema GN, Nibbering PH. Complement activation and inhibition in wound healing. Clin Dev Immunol 2012;2012:1–14.
[42] Sabino F, Hermes O, Egli FE, Kockmann T, Schlage P, Croizat P, et al. In vivo assessment of protease dynamics in cutaneous wound healing by degradomics analysis of porcine wound exudates. Mol Cell Proteomics 2015 Feb;14(2):354–70.
[43] Barrett AJ. Proteases. Curr Protoc Protein Sci [Internet] 2000 Sep;21(1). [cited 2021 Jan 26]. Available from: https://onlinelibrary.wiley.com/doi/10.1002/0471140864.ps2101s21.
[44] López-Otín C, Bond JS. Proteases: multifunctional enzymes in life and disease. J Biol Chem 2008 Nov;283(45):30433–7.
[45] Neurath H, Walsh KA. Role of proteolytic enzymes in biological regulation (a review). Proc Natl Acad Sci 1976 Nov 1;73(11):3825–32.
[46] Lichtenthaler SF, Lemberg MK, Fluhrer R. Proteolytic ectodomain shedding of membrane proteins in mammals—hardware, concepts, and recent developments. EMBO J [Internet] 2018 Aug;37(15). [cited 2021 Jan 26]. Available from: https://onlinelibrary.wiley.com/doi/abs/10.15252/embj.201899456.
[47] Klein T, Eckhard U, Dufour A, Solis N, Overall CM. Proteolytic cleavage—mechanisms, function, and "omic" approaches for a near-ubiquitous posttranslational modification. Chem Rev 2018 Feb 14;118(3):1137–68.
[48] Düsterhöft S, Lokau J, Garbers C. The metalloprotease ADAM17 in inflammation and cancer. Pathol Res Pract 2019 Jun;215(6):152410.
[49] ten Cate H, Hackeng TM, de Frutos PG. Coagulation factor and protease pathways in thrombosis and cardiovascular disease. Thromb Haemost 2017;117(07):1265–71.
[50] Agbowuro AA, Huston WM, Gamble AB, Tyndall JDA. Proteases and protease inhibitors in infectious diseases. Med Res Rev 2018 Jul;38(4):1295–331.
[51] Quintero-Fabián S, Arreola R, Becerril-Villanueva E, Torres-Romero JC, Arana-Argáez V, Lara-Riegos J, et al. Role of matrix metalloproteinases in angiogenesis and cancer. Front Oncol 2019 Dec 6;9:1370.
[52] Boon L, Ugarte-Berzal E, Vandooren J, Opdenakker G. Protease propeptide structures, mechanisms of activation, and functions. Crit Rev Biochem Mol Biol 2020 Apr;55(2):111–65.
[53] Verhamme IM, Leonard SE, Perkins RC. Proteases: pivot points in functional proteomics. In: Wang X, Kuruc M, editors. Functional proteomics [Internet]. New York, NY: Springer New York; 2019. p. 313–92. [cited 2020 Jan 31]. Available from: http://link.springer.com/10.1007/978-1-4939-8814-3_20.
[54] Pahwa S, Bhowmick M, Amar S, Cao J, Strongin AY, Fridman R, et al. Characterization and regulation of MT1-MMP cell surface-associated activity. Chem Biol Drug Des 2019 Jun;93(6):1251–64.
[55] Van Doren SR. Matrix metalloproteinase interactions with collagen and elastin. Matrix Biol 2015 May;44–46:224–31.
[56] Kaysser L. Built to bind: biosynthetic strategies for the formation of small-molecule protease inhibitors. Nat Prod Rep 2019 Dec 11;36(12):1654–86.

[57] Gupta SP, Patil VM. Specificity of binding with matrix metalloproteinases. Exp Suppl 2012 2012;103:35–56.
[58] Wolfe MS. Substrate recognition and processing by γ-secretase. Biochim Biophys Acta Biomembr 2020 Jan 1;1862(1):183016. 2019/07/08 ed.
[59] Song J, Wang Y, Li F, Akutsu T, Rawlings ND, Webb GI, et al. iProt-Sub: a comprehensive package for accurately mapping and predicting protease-specific substrates and cleavage sites. Brief Bioinform 2019 Mar 25;20(2):638–58.
[60] Rawlings ND, Barrett AJ, Thomas PD, Huang X, Bateman A, Finn RD. The MEROPS database of proteolytic enzymes, their substrates and inhibitors in 2017 and a comparison with peptidases in the PANTHER database. Nucleic Acids Res 2018 Jan 4;46(D1):D624–32.
[61] Di Cera E. Serine proteases. IUBMB Life 2009 May;61(5):510–5.
[62] Sim RB, Laich A. Serine proteases of the complement system. Biochem Soc Trans 2000 Oct 1;28(5):A488.
[63] Sidhu G, Soff GA. The coagulation system and angiogenesis. In: Kwaan HC, Green D, editors. Coagulation in cancer [Internet]. Boston, MA: Springer US; 2009. p. 67–80. [cited 2021 Jan 27]. (Cancer Treatment and Research; vol. 148). Available from: http://link.springer.com/10.1007/978-0-387-79962-9_5.
[64] Gajsiewicz J, Morrissey J. Structure–function relationship of the interaction between tissue factor and factor VIIa. Semin Thromb Hemost 2015 Sep 26;41(07):682–90.
[65] Yamamoto M, Nakagaki T, Kisiel W. Tissue factor-dependent autoactivation of human blood coagulation factor VII. J Biol Chem 1992 Sep;267(27):19089–94.
[66] Römisch J, Feussner A, Vermöhlen S, Stöhr HA. A protease isolated from human plasma activating factor VII independent of tissue factor. Blood Coagul Fibrinolysis Int J Haemost Thromb 1999 Dec;10(8):471–9.
[67] Römisch J. Factor VII activating protease (FSAP): a novel protease in hemostasis. Biol Chem 2002 Aug;383(7–8):1119–24.
[68] Wood JP, Ellery PER, Maroney SA, Mast AE. Biology of tissue factor pathway inhibitor. Blood 2014 May;123(19):2934–43.
[69] Ruf W, Rehemtulla A, Morrissey JH, Edgington TS. Phospholipid-independent and -dependent interactions required for tissue factor receptor and cofactor function. J Biol Chem 1991 Feb;266(4):2158–66.
[70] Richter G, Schwarz HP, Dorner F, Turecek PL. Activation and inactivation of human factor X by proteases derived from *Ficus carica*: *Activation and Inactivation of Human FX*. Br J Haematol 2002 Dec;119(4):1042–51.
[71] Zögg T, Brandstetter H. Activation mechanisms of coagulation factor IX. Biol Chem [Internet] 2009 Jan 1;390(5/6). [cited 2021 Jan 27]. Available from: http://www.degruyter.com/view/j/bchm.2009.390.issue-5-6/bc.2009.057/bc.2009.057.xml.
[72] Duga S, Asselta R, Tenchini ML. Coagulation factor V. Int J Biochem Cell Biol 2004 Aug;36(8):1393–9.
[73] Spencer FA, Becker RC. The prothrombinase complex: assembly and function. J Thromb Thrombolysis 1997;4(3/4):357–64.
[74] Davie E, Kulman J. An overview of the structure and function of thrombin. Semin Thromb Hemost 2006 Feb;32(S1):3–15.
[75] Chinnaraj M, Chen Z, Pelc LA, Grese Z, Bystranowska D, Di Cera E, et al. Structure of prothrombin in the closed form reveals new details on the mechanism of activation. Sci Rep 2018 Dec;8(1):2945.
[76] Haynes LM, Bouchard BA, Tracy PB, Mann KG. Prothrombin activation by platelet-associated prothrombinase proceeds through the prethrombin-2 pathway via a concerted mechanism. J Biol Chem 2012 Nov;287(46):38647–55.
[77] Wood JP, Silveira JR, Maille NM, Haynes LM, Tracy PB. Prothrombin activation on the activated platelet surface optimizes expression of procoagulant activity. Blood 2011 Feb;117(5):1710–8.
[78] Bradford HN, Orcutt SJ, Krishnaswamy S. Membrane binding by prothrombin mediates its constrained presentation to prothrombinase for cleavage. J Biol Chem 2013 Sep;288(39):27789–800.
[79] Amiral J, Seghatchian J. Revisiting antithrombin in health and disease, congenital deficiencies and genetic variants, and laboratory studies on α and β forms. Transfus Apher Sci 2018 Apr;57(2):291–7.

[80] Blombäck B, Hessel B, Hogg D, Therkildsen L. A two-step fibrinogen–fibrin transition in blood coagulation. Nature 1978 Oct;275(5680):501–5.
[81] Mosesson MW. Fibrinogen and fibrin structure and functions. J Thromb Haemost 2005 Aug;3(8):1894–904.
[82] Litvinov RI, Pieters M, de Lange-Loots Z, Weisel JW. Fibrinogen and fibrin. In: Harris JR, Marles-Wright J, editors. Macromolecular protein complexes III: structure and function [Internet]. Cham: Springer International Publishing; 2021. p. 471–501. Available from https://doi.org/10.1007/978-3-030-58971-4_15.
[83] Gupta S, Biswas A, Akhter MS, Krettler C, Reinhart C, Dodt J, et al. Revisiting the mechanism of coagulation factor XIII activation and regulation from a structure/functional perspective. Sci Rep 2016 Sep;6(1):30105.
[84] Muszbek L, Bereczky Z, Bagoly Z, Komáromi I, Katona É. Factor XIII: a coagulation factor with multiple plasmatic and cellular functions. Physiol Rev 2011 Jul;91(3):931–72.
[85] Davie EW, Ratnoff OD. Waterfall sequence for intrinsic blood clotting. Science 1964 Sep 18;145(3638):1310–2.
[86] Macfarlane RG. An enzyme cascade in the blood clotting mechanism, and its function as a biochemical amplifier. Nature 1964 May;202(4931):498–9.
[87] Naudin C, Burillo E, Blankenberg S, Butler L, Renné T. Factor XII contact activation. Semin Thromb Hemost 2017 Nov;43(08):814–26.
[88] van der Graaf F, Keus F, Vlooswijk R, Bouma B. The contact activation mechanism in human plasma: activation induced by dextran sulfate. Blood 1982 Jun 1;59(6):1225–33.
[89] Grover SP, Mackman N. Intrinsic pathway of coagulation and thrombosis: insights from animal models. Arterioscler Thromb Vasc Biol 2019 Mar;39(3):331–8.
[90] Mohammed BM, Matafonov A, Ivanov I, Sun M, Cheng Q, Dickeson SK, et al. An update on factor XI structure and function. Thromb Res 2018 Jan;161:94–105.
[91] Gailani D. Activation of factor IX by factor XIa. Trends Cardiovasc Med 2000 Jul;10(5):198–204.
[92] Fay PJ. Activation of factor VIII and mechanisms of cofactor action. Blood Rev 2004 Mar;18(1):1–15.
[93] Mazurkiewicz-Pisarek A, Płucienniczak G, Ciach T, Płucienniczak A, et al. Acta Biochim Pol [Internet] 2016 Jan 28;63(1). [cited 2021 Jan 27]. Available from: https://ojs.ptbioch.edu.pl/index.php/abp/article/view/1672.
[94] Soeda T, Nogami K, Matsumoto T, Ogiwara K, Shima M. Mechanisms of factor VIIa-catalyzed activation of factor VIII: mechanism of FVIIa-catalyzed FVIII activation. J Thromb Haemost 2010 Nov;8(11):2494–503.
[95] Hertzberg M. Biochemistry of factor X. Blood Rev 1994 Mar;8(1):56–62.
[96] Walker CPR, Royston D. Thrombin generation and its inhibition: a review of the scientific basis and mechanism of action of anticoagulant therapies. Br J Anaesth 2002;88(6):848–63.
[97] Mann KG, Lawson JH. The role of the membrane in the expression of the vitamin K-dependent enzymes. Arch Pathol Lab Med 1992 Dec;116(12):1330–6.
[98] Furie B, Bouchard BA, Furie BC. Vitamin K-dependent biosynthesis of γ-carboxyglutamic acid. Blood 1999 Mar 15;93(6):1798–808.
[99] Rawala-Sheikh R, Ahmad SS, Monroe D, Roberts H, Walsh P. Role of y-carboxyglutamic acid residues in the binding of factor IXa to platelets and in factor-X activation. Blood 1992;79:398–405.
[100] van Dieijen G, Tans G, Rosing J, Hemker HC. The role of phospholipid and factor VIIIa in the activation of bovine factor X. J Biol Chem 1981 Apr;256(7):3433–42.
[101] Banks R, Forbes M, Kinsey S, Stanley A, Ingham E, Walters C, et al. Release of the angiogenic cytokine vascular endothelial growth factor (VEGF) from platelets: significance for VEGF measurements and cancer biology. Br J Cancer 1998 Mar;77(6):956–64.
[102] Coughlin SR. Thrombin signalling and protease-activated receptors. Nature 2000 Sep;407(6801):258–64.
[103] Hattori R, Hamilton KK, Fugate RD, McEver RP, Sims PJ. Stimulated secretion of endothelial von Willebrand factor is accompanied by rapid redistribution to the cell surface of the intracellular granule membrane protein GMP-140. J Biol Chem 1989 May;264(14):7768–71.

[104] Esmon CT, Fukudome K, Mather T, Bode W, Regan LM, Stearns-Kurosawa DJ, et al. Inflammation, sepsis, and coagulation. Haematologica 1999 Mar;84(3):254–9.
[105] Vu T-KH, Hung DT, Wheaton VI, Coughlin SR. Molecular cloning of a functional thrombin receptor reveals a novel proteolytic mechanism of receptor activation. Cell 1991 Mar;64(6):1057–68.
[106] Xu W-F, Andersen H, Whitmore TE, Presnell SR, Yee DP, Ching A, et al. Cloning and characterization of human protease-activated receptor 4. Proc Natl Acad Sci 1998 Jun 9;95(12):6642–6.
[107] Rezaie A. Protease-activated receptor signalling by coagulation proteases in endothelial cells. Thromb Haemost 2014;112(11):876–82.
[108] Nieman MT. Protease-activated receptors in hemostasis. Blood 2016 Jul 14;128(2):169–77. 2016/04/28 ed.
[109] Posma JJN, Posthuma JJ, Spronk HMH. Coagulation and non-coagulation effects of thrombin. J Thromb Haemost 2016 Oct;14(10):1908–16.
[110] van den Biggelaar M, Hernández-Fernaud JR, van den Eshof BL, Neilson LJ, Meijer AB, Mertens K, et al. Quantitative phosphoproteomics unveils temporal dynamics of thrombin signaling in human endothelial cells. Blood 2014 Mar 20;123(12):e22–36.
[111] Flaumenhaft R, De Ceunynck K. Targeting PAR1: now what? Trends Pharmacol Sci 2017 Aug;38(8):701–16.
[112] Flaumenhaft R. Protease-activated receptor-1 signaling: the big picture. Arterioscler Thromb Vasc Biol 2017 Oct;37(10):1809–11.
[113] Walsh SW, Nugent WH, Solotskaya AV, Anderson CD, Grider JR, Strauss JF. Matrix metalloprotease-1 and elastase are novel uterotonic agents acting through protease-activated receptor 1. Reprod Sci 2018 Jul;25(7):1058–66.
[114] Zhao P, Lieu T, Barlow N, Sostegni S, Haerteis S, Korbmacher C, et al. Neutrophil elastase activates protease-activated receptor-2 (PAR2) and transient receptor potential vanilloid 4 (TRPV4) to cause inflammation and pain. J Biol Chem 2015 May;290(22):13875–87.
[115] Amara U, Flierl MA, Rittirsch D, Klos A, Chen H, Acker B, et al. Molecular intercommunication between the complement and coagulation systems. J Immunol 2010 Nov 1;185(9):5628–36.
[116] Stavrou EX. Factor XII in inflammation and wound healing. Curr Opin Hematol 2018 Sep;25(5):403–9.
[117] Weiler H, Isermann BH. Thrombomodulin. J Thromb Haemost 2003 Jul;1(7):1515–24.
[118] Rezaie AR, Cooper ST, Church FC, Esmon CT. Protein C inhibitor is a potent inhibitor of the thrombin-thrombomodulin complex. J Biol Chem 1995 Oct;270(43):25336–9.
[119] Esmon CT. The protein C pathway. Chest 2003 Sep;124(3 Suppl):26S–32S.
[120] Griffin JH. 2007 Activated protein C. 8.
[121] Dahlbäck B, Villoutreix BO. Regulation of blood coagulation by the protein C anticoagulant pathway: novel insights into structure-function relationships and molecular recognition. Arterioscler Thromb Vasc Biol 2005 Jul;25(7):1311–20.
[122] Joyce DE, Gelbert L, Ciaccia A, DeHoff B, Grinnell BW. Gene expression profile of antithrombotic protein C defines new mechanisms modulating inflammation and apoptosis. J Biol Chem 2001 Apr;276(14):11199–203.
[123] Riewald M. Activation of endothelial cell protease activated receptor 1 by the protein C pathway. Science 2002 Jun 7;296(5574):1880–2.
[124] Zhao R, Lin H, Bereza-Malcolm L, Clarke E, Jackson C, Xue M. Activated protein C in cutaneous wound healing: from bench to bedside. Int J Mol Sci 2019 Feb 19;20(4):903.
[125] Meijers JCM, Herwald H. Protein C inhibitor. Semin Thromb Hemost 2011 Jun;37(4):349–54.
[126] Wang X, Terzyan S, Tang J, Loy JA, Lin X, Zhang XC. Human plasminogen catalytic domain undergoes an unusual conformational change upon activation. J Mol Biol 2000 Jan;295(4):903–14.
[127] Schaller J, Gerber SS. The plasmin-antiplasmin system: structural and functional aspects. Cell Mol Life Sci 2011 Mar;68(5):785–801.
[128] Cesarman-Maus G, Hajjar KA. Molecular mechanisms of fibrinolysis. Br J Haematol 2005 May;129(3):307–21.
[129] Crippa MP. Urokinase-type plasminogen activator. Int J Biochem Cell Biol 2007;39(4):690–4.

[130] Thelwell C, Longstaff C. The regulation by fibrinogen and fibrin of tissue plasminogen activator kinetics and inhibition by plasminogen activator inhibitor 1. J Thromb Haemost 2007 Apr;5(4):804–11.
[131] Hoylaerts M, Rijken DC, Lijnen HR, Collen D. Kinetics of the activation of plasminogen by human tissue plasminogen activator. Role of fibrin. J Biol Chem 1982 Mar;257(6):2912–9.
[132] Moali C, Hulmes DJ. Extracellular and cell surface proteases in wound healing: new players are still emerging. Eur J Dermatol 2009 Nov;19(6):552–64.
[133] Bouma BN, Mosnier LO. Thrombin activatable fibrinolysis inhibitor (TAFI)—how does thrombin regulate fibrinolysis? Ann Med 2006 Jan;38(6):378–88.
[134] Miljic P, Heylen E, Willemse J, Djordjevic V, Radojkovic D, Colovic M, et al. Thrombin activatable fibrinolysis inhibitor (TAFI): a molecular link between coagulation and fibrinolysis. Srp Arh Celok Lek 2010;138(suppl. 1):74–8.
[135] Gils A, Declerck PJ. Plasminogen activator inhibitor-1. Curr Med Chem 2004 Sep;11(17):2323–34.
[136] Medcalf RL, Stasinopoulos SJ. The undecided serpin: the ins and outs of plasminogen activator inhibitor type 2. FEBS J 2005 Sep 1;272(19):4858–67.
[137] Gebbink MFBG. Tissue-type plasminogen activator-mediated plasminogen activation and contact activation, implications in and beyond haemostasis: activation of tPA by cross-β structures. J Thromb Haemost 2011 Jul;9:174–81.
[138] Korkmaz B, Horwitz MS, Jenne DE, Gauthier F. Neutrophil elastase, proteinase 3, and cathepsin G as therapeutic targets in human diseases. Sibley D, editor. Pharmacol Rev 2010 Dec;62(4):726–59.
[139] Padrines M, Wolf M, Walz A, Baggiolini M. Interleukin-8 processing by neutrophil elastase, cathepsin G and proteinase-3. FEBS Lett 1994 Sep 26;352(2):231–5.
[140] Owen CA, Campbell EJ. The cell biology of leukocyte-mediated proteolysis. J Leukoc Biol 1999 Feb;65(2):137–50.
[141] Hermant B, Bibert S, Concord E, Dublet B, Weidenhaupt M, Vernet T, et al. Identification of proteases involved in the proteolysis of vascular endothelium cadherin during neutrophil transmigration. J Biol Chem 2003 Apr;278(16):14002–12.
[142] Groutas WC, Dou D, Alliston KR. Neutrophil elastase inhibitors. Expert Opin Ther Pat 2011 Mar;21(3):339–54.
[143] Grinnell F, Zhu M. Identification of neutrophil elastase as the proteinase in burn wound fluid responsible for degradation of fibronectin. J Invest Dermatol 1994 Aug;103(2):155–61.
[144] Eming SA, Koch M, Krieger A, Brachvogel B, Kreft S, Bruckner-Tuderman L, et al. Differential proteomic analysis distinguishes tissue repair biomarker signatures in wound exudates obtained from normal healing and chronic wounds. J Proteome Res 2010 Sep 3;9(9):4758–66.
[145] Afshar-Kharghan V, Thiagarajan P. Leukocyte adhesion and thrombosis. Curr Opin Hematol 2006 Jan;13(1):34–9.
[146] Reilly CF, Tewksbury DA, Schechter NM, Travis J. Rapid conversion of angiotensin I to angiotensin II by neutrophil and mast cell proteinases. J Biol Chem 1982 Aug;257(15):8619–22.
[147] Gao S, Zhu H, Zuo X, Luo H. Cathepsin G and its role in inflammation and autoimmune diseases. Archiv Rheumatol 2018;33:748–9.
[148] Hiebert PR, Granville DJ. Granzyme B in injury, inflammation, and repair. Trends Mol Med 2012 Dec;18(12):732–41.
[149] Afonina IS, Tynan GA, Logue SE, Cullen SP, Bots M, Lüthi AU, et al. Granzyme B-dependent proteolysis acts as a switch to enhance the proinflammatory activity of IL-1α. Mol Cell 2011 Oct;44(2):265–78.
[150] Omoto Y, Yamanaka K, Tokime K, Kitano S, Kakeda M, Akeda T, et al. Granzyme B is a novel interleukin-18 converting enzyme. J Dermatol Sci 2010 Aug;59(2):129–35.
[151] Buzza MS, Zamurs L, Sun J, Bird CH, Smith AI, Trapani JA, et al. Extracellular matrix remodeling by human granzyme B via cleavage of vitronectin, fibronectin, and laminin. J Biol Chem 2005 Jun;280(25):23549–58.

[152] Choy JC, Hung VHY, Hunter AL, Cheung PK, Motyka B, Goping IS, et al. Granzyme B induces smooth muscle cell apoptosis in the absence of perforin: involvement of extracellular matrix degradation. Arterioscler Thromb Vasc Biol 2004 Dec;24(12):2245–50.
[153] Pampalakis G, Sotiropoulou G. Tissue kallikrein proteolytic cascade pathways in normal physiology and cancer. Biochim Biophys Acta BBA Rev Cancer 2007 Sep;1776(1):22–31.
[154] Kalinska M, Meyer-Hoffert U, Kantyka T, Potempa J. Kallikreins—the melting pot of activity and function. Biochimie 2016 Mar;122:270–82.
[155] Borgoño CA, Diamandis EP. The emerging roles of human tissue kallikreins in cancer. Nat Rev Cancer 2004 Nov;4(11):876–90.
[156] Yoon H, Laxmikanthan G, Lee J, Blaber SI, Rodriguez A, Kogot JM, et al. Activation profiles and regulatory cascades of the human kallikrein-related peptidases. J Biol Chem 2007 Nov;282(44):31852–64.
[157] Rojkjaer R, Schmaier AH. Activation of the plasma kallikrein/kinin system on endothelial cells. Proc Assoc Am Physicians 1999 Jun;111(3):220–7.
[158] Campbell DJ. The renin–angiotensin and the kallikrein–kinin systems. Int J Biochem Cell Biol 2003 Jun;35(6):784–91.
[159] Matus CE, Bhoola KD, Figueroa CD. Kinin B1 receptor signaling in skin homeostasis and wound healing. Yale J Biol Med 2020 Mar 27;93(1):175–85.
[160] Chao J, Shen B, Gao L, Xia C-F, Bledsoe G, Chao L. Tissue kallikrein in cardiovascular, cerebrovascular and renal diseases and skin wound healing. Biol Chem 2010 Apr;391(4):345–55.
[161] Schmaier AH. The contact activation and kallikrein/kinin systems: pathophysiologic and physiologic activities. J Thromb Haemost 2016 Jan;14(1):28–39.
[162] Egelrud TR. Desquamation in the stratum corneum. Acta Derm Venereol 2000 Sep 20;208:44–5.
[163] Gao L, Chao L, Chao J. A novel signaling pathway of tissue kallikrein in promoting keratinocyte migration: activation of proteinase-activated receptor 1 and epidermal growth factor receptor. Exp Cell Res 2010 Feb;316(3):376–89.
[164] Nylander-Lundqvist E, Bäck O, Egelrud T. IL-1 beta activation in human epidermis. J Immunol 1996 Aug 15;157(4):1699–704.
[165] Mancek-Keber M. Inflammation-mediating proteases: structure, function in (patho) physiology and inhibition. Protein Pept Lett 2014;21(12):1209–29.
[166] Wachtfogel YT, Kucich U, James HL, Scott CF, Schapira M, Zimmerman M, et al. Human plasma kallikrein releases neutrophil elastase during blood coagulation. J Clin Invest 1983 Nov 1;72(5):1672–7.
[167] Markiewski MM, Daugherity E, Reese B, Karbowniczek M. The role of complement in angiogenesis. Antibodies 2020 Dec 1;9(4):67.
[168] Rafail S, Kourtzelis I, Foukas PG, Markiewski MM, DeAngelis RA, Guariento M, et al. Complement deficiency promotes cutaneous wound healing in mice. J Immunol 2015 Feb 1;194(3):1285–91.
[169] Sim RB, Tsiftsoglou SA. Proteases of the complement system. Biochem Soc Trans 2004 Feb 1;32(1):21–7.
[170] Dobó J, Pál G, Cervenak L, Gál P. The emerging roles of mannose-binding lectin-associated serine proteases (MASPs) in the lectin pathway of complement and beyond. Immunol Rev 2016 Nov;274(1):98–111.
[171] Matsushita M, Endo Y, Fujita T. Cutting edge: complement-activating complex of ficolin and mannose-binding lectin-associated serine protease. J Immunol 2000 Mar 1;164(5):2281–4.
[172] Bond JS. Proteases: history, discovery, and roles in health and disease. J Biol Chem 2019 Feb 1;294(5):1643–51.
[173] Ratnikov BI, Cieplak P, Gramatikoff K, Pierce J, Eroshkin A, Igarashi Y, et al. Basis for substrate recognition and distinction by matrix metalloproteinases. Proc Natl Acad Sci 2014 Oct 7;111(40):E4148–55.
[174] Martins VL, Caley M, O'Toole EA. Matrix metalloproteinases and epidermal wound repair. Cell Tissue Res 2013 Feb;351(2):255–68.
[175] Shay G, Lynch CC, Fingleton B. Moving targets: emerging roles for MMPs in cancer progression and metastasis. Matrix Biol 2015 May;44–46:200–6.

[176] Kessenbrock K, Plaks V, Werb Z. Matrix metalloproteinases: regulators of the tumor microenvironment. Cell 2010 Apr;141(1):52–67.

[177] Wang X, Khalil RA. Matrix metalloproteinases, vascular remodeling, and vascular disease. In: Advances in pharmacology [Internet]. Elsevier; 2018. p. 241–330. [cited 2021 Feb 1]. Available from: https://linkinghub.elsevier.com/retrieve/pii/S1054358917300571.

[178] Seizer P, May AE. Platelets and matrix metalloproteinases. Thromb Haemost 2013;110(11):903–9.

[179] Saito S, Trovato MJ, You R, Lal BK, Fasehun F, Padberg FT, et al. Role of matrix metalloproteinases 1, 2, and 9 and tissue inhibitor of matrix metalloproteinase-1 in chronic venous insufficiency. J Vasc Surg 2001 Nov;34(5):930–8.

[180] Nagase H, Visse R, Murphy G. Structure and function of matrix metalloproteinases and TIMPs. Cardiovasc Res 2006 Feb 15;69(3):562–73.

[181] Murphy G, Stanton H, Cowell S, Butler G, Knäuper V, Atkinson S, et al. Mechanisms for pro matrix metalloproteinase activation. APMIS 1999 Mar;107(1–6):38–44.

[182] Nagase H, Enghild JJ, Suzuki K, Salvesen G. Stepwise activation mechanisms of the precursor of matrix metalloproteinase 3 (stromelysin) by proteinases and (4-aminophenyl)mercuric acetate. Biochemistry 1990 Jun 19;29(24):5783–9.

[183] Nagase H. Activation mechanisms of matrix metalloproteinases. Biol Chem 1997 Apr;378(3–4):151–60.

[184] Cui N, Hu M, Khalil RA. Biochemical and biological attributes of matrix metalloproteinases. In: Progress in molecular biology and translational science [Internet]. Elsevier; 2017. p. 1–73. [cited 2021 Feb 1]. Available from: https://linkinghub.elsevier.com/retrieve/pii/S1877117317300327.

[185] Parks WC, Wilson CL, López-Boado YS. Matrix metalloproteinases as modulators of inflammation and innate immunity. Nat Rev Immunol 2004 Aug;4(8):617–29.

[186] Beidler SK, Douillet CD, Berndt DF, Keagy BA, Rich PB, Marston WA. Multiplexed analysis of matrix metalloproteinases in leg ulcer tissue of patients with chronic venous insufficiency before and after compression therapy: multiplexed analysis of MMPs in chronic leg ulcers. Wound Repair Regen 2008 Sep;16(5):642–8.

[187] Lobmann R, Ambrosch A, Schultz G, Waldmann K, Schiweck S, Lehnert H. Expression of matrix-metalloproteinases and their inhibitors in the wounds of diabetic and non-diabetic patients. Diabetologia 2002 Jun;45(7):1011–6.

[188] Vaalamo M, Mattila L, Johansson N, Kariniemi AL, Karjalainen-Lindsberg ML, Kähäri VM, et al. Distinct populations of stromal cells express collagenase-3 (MMP-13) and collagenase-1 (MMP-1) in chronic ulcers but not in normally healing wounds. J Invest Dermatol 1997 Jul;109(1):96–101.

[189] Nwomeh BC, Liang H-X, Cohen IK, Yager DR. MMP-8 is the predominant collagenase in healing wounds and nonhealing ulcers. J Surg Res 1999 Feb;81(2):189–95.

[190] Pirilä E, Korpi JT, Korkiamäki T, Jahkola T, Gutierrez-Fernandez A, Lopez-Otin C, et al. Collagenase-2 (MMP-8) and matrilysin-2 (MMP-26) expression in human wounds of different etiologies. Wound Repair Regen 2007 Jan;15(1):47–57.

[191] Gutiérrez-Fernández A, Inada M, Balbín M, Fueyo A, Pitiot AS, Astudillo A, et al. Increased inflammation delays wound healing in mice deficient in collagenase-2 (MMP-8). FASEB J 2007 Aug;21(10):2580–91.

[192] Hartenstein B, Dittrich BT, Stickens D, Heyer B, Vu TH, Teurich S, et al. Epidermal development and wound healing in matrix metalloproteinase 13-deficient mice. J Invest Dermatol 2006 Feb;126(2):486–96.

[193] Yager DR, Zhang LY, Liang HX, Diegelmann RF, Cohen IK. Wound fluids from human pressure ulcers contain elevated matrix metalloproteinase levels and activity compared to surgical wound fluids. J Invest Dermatol 1996 Nov;107(5):743–8.

[194] Hattori N, Mochizuki S, Kishi K, Nakajima T, Takaishi H, D'Armiento J, et al. MMP-13 plays a role in keratinocyte migration, angiogenesis, and contraction in mouse skin wound healing. Am J Pathol 2009 Aug;175(2):533–46.

[195] Kuivanen TT, Jeskanen L, Kyllönen L, Impola U, Saarialho-Kere UK. Transformation-specific matrix metalloproteinases, MMP-7 and MMP-13, are present in epithelial cells of keratoacanthomas. Mod Pathol 2006 Sep;19(9):1203–12.

[196] Chang M. Restructuring of the extracellular matrix in diabetic wounds and healing: a perspective. Pharmacol Res 2016 May;107:243–8.

[197] Cullen B, Smith R, McCulloch E, Silcock D, Morrison L. Mechanism of action of PROMOGRAN, a protease modulating matrix, for the treatment of diabetic foot ulcers. Wound Repair Regen Off Publ Wound Heal Soc Eur Tissue Repair Soc 2002 Feb;10(1):16–25.

[198] Allan JA, Docherty AJ, Barker PJ, Huskisson NS, Reynolds JJ, Murphy G. Binding of gelatinases A and B to type-I collagen and other matrix components. Biochem J 1995 Jul 1;309(Pt 1):299–306.

[199] Giannelli G, Falk-Marzillier J, Schiraldi O, Stetler-Stevenson WG, Quaranta V. Induction of cell migration by matrix metalloprotease-2 cleavage of laminin-5. Science 1997 Jul 11;277(5323):225–8.

[200] Mohan R, Chintala SK, Jung JC, Villar WVL, McCabe F, Russo LA, et al. Matrix metalloproteinase gelatinase B (MMP-9) coordinates and effects epithelial regeneration. J Biol Chem 2002 Jan 18;277(3):2065–72.

[201] Lechapt-Zalcman E, Prulière-Escabasse V, Advenier D, Galiacy S, Charrière-Bertrand C, Coste A, et al. Transforming growth factor-beta1 increases airway wound repair via MMP-2 upregulation: a new pathway for epithelial wound repair? Am J Physiol Lung Cell Mol Physiol 2006 Jun;290(6):L1277–82.

[202] Cook H, Davies KJ, Harding KG, Thomas DW. Defective extracellular matrix reorganization by chronic wound fibroblasts is associated with alterations in TIMP-1, TIMP-2, and MMP-2 activity. J Invest Dermatol 2000 Aug;115(2):225–33.

[203] Bullen EC, Longaker MT, Updike DL, Benton R, Ladin D, Hou Z, et al. Tissue inhibitor of metalloproteinases-1 is decreased and activated gelatinases are increased in chronic wounds. J Invest Dermatol 1995 Feb;104(2):236–40.

[204] Utz ER, Elster EA, Tadaki DK, Gage F, Perdue PW, Forsberg JA, et al. Metalloproteinase expression is associated with traumatic wound failure. J Surg Res 2010 Apr;159(2):633–9.

[205] Rayment EA, Upton Z, Shooter GK. Increased matrix metalloproteinase-9 (MMP-9) activity observed in chronic wound fluid is related to the clinical severity of the ulcer. Br J Dermatol 2008 May;158(5):951–61.

[206] Serra R, Buffone G, Falcone D, Molinari V, Scaramuzzino M, Gallelli L, et al. Chronic venous leg ulcers are associated with high levels of metalloproteinases-9 and neutrophil gelatinase-associated lipocalin. Wound Repair Regen Off Publ Wound Heal Soc Eur Tissue Repair Soc 2013 Jun;21(3):395–401.

[207] Reiss MJ, Han Y-P, Garcia E, Goldberg M, Yu H, Garner WL. Matrix metalloproteinase-9 delays wound healing in a murine wound model. Surgery 2010 Feb;147(2):295–302.

[208] Fray MJ, Dickinson RP, Huggins JP, Occleston NL. A potent, selective inhibitor of matrix metalloproteinase-3 for the topical treatment of chronic dermal ulcers. J Med Chem 2003 Jul 1;46(16):3514–25.

[209] Ladwig GP, Robson MC, Liu R, Kuhn MA, Muir DF, Schultz GS. Ratios of activated matrix metalloproteinase-9 to tissue inhibitor of matrix metalloproteinase-1 in wound fluids are inversely correlated with healing of pressure ulcers. Wound Repair Regen 2002 Jan;10(1):26–37.

[210] Snyder RJ, Driver V, Fife CE, Lantis J, Peirce B, Serena T, et al. Using a diagnostic tool to identify elevated protease activity levels in chronic and stalled wounds: a consensus panel discussion. Ostomy Wound Manage 2011 Dec;57(12):36–46.

[211] Visse R, Nagase H. Matrix metalloproteinases and tissue inhibitors of metalloproteinases: structure, function, and biochemistry. Circ Res 2003 May 2;92(8):827–39.

[212] Krampert M, Bloch W, Sasaki T, Bugnon P, Rülicke T, Wolf E, et al. Activities of the matrix metalloproteinase stromelysin-2 (MMP-10) in matrix degradation and keratinocyte organization in wounded skin. Mol Biol Cell 2004 Dec;15(12):5242–54.

[213] Saarialho-Kere UK. Patterns of matrix metalloproteinase and TIMP expression in chronic ulcers. Arch Dermatol Res 1998 Jul;290(Suppl):S47–54.

[214] Krishnaswamy VR, Mintz D, Sagi I. Matrix metalloproteinases: the sculptors of chronic cutaneous wounds. Biochim Biophys Acta BBA Mol Cell Res 2017 Nov;1864(11):2220–7.

[215] Schlage P, Kockmann T, Sabino F, Kizhakkedathu Jayachandran N, auf dem Keller U. Matrix metalloproteinase 10 degradomics in keratinocytes and epidermal tissue identifies bioactive substrates with pleiotropic functions. Mol Cell Proteomics 2015 Dec;14(12):3234–46.

[216] Vaalamo M, Weckroth M, Puolakkainen P, Kere J, Saarinen P, Lauharanta J, et al. Patterns of matrix metalloproteinase and TIMP-1 expression in chronic and normally healing human cutaneous wounds. Br J Dermatol 1996 Jul;135(1):52–9.
[217] Bullard KM, Lund L, Mudgett JS, Mellin TN, Hunt TK, Murphy B, et al. Impaired wound contraction in stromelysin-1-deficient mice. Ann Surg 1999 Aug;230(2):260–5.
[218] Bullard KM, Mudgett J, Scheuenstuhl H, Hunt TK, Banda MJ. Stromelysin-1-deficient fibroblasts display impaired contraction in vitro. J Surg Res 1999 Jun 1;84(1):31–4.
[219] Gearing AJH, Thorpe SJ, Miller K, Mangan M, Varley PG, Dudgeon T, et al. Selective cleavage of human IgG by the matrix metalloproteinases, matrilysin and stromelysin. Immunol Lett 2002 Apr 1;81(1):41–8.
[220] Parks WC. Matrix metalloproteinases in repair. Wound Repair Regen Off Publ Wound Heal Soc Eur Tissue Repair Soc 1999 Dec;7(6):423–32.
[221] Rohani MG, McMahan RS, Razumova MV, Hertz AL, Cieslewicz M, Pun SH, et al. MMP-10 regulates collagenolytic activity of alternatively activated resident macrophages. J Invest Dermatol 2015 Oct;135(10):2377–84.
[222] Chen P, Abacherli LE, Nadler ST, Wang Y, Li Q, Parks WC. MMP7 shedding of syndecan-1 facilitates re-epithelialization by affecting α2β1 integrin activation. PLoS One 2009 Aug 10;4(8), e6565.
[223] McGuire JK, Li Q, Parks WC. Matrilysin (matrix metalloproteinase-7) mediates E-cadherin ectodomain shedding in injured lung epithelium. Am J Pathol 2003 Jun;162(6):1831–43.
[224] Swee M, Wilson CL, Wang Y, McGuire JK, Parks WC. Matrix metalloproteinase-7 (matrilysin) controls neutrophil egress by generating chemokine gradients. J Leukoc Biol 2008 Jun;83(6):1404–12. 2008/03/11 ed.
[225] Shapiro SD, Kobayashi DK, Ley TJ. Cloning and characterization of a unique elastolytic metalloproteinase produced by human alveolar macrophages. J Biol Chem 1993 Nov 15;268(32):23824–9.
[226] Madlener M, Parks WC, Werner S. Matrix metalloproteinases (MMPs) and their physiological inhibitors (TIMPs) are differentially expressed during excisional skin wound repair. Exp Cell Res 1998 Jul 10;242(1):201–10.
[227] Cornelius LA, Nehring LC, Harding E, Bolanowski M, Welgus HG, Kobayashi DK, et al. Matrix metalloproteinases generate angiostatin: effects on neovascularization. J Immunol Baltim MD 1950 1998 Dec 15;161(12):6845–52.
[228] Muller M, Trocme C, Lardy B, Morel F, Halimi S, Benhamou PY. Matrix metalloproteinases and diabetic foot ulcers: the ratio of MMP-1 to TIMP-1 is a predictor of wound healing. Diabet Med 2008 Apr;25(4):419–26.
[229] Chen C, Schultz GS, Bloch M, Edwards PD, Tebes S, Mast BA. Molecular and mechanistic validation of delayed healing rat wounds as a model for human chronic wounds. Wound Repair Regen Off Publ Wound Heal Soc Eur Tissue Repair Soc 1999 Dec;7(6):486–94.
[230] Loffek S, Schilling O, Franzke C-W. Biological role of matrix metalloproteinases: a critical balance. Eur Respir J 2011 Jul 1;38(1):191–208.
[231] Austin KM, Covic L, Kuliopulos A. Matrix metalloproteases and PAR1 activation. Blood 2013 Jan 17;121(3):431–9. 2012/10/18 ed.
[232] Trivedi V, Boire A, Tchernychev B, Kaneider NC, Leger AJ, O'Callaghan K, et al. Platelet matrix metalloprotease-1 mediates thrombogenesis by activating PAR1 at a cryptic ligand site. Cell 2009 Apr;137(2):332–43.
[233] Tousseyn T, Jorissen E, Reiss K, Hartmann D. (Make) Stick and cut loose—disintegrin metalloproteases in development and disease. Birth Defects Res Part C Embryo Today Rev 2006 Mar;78(1):24–46.
[234] Yamamoto S, Higuchi Y, Yoshiyama K, Shimizu E, Kataoka M, Hijiya N, et al. ADAM family proteins in the immune system. Immunol Today 1999 Jun;20(6):278–84.
[235] Charrier L, Yan Y, Driss A, Laboisse CL, Sitaraman SV, Merlin D. ADAM-15 inhibits wound healing in human intestinal epithelial cell monolayers. Am J Physiol Gastrointest Liver Physiol 2005 Feb;288(2):G346–53.
[236] Zunke F, Rose-John S. The shedding protease ADAM17: physiology and pathophysiology. Biochim Biophys Acta BBA Mol Cell Res 2017 Nov;1864(11):2059–70.

[237] Hirano M, Ma BY, Kawasaki N, Okimura K, Baba M, Nakagawa T, et al. Mannan-binding protein blocks the activation of metalloproteases meprin α and β. J Immunol 2005 Sep 1;175(5):3177–85.
[238] Vadon-Le Goff S, Hulmes DJS, Moali C. BMP-1/tolloid-like proteinases synchronize matrix assembly with growth factor activation to promote morphogenesis and tissue remodeling. Met Extracell Matrix Biol 2015 May 1;44–46:14–23.
[239] Amano S, Scott IC, Takahara K, Koch M, Champliaud MF, Gerecke DR, et al. Bone morphogenetic protein 1 is an extracellular processing enzyme of the laminin 5 gamma 2 chain. J Biol Chem 2000 Jul 28;275(30):22728–35.
[240] Troilo H, Bayley CP, Barrett AL, Lockhart-Cairns MP, Jowitt TA, Baldock C. Mammalian tolloid proteinases: role in growth factor signalling. FEBS Lett 2016 Aug;590(15):2398–407. 2016/07/22 ed.
[241] Ge G, Fernández CA, Moses MA, Greenspan DS. Bone morphogenetic protein 1 processes prolactin to a 17-kDa antiangiogenic factor. Proc Natl Acad Sci U S A 2007 Jun 12;104(24):10010–5.
[242] Mason SD, Joyce JA. Proteolytic networks in cancer. Trends Cell Biol 2011 Apr;21(4):228–37.
[243] Vidak E, Javoršek U, Vizovišek M, Turk B. Cysteine cathepsins and their extracellular roles: shaping the microenvironment. Cell 2019 Mar 20;8(3):264.
[244] Brömme D, Wilson S. Role of cysteine cathepsins in extracellular proteolysis. In: Parks WC, Mecham RP, editors. Extracellular matrix degradation [Internet]. Berlin, Heidelberg: Springer Berlin Heidelberg; 2011. p. 23–51. https://doi.org/10.1007/978-3-642-16861-1_2.
[245] Vizovišek M, Fonović M, Turk B. Cysteine cathepsins in extracellular matrix remodeling: extracellular matrix degradation and beyond. Matrix Model Remodel 2019 Jan 1;75–76:141–59.
[246] Garnero P, Borel O, Byrjalsen I, Ferreras M, Drake FH, McQueney MS, et al. The collagenolytic activity of cathepsin K is unique among mammalian proteinases. J Biol Chem 1998 Nov;273(48):32347–52.
[247] Yasuda Y, Li Z, Greenbaum D, Bogyo M, Weber E, Brömme D. Cathepsin V, a novel and potent elastolytic activity expressed in activated macrophages. J Biol Chem 2004 Aug 27;279(35):36761–70.
[248] Ohashi K, Naruto M, Nakaki T, Sano E. Identification of interleukin-8 converting enzyme as cathepsin L. Biochim Biophys Acta 2003 Jun 26;1649(1):30–9.
[249] Repnik U, Starr AE, Overall CM, Turk B. Cysteine cathepsins activate ELR chemokines and inactivate non-ELR chemokines. J Biol Chem 2015 May 29;290(22):13800–11. 2015/04/01 ed.
[250] Lindsay S, Oates A, Bourdillon K. The detrimental impact of extracellular bacterial proteases on wound healing: bacterial proteases. Int Wound J 2017 Dec;14(6):1237–47.
[251] Okamoto T, Akaike T, Suga M, Tanase S, Horie H, Miyajima S, et al. Activation of human matrix metalloproteinases by various bacterial proteinases. J Biol Chem 1997 Feb;272(9):6059–66.
[252] Serra R, Grande R, Buffone G, Molinari V, Perri P, Perri A, et al. Extracellular matrix assessment of infected chronic venous leg ulcers: role of metalloproteinases and inflammatory cytokines: ECM, CVU, MMPs and cytokines. Int Wound J 2016 Feb;13(1):53–8.
[253] Sabino F, Egli FE, Savickas S, Holstein J, Kaspar D, Rollmann M, et al. Comparative degradomics of porcine and human wound exudates unravels biomarker candidates for assessment of wound healing progression in trauma patients. J Invest Dermatol 2018 Feb;138(2):413–22.
[254] Schlage P, Egli FE, Nanni P, Wang LW, Kizhakkedathu JN, Apte SS, et al. Time-resolved analysis of the matrix metalloproteinase 10 substrate degradome. Mol Cell Proteomics 2014 Feb;13(2):580–93.
[255] Sabino F, Madzharova E, auf dem Keller U. Cell density-dependent proteolysis by HtrA1 induces translocation of zyxin to the nucleus and increased cell survival. Cell Death Dis 2020 Aug 21;11(8):674.

CHAPTER 9

Proteinase imbalance in oral cancer and other diseases

Luciana D. Trino[a], Daniela C. Granato[a], Leandro X. Neves[a], Hinrich P. Hansen[b], and Adriana F. Paes Leme[a]
[a]*Laboratório Nacional de Biociências, LNBio, Centro Nacional de Pesquisa em Energia e Materiais, CNPEM, Campinas, SP, Brazil,* [b]*Department of Internal Medicine I, University Hospital Cologne, CECAD Research Center, Cologne, Germany*

Proteolysis in oral cancer

Overview on proteolysis

Proteolysis is a ubiquitous cellular process implicated in a range of physiological events. From functionalization of pre- and proproteins to the maintenance of proteostasis via degradation pathways, several proteolytic systems are required to sustain cellular homeostasis. Those systems are nothing but complex, displaying the ability to target specific substrates (e.g., polyubiquitinated proteins, signal sequences, and pro/ectodomain shedding) and operate at distinct cellular compartments (e.g., lysosomes, mitochondria, cytosol, plasma membrane, and extracellular space) in tightly regulated fashion. Additionally, proteinase-protein inhibitor interactions are critical for the maintenance of the physiological balance in the protease web (Fig. 1).

Uncontrolled proteolysis may lead to deleterious effects on health since it plays an important role in many signaling processes and constitutes an irreversible step in signaling cascades—in contrast to other reversible posttranslational protein modifications such as phosphorylation. In fact, perturbations on the proteostasis have been observed in a myriad of diseases such as neurodegeneration, inflammation, rheumatoid arthritis, fibrosis, and tumorigenesis [1]. For instance, proteinase activity is usually found augmented in the tumor microenvironment and seems closely related to key processes of cancer progression such as extracellular matrix modeling, invasion, angiogenesis, and metastasis [2].

Proteinases and inhibitors in oral cancer

The knowledge accumulated from previous work developed by our and other research groups converged to multiple observations of changes in proteolytic processes in oral squamous

Proteolytic Signaling in Health and Disease. https://doi.org/10.1016/B978-0-323-85696-6.00002-0

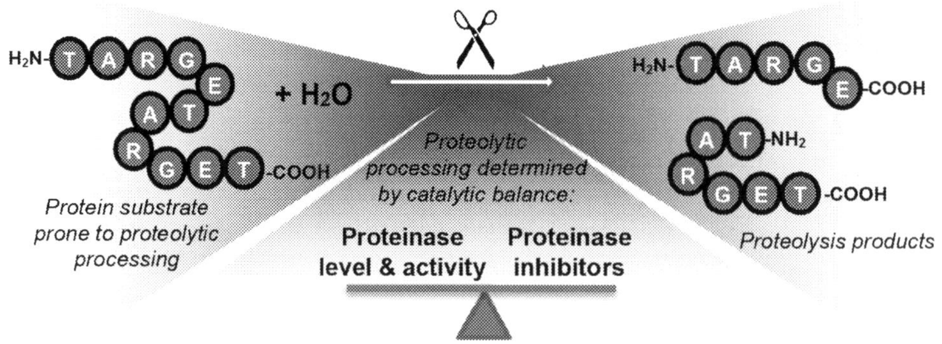

Fig. 1
Schematic representation of the proteolytic web in homeostasis.

cell carcinomas (OSCC). First, a proteomic analysis of saliva revealed reduced levels of peptidase inhibitors—including cystatins B/SA/SN/S and serine proteinase inhibitor Kazal type 5—in the saliva of OSCC patients compared to healthy individuals [3]. Furthermore, it was observed increased levels of matrix metalloproteinase MMP-9 in the saliva of patients with active tumor lesions compared to those submitted to surgical resection. Notably, another study investigated the proteinase spectrum of OSCC saliva patients and found increased levels of matrix metalloproteinases MMP-1, MMP-2, MMP-10, and MMP-12, together with ADAMST13, ADAM9, cathepsin V, and kallikrein-5 [3]. In the same study, the authors proposed a panel composed by ADAM9, cathepsin V, and kallikrein-5 as potentially useful biomarkers in the screening and diagnosis of OSCC. The results of both studies highlight the connection between proteinases and antiproteinase activity in OSCC.

The verification of perturbations in proteinase inhibitor levels has proven relevant not only for the understanding of OSCC biology but can also be promising in clinical management, therapeutics, and discovery of prognostic/diagnostic signatures. We have demonstrated that reduced levels of cystatin (CST) B—a cysteine-proteinase inhibitor—in neoplastic islands from the invasive tumor front performs as an independent marker of local recurrence while its lower saliva levels correlate with the occurrence of cervical lymph node metastasis [4].

In addition to cystatin B, lower levels of other cysteine-peptidase inhibitors such as cystatin M and C have been associated with increased proliferation, invasion of metastatic oral cancer cell lines [5], and lower survival rates of head and neck squamous cell carcinoma (HNSCC) patients [6]. On the other side of the balance, increased proteolytic activity in the tumor microenvironment is often reported and may lead to a poor clinical outcome. For instance, elevated expression of matrix metalloproteinases correlates with OSCC tumor invasion and metastasis [7], while cathepsin K associates with lymph node metastasis and poor prognosis [8], similarly to cathepsins B and D [9].

Saliva peptidomics in oral cancer

Several studies have addressed proteolysis by monitoring the human degradome, comprised of 588 proteinases, plus nearly 160 proteinase inhibitors encoded in the human genome [10]. Information about proteinase and antiproteinase expression levels can be readily assessed using large-scale approaches of different organs, tissues, and cancer cell lineages [11]. However, little information is available about proteins targeted by proteolytic processing as well as the resulting proteolytically cleaved substrates in pathological conditions.

In the context of oral squamous cell carcinoma, it is noteworthy to consider the proximity and constant contact of the saliva with tumor lesions, and how this could eventually affect saliva composition. Compared to other clinically relevant body fluids, low molecular weight peptides and degradation products account for 20%–30% of the saliva proteinaceous content of saliva [12]. That makes saliva a promising fluid for the development of noninvasive studies of proteolytic dynamics in OSCC, or other oral cavity pathologies.

Recently, we developed a mass spectrometry (MS)-based pipeline for integrated peptidomics and proteomics analyses that enabled the identification of degradation products in OSCC patients' saliva, and simultaneous evaluation of the levels of proteinases and proteinase inhibitors [13]. The compositional information and differences were evaluated between patients pathologically diagnosed with (pN+) and without (pN0) lymph node metastasis. Compositional differences were interrogated using bioinformatics tools for the prediction of active proteinases and tested for correlation with OSCC prognostic factors.

The comparison between the saliva peptidomes revealed 1720 peptides exclusively detected in pN+ patients, 1001 unique to pN0, and 1628 shared (~ 37% of all 4349 peptide identifications) between the groups. This relatively small fraction of commonalities between the two groups revealed that the saliva peptidome is remarkably dynamic. Quantitative analysis of endogenous peptides pointed out 77 sequences differentially abundant between the groups with the striking majority (75 peptides) exhibiting higher levels in pN+. This evidences an exacerbated proteolytic activity in the saliva of patients with advanced tumor stages diagnosed with lymph node metastasis.

Interestingly, at the global protein levels, only a few differences were detected between pN+ and pN0 patients. More specifically, a group comparison of saliva proteome revealed that 95% of all proteins identified were shared between pN0 and pN+ saliva samples, while 18 proteins were differentially abundant. Importantly, none of the proteins assigned to differential peptidome constituents showed significant group differences in the proteomic analysis. This indicates that major changes resulting from accentuated proteolysis—detected via peptidomics—are off the radar of conventional bottom-up proteomics.

However, the proteomic analysis did reveal a reduction of proteinase inhibitors in pN+ saliva that might favor proteolysis, even though no remarkable differences could be detected for proteinase levels in patients' saliva. Therefore, proteomics and peptidomics data corroborated that the increase in proteolytic activity is likely modulated via reduced inhibitory capacity. More specifically, 32 proteinase inhibitors present in patients' saliva exhibited an average reduction of 1.7-fold in pN+ patients, whereas over 40 distinct proteinases were found closer to equivalence (1.1-fold decrease in pN+). Among peptidase inhibitors significantly reduced in pN+ saliva are: fetuin A, interalpha-trypsin inhibitor heavy chain H_2, lipocalin-1, and serine proteinase inhibitor Kazal type 5—the latter corroborating previous findings of our group [14].

Peptidome differences could also be explored for correlation with clinical-pathological data to evaluate their prognostic utility. For instance, saliva levels of 10 out of 77 differential endogenous peptides correlated with the occurrence of extracapsular extension and perineural invasion, in addition to lymph node metastasis. Furthermore, a panel of five peptides displayed area under the ROC (receiver operating characteristic) curves of 86% for classification of pN+ and pN0 patients. This highlights the impact that studies on proteolytic dynamics in OSCC might have in clinical management. Notably, other studies have reported cellular and plasma peptidomics features with importance in the context of breast, ovarian, and bladder cancer [15, 16].

Deeper sequence analysis of endogenous peptides may hold important information on proteolytic activity and specificity in a biological context. By deriving N- and C-terminal amino acid residues from full-length protein sequences, we reconstructed the putative cleavage sites of differential peptides in the OSCC saliva peptidome (Fig. 2). This permitted the prediction of 19 active proteinases, including cathepsins with lysosomal (cathepsins D, L, K, S) or vacuole (cathepsin E) origin, and membrane metalloproteinase 2 and 25, mainly involved in structural modeling of the extracellular matrix. Together, cathepsins L, K, and S; calpains 1 and 2; and caspase 6 exhibit cysteine-type peptidase activity.

A remarkable enrichment of proline residues has been observed in the sequence logo of differential peptides highlighting a major contribution of breakdown products derived from salivary proline-rich protein (PRP) in the saliva of pN+ patients. Together with histatins and statherins, PRP fragments are the most abundant products of oral cavity proteolysis. [17] PRP peptides are often released upon the cleavage of the Gln-Gly (P1–P1′) motif likely as a result of microbial glutamine endopeptidase activity [18], corroborating our observations in OSCC saliva peptidome (Fig. 2). Importantly, the role of oral microbiota in cancer development [19] and OSCC aggressiveness [20] has been recently addressed and could serve as a diagnostic indicator [21, 22], reinforcing that oral microbiome can be intimately linked to OSCC biology. Noteworthy, four out of five endogenous peptides with the highest AUC-ROC are derived from basic salivary proline-rich protein 1 upon cleavage on the Gln-Gly motif. This implies that the activity of microbial proteinases in the oral cavity provided unique features relevant for patient classification.

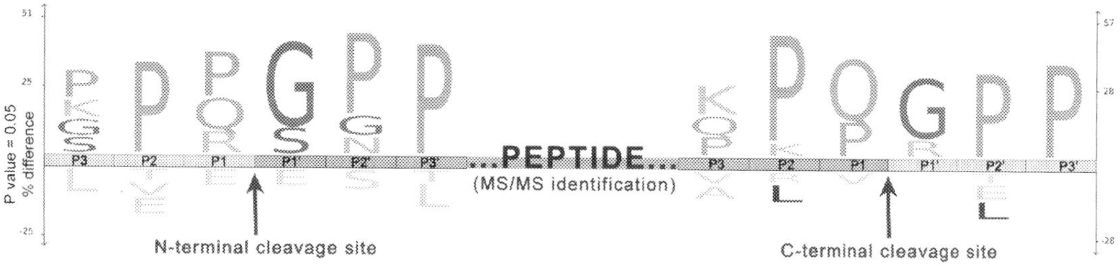

Fig. 2

Sequence analysis of saliva peptides differentially abundant between pN+ and pN0 OSCC patients highlighting the enrichment of glutamine-glycine motif in cleavage sites. *Adapted from Neves LX, Granato DC, Busso-Lopes AF, et al. Peptidomics-driven strategy reveals peptides and predicted proteases associated with oral cancer prognosis. Mol Cell Proteomics 11, 2020. https://doi.org/10.1074/mcp.RA120.002227*

Despite the lack of evidence that the saliva proteinase levels predicted to cleave differential peptides are being modulated, we found that their expression pattern in tumor tissues is associated with prognosis in HNSCC. Using data available from public repositories, we found out that increased tissue expression of calpains 1 and 2, cathepsin B, and MMP11 decreases overall patient survival, while higher levels of meprin 1A are associated with recurrence and tumor size. In addition, upregulation of most predicted cathepsins correlated with poor prognostic features such as recurrence, perineural invasion, nodal extracapsular extension, and tumor size.

The multifaceted approach for characterization of proteolytic events via peptidomics, bottom-up proteomics, and in silico analysis of cleavage sites provided a better understanding of OSCC biology, with implications for future translational studies and discovery of new therapeutic routes. The salivary peptidome is differentially regulated in OSCC patients with nodal metastasis as a result of the increased proteinase activity from oral microflora or host cell origin. This accentuated proteolysis in pN+ patients' saliva concurs with a reduction of inhibitory function suggesting an imbalance in the protease webs (Fig. 3).

Zinc metalloproteinases: A family of proteinases involved in the balance between health and disease

Several proteinases are involved in the balance between health and disease, especially the zinc metalloproteinases. The zinc metalloproteinases comprise a family of zinc proteinases that have catalytic zinc-binding motif with two histidines nearby the catalytic glutamate (Fig. 4). The metalloproteinase or catalytic domain is grouped into the subfamily of metalloproteinases, metzincins, which present a methionine residue on the C-terminal side of the HEXXHXXGXXH sequence (motif containing three histidines) involved in coordination

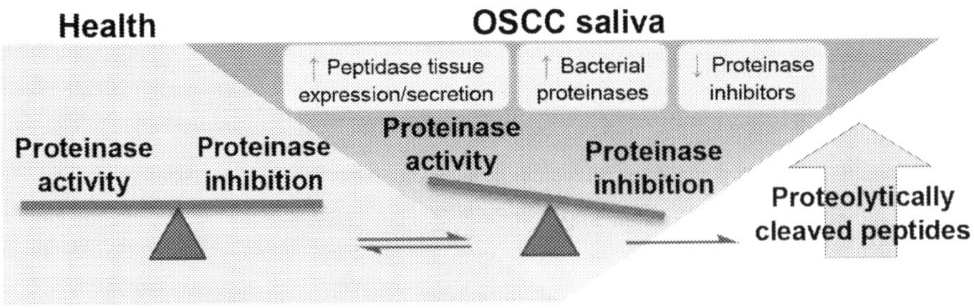

Fig. 3

Protease web modulation in the saliva of oral squamous cell carcinoma patients.

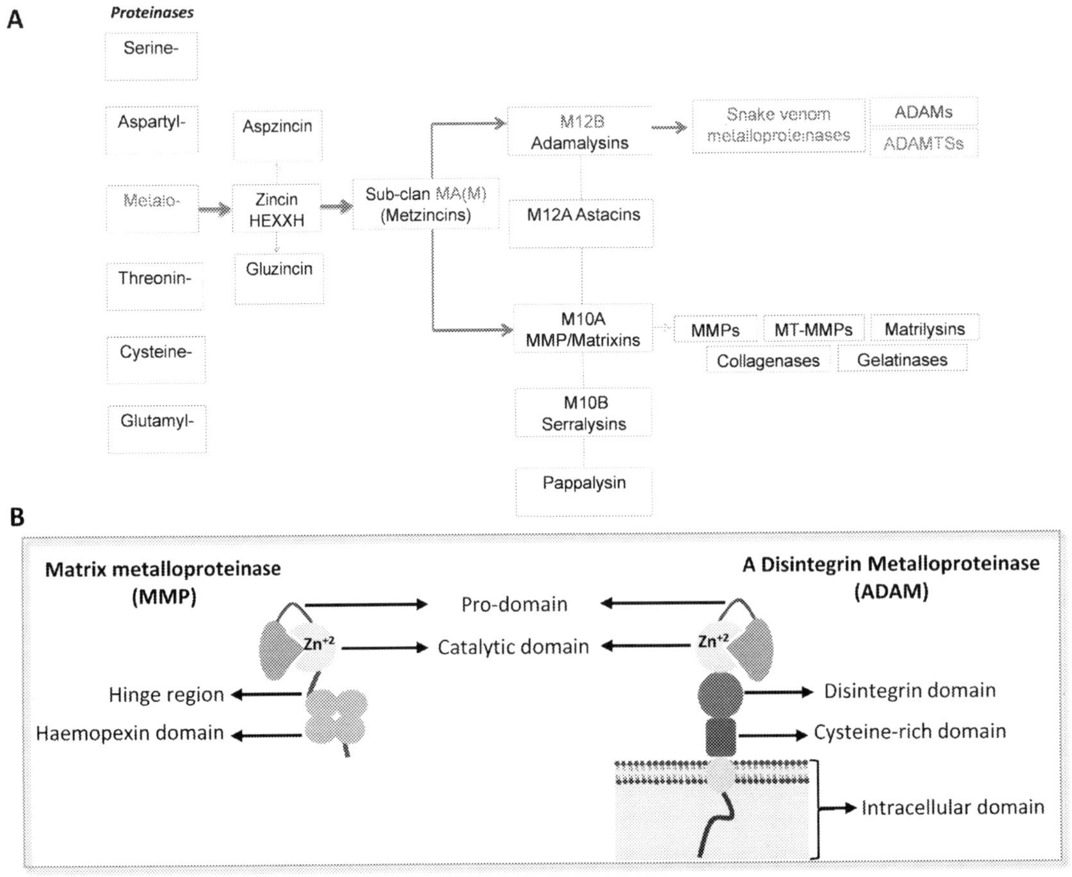

Fig. 4

Family of proteinases. (A) Most proteinases and their proteolytic sites are deposited in the repository named MEROPS: http://merops.sanger.ac.uk/index.htm. Matrix metalloproteinases (MMPs) and disintegrin metalloproteinases (ADAMs) structure domains. (B) MMPs and ADAMs have in common the zinc-containing catalytic domain, which is inactive in the presence of the prodomain. The metalloproteinases have other domains responsible for specific biological functions.

with a zinc atom. Among the proteinases that belong to the metzincin superfamily are the matrix metalloproteinases (MMPs) and adamalysins. The MMPs, also known as matrixins, are calcium-dependent zinc-containing endopeptidases and are capable of degrading all kinds of extracellular matrix proteins [23]. In addition, among the adamalysins, which are a zinc-dependent family of metalloproteinases, we encounter the ADAMs (*a disintegrin and metalloproteinase*) and ADAMTS (*a disintegrin and metalloproteinase with thrombospondin* motif), also characterized matrix metalloproteinases involved in health and disease [24]. ADAM17, also known as TNFα-converting enzyme, is one of the main components of the ADAM family of proteinases and will be the main focus of our discussion from now on.

ADAM17

Our group has been focusing on studying the ADAM17 signaling pathways to understand ADAM17 function through its substrates and interaction binding partners [25–30]. Therefore, the next part of this chapter is focused on ADAM17 history, structure, function in health and disease, and, finally, on the studies of our group that contributed to the ADAM17 history.

ADAM17 history

In 1988, it was discovered that the activity of tumor necrosis factor-alpha (TNFα), a proinflammatory mediator, is dependent on its proteolytic cleavage [31, 32], which was identified by Kriegler et al. between the amino acid residues Ala 76/Val 77 [32]. Once TNFα was recognized as a key regulator of the immune system, being able to induce effects ranging from inflammation to apoptotic cell death [33], a lot of effort was employed to identify the responsible proteinase for TNFα cleavage due to its possible application as a therapeutic target (see Section "ADAM17: A potential therapeutic biomarker"). The first advances in its identification were achieved in 1994, in which it was reported that the enzyme that cleaves TNFα was a metalloproteinase [34]. After three years, in 1997, the clone of cDNAs coding for human and murine TNFα-converting enzyme was reported [35, 36]. The research from Black et al. also identified that this enzyme was a new member of the mammalian adamalysin family or a disintegrin and metalloproteinase (ADAMs) [35], renaming it as ADAM17. The crystal structure of the catalytic domain from ADAM17 was revealed in 1998, in which the catalytic domain of ADAM17 was cocrystallized with a hydroxamic acid inhibitor making it possible to identify a polypeptide fold and a catalytic zinc region [37], confirming its classification in the ADAM family.

Over time, many researchers focused on identifying the role of ADAM17 as a therapeutic target. However, the first knockout of ADAM17 in mice was not viable [38]. The animals did not present the zinc-binding site on the metalloenzyme domain, which caused severe dysregulation of epithelial development. The study emphasized the essential role of

ADAM17-dependent ectodomain shedding in mammalian development and also identified other substrates of ADAM17 besides TNFα, such as L-selectin adhesion molecule, and transforming growth factor-alpha (TGFα) [38]. As TGFα was known as a ligand of the epidermal growth factor receptor (EGF-R), it was considered that ligands of EGF-R were substrates of ADAM17. This was hypothesized because EGF-R ligands are transmembrane proteins that need to be cleaved in order to be active [31].

Another process in which ADAM17 is involved in was discovered in 2000, known as RIP (Regulated Intramembrane Proteolysis) process. This process is known as the cleavage of certain surface proteins by ADAM17, on which the protein ectodomain is released, followed by a second cut inside the membrane. The second cleavage is performed by proteinases that are active inside the lipid environment of membranes, known as intramembrane cleaving proteinases [39]. The released fragment, generally the cytoplasmic tail, diffuses through the cell performing signaling activity [39, 40]. RIPping activity was identified when Brou et al. recognized the role of ADAM17 in the activation of the notch pathway [41]. The notch signaling transduction is involved in many processes, such as cellular specification, differentiation, proliferation, and survival [42]. First, the notch receptor cleavage by ADAM17 on its ectodomain occurs, which facilitates an intermembrane cleavage by another proteinase, like presenilin-dependent γ-secretase [42]. Then, the cytoplasmic domain of notch is transported to the nucleus where it acts regulating the transcription of various genes, such as HES1 [42, 43].

In addition to the shedding activity, ADAM17 has historically had several important ligands characterized. Among them is thioredoxin-1 (Trx-1) characterized in 2012 by our group as a novel interaction partner with the cytoplasmic domain of ADAM17 [25]. A more detailed information regarding the findings given by this study is described in Section "ADAM17: A potential therapeutic biomarker".

Other relevant ligands identified in 2012 were the proteins belonging to the iRhom family. Two important publications highlighted the association between ADAM17 and iRhoms in mammals, identifying the critical role of iRhoms in regulating ADAM17 trafficking and, consequently, in its maturation and surface expression [44, 45]. Furthermore, genetic mutations in the ADAM17 and iRhom2 genes were discovered in human patients, accentuating the importance of this interaction in immunity, host defense, and diseases like cancer [46].

Structural domains of ADAM17 have been well characterized in the past due to the similarity with other MMPs. Recently, however, a new domain was identified. A conserved stalk region, named as "conserved ADAM seventeen dynamic interaction sequence" (CANDIS), was discovered in 2014 [47]. This region is responsible for the recognition of substrates such as the IL-6 receptor and the IL-1 receptor 2 [48].

Other studies from our group regarding the role of ADAM17 in oral cancer development, cell proliferation, and migration was also investigated in 2014 [30, 49]. Using mass spectrometry, Dr. Paes Leme's group also identified glypican-1 as a novel ADAM17 substrate in 2017 [26]. Additionally, interaction studies identified the ADAM17 cytoplasmic domain as a novel interaction partner of Trx-1, which led to the discovery that Trx-1 directly and indirectly modulates ADAM17 activity (2018) [27], but also that ADAM17 plays a role in the conformation and activation of Trx-1 (2020) [28]. These findings are better explained in Section "ADAM17".

Furthermore, we can highlight some studies that indicated associations between ADAM17 and the coronavirus disease in 2019 (COVID-19), caused by the severe acute respiratory syndrome coronavirus 2 (SARS-CoV-2). It has been speculated that angiotensin-converting enzyme 2 (ACE2) is the receptor for the spike (S) protein of SARS-CoV-2 [50–54], on which the viral S protein receptor-binding domain binds to ACE2 to enter the host cell. ACE2 is a surface protein expressed on different tissues and organs, like lungs and the myocardium, being located in the inner track of the respiratory system [52]. In cell lines (BHK-21 fibroblast and HEK293), it was demonstrated that the ACE2 ectopic expression conferred permissiveness to pseudovirions expressing the SARS-CoV S protein (VSV-SARS-S) [55]. Also, other findings suggest that the receptor-binding domain (RBD), from S-glycoprotein, acts as the binding interface between the S-glycoprotein and the ACE2 receptor. The importance of this interaction remains in the fact that the S-glycoprotein on the virion surface mediates receptor recognition and membrane fusion [56]. ADAM17 is responsible together with a transmembrane serine proteinase 2 (TMPRSS2) for the ectodomain shedding of ACE2 [51, 53, 57]. In this way, it was hypothesized that ACE2 cleavage by ADAM17 might facilitate virus entry into host cells [54]. Other studies speculate that ADAM17 cleavage offers protection to organs while the shedding by TMPRSS2 is associated with the respiratory diseases [58]. Therefore, more detailed clinical studies need to be performed in order to clarify the role of ADAM17 in COVID-19.

Fig. 5 summarizes the most relevant events associated with the discovery, characterization, and biological function of ADAM17.

ADAM17 structure and biological function

ADAM17 is a type I transmembrane protein and was the first shedding proteinase to be molecularly characterized [31]. The extracellular part of the proteinase structure consists of an N-terminal signal peptide followed by four functionally distinctive domains: a prodomain, a metalloproteinase or catalytic domain, a disintegrin-like domain, and a cysteine-rich or EGF-like domain [59, 60]. This extracellular part is connected to an intracellular region by a transmembrane domain followed by a cytoplasmic portion [48]. The general structure of ADAM17 is shown in Fig. 6.

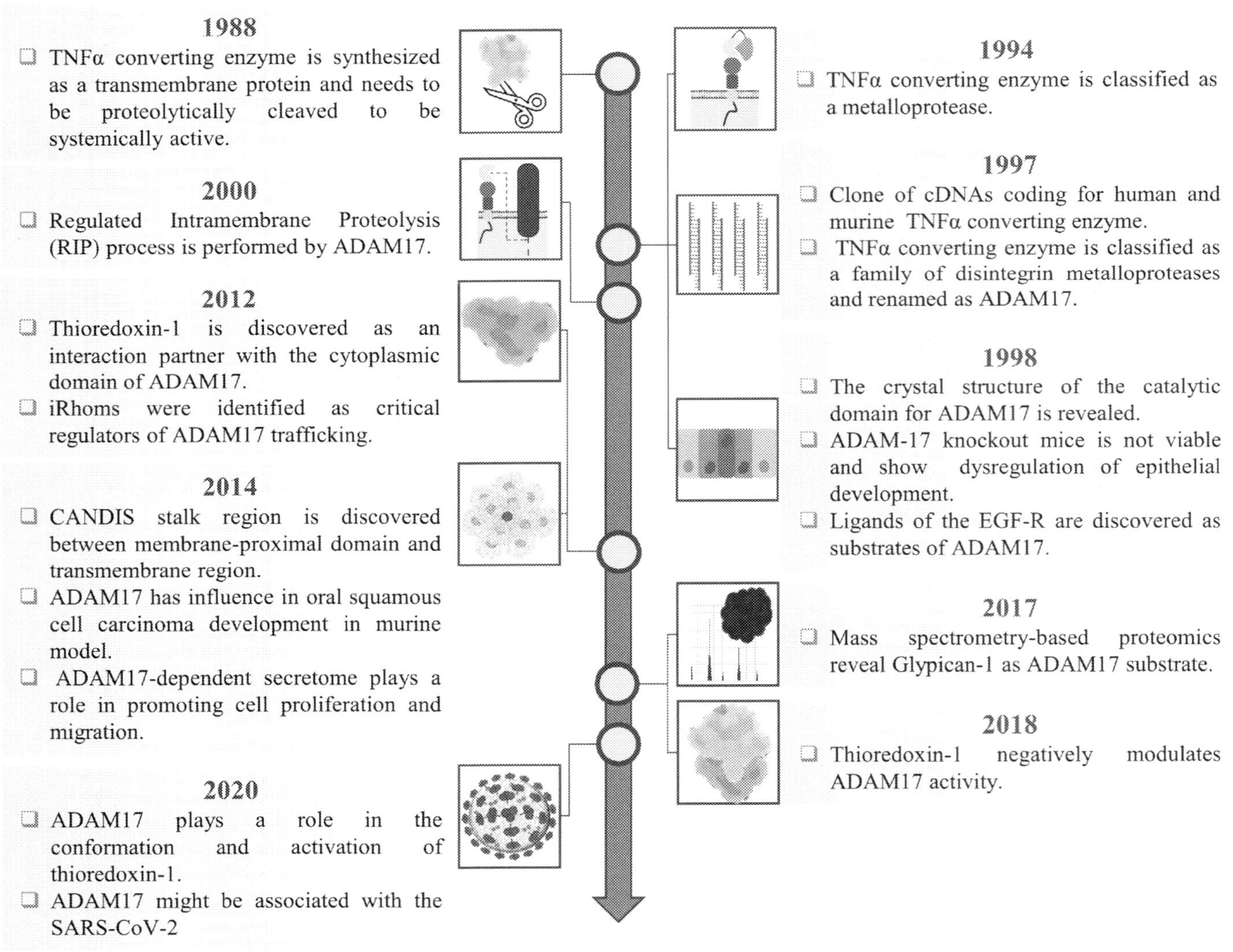

Fig. 5

Timeline of ADAM17 history concerning the key events associated with its discovery, characterization, and biological function. The research developed in the group of Dr. Paes-Leme has been included in the timeline. The year correlates to the publication date of the first article associated with the specific event.

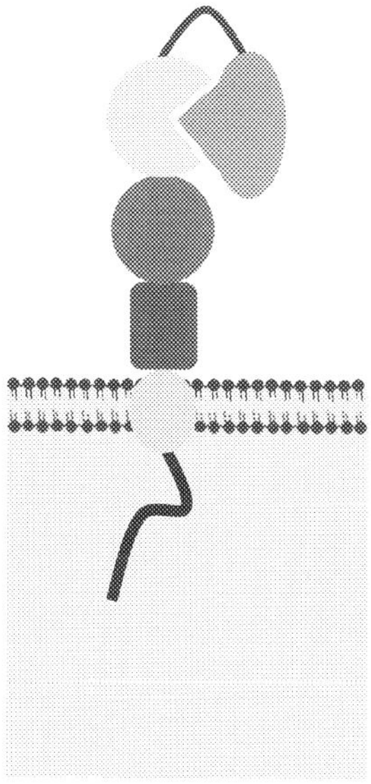

ADAM17 domain	Position	Biological function
Pro-domain	18 – 214	Inhibitory and chaperone function
Metalloprotease or catalytic domain	215 – 473	Ectodomain shedding
Disintegrin-like domain	474 – 572	Interaction with integrins
Cysteine-rich domain	603 – 671	Recognition of substrates, maturation of ADAM17, and support in adhesion of disintegrin-like domain
Transmembrane domain	672 – 694	Subcellular localization
Cytoplasmic domain	695 – 824	Signaling function, activation by phosphorylation, regulatory mechanism, and trafficking

Fig. 6

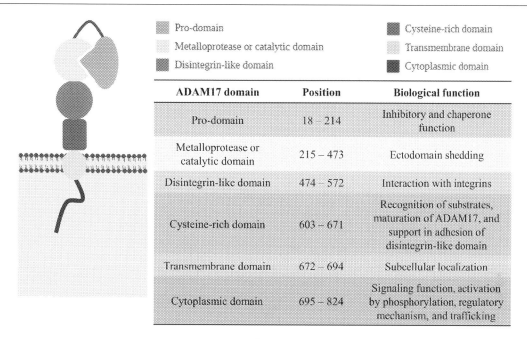

Schematic illustration of the general structure of ADAM17 showing the different domains, their position in the amino acid sequence of the protein, and their biological functions. The extracellular portion of ADAM17 consists of an N-terminal signal peptide (1–17 aa) followed by four functionally distinctive domains linked to an intracellular region by a transmembrane domain and followed by a cytoplasmic domain. *Adapted from Zunke F, Rose-John S. The shedding protease ADAM17: physiology and pathophysiology. Biochim Biophys Acta (BBA) Mol Cell Res 2017;1864(11, Pt B):2059–2070. https://doi.org/10.1016/j.bbamcr.2017.07.001*

The prodomain is constituted by a cysteine motif that can interact with a zinc ion in the catalytic domain of ADAM17 and inactivates the metalloproteinase catalytic site from the endoplasmic reticulum to the Golgi apparatus, known as a cysteine-switch mechanism to maintain enzyme latency [40, 60, 61]. The prodomain is removed by the proprotein convertase furin or by autocatalytic removal in the trans-Golgi network [31, 62]. Besides the inhibitory function, the prodomain acts as a chaperone assisting in protein folding and presents a protective role avoiding the enzyme degradation throughout transport via the secretory pathway [60].

In terms of biological function, the metalloproteinase domain is the most important due to its role in ectodomain shedding, which consists of a cleavage mechanism that releases the ectodomain to the extracellular environment and affects structurally and functionally several substrate proteins [62]. It has been reported more than 90 substrates for ADAM17 [31, 59, 63–65], including cytokines (TNFα, IL6R) [60, 62, 66], ligands of the epidermal growth factor receptor (EGF-R) like TNF receptor-I and TNF receptor-II; ErbB2;

ErbB4 [62]; adhesion molecules such as L-selectin, ICAM-1, and VCAM1 [67–69]; as well as chemokines like CX3CL1 [70–72]. Usually, the ADAM17 substrates are cleaved in a cis manner with higher preselection for alanine, leucine, and valine residues, and a lower tendency to cleave proline residue [73]. Furthermore, the ectodomain shedding mechanism tends to occur with an α-helical conformation and appropriate physical length, on which the specific site of cleavage will be dependent on the interaction between the substrate and the membrane [74–76]. The regulation of ectodomain shedding is based on gene expression, spatial redistribution of the sheddase and its substrates within the plasma membrane, proenzyme conversion, enzyme inhibition, and allosteric control [73]. These properties can be influenced by the cell and substrate type. The consequences of ectodomain shedding will be dependent on the function of the substrate protein, in which it can allow the participation of cytokines in paracrine signaling and can inactivate or activate receptors leading to inflammatory responses [62]. Therefore, the ectodomain shedding can act as a posttranslational regulator of membrane proteins and it is important to note that for each protein substrate, an experimental assay needs to be performed in order to better understand the consequences of ADAM17 catalytic activity.

The disintegrin-like domain in ADAMs was named after snake venom disintegrins, due to the specificity in binding some integrins [60, 62]. Integrins are part of heterodimeric (αβ) cellular receptors that can bind to specific extracellular ligands and consequently regulate cell adhesion, cytoskeleton reorganization, and cell polarity [60, 77]. In addition, these aspects of the cellular behavior regulated by the interaction between integrins and the ADAM17 disintegrin-like domain show critical application in a variety of physiological and pathophysiological processes. For instance, the interaction between ADAM17 and α5β1 integrin has been investigated for therapeutic use in chronic kidney diseases [78].

The cysteine-rich domain plays a role in the recognition of substrates as well as in the maturation of ADAM17 [79, 80]. It is also possible that the cysteine-rich domain supports the adhesion or binding capacity of the disintegrin-like domain including specificity to the mediated interactions with integrins and heparan sulfate proteoglycans [40]. Besides that, some studies reported that the cysteine-rich domain is crucial for substrate recognition and selection by recruiting accessory proteins or through direct contact with the substrates [60, 81].

The CANDIS (conserved ADAM17 dynamic interaction sequence) region is a short juxtamembrane segment located between the membrane-proximal domain and the transmembrane region. The CANDIS segment is a regulatory element involved in substrate recognition of human IL-6R and IL-1RII (transmembrane type I proteins) [47]. It has been shown that this segment cannot bind to type 2 transmembrane protein pro-TNFα [61]. However, more studies are necessary before we can generalize the fact that the CANDIS recognizes only transmembrane type I proteins. Other studies also revealed that CANDIS can interact with lipid bilayers in vitro, which can regulate ADAM17 shedding activity [61, 82].

This interaction is enhanced in the presence of phosphatidylserine on the outer leaflet of the membrane [83], while the presence of cholesterol blocks the binding of CANDIS to the membrane [82].

The role of the transmembrane domain is to bind the ADAM17 in the cell membrane allowing it to perform most of its physiological functions. Additionally, some findings suggest that the transmembrane domain of ADAM17 plays a role in regulating its activation, such as in the interaction with Rhbdf2 that is essential for the regulation of Rhbdf2–ADAM17-dependent shedding events [83–85].

The C-terminal cytoplasmic domain contains an SRC homology 3 domain-binding region as well as phosphorylation sites that can interact with a variety of intracellular signaling molecules. Some studies suggest that the cytoplasmic tail is important in negatively regulating ADAM17 [27, 86]. The cytoplasmic tail promotes dimerization of cell surface ADAM17, which is reduced by its phosphorylation through mitogen-activated protein kinases (MAPKs), extracellular signal-regulated kinase (Erk), and p38 [86]. As a consequence of the phosphorylation, it decreases tissue inhibitor of metalloproteinase-3 (TIMP3) interaction with the extracellular region of ADAM17 and promotes activation [86]. Recently, another study suggests that Trx-1 is also capable of regulating ADAM17 activity in a direct and indirect manner by interacting with the ADAM17 cytoplasmic domain (ADAM17cyto) [27]. The Trx-1 or its peptide [78],QFFKKGQ [84], interacts with ADAM17cyto leading to an inhibition of the metalloproteinase activity upon the presence of pro-HB-EGF [27]. On the contrary, Trx-$1^{C32/35S}$ recovers ADAM17 activity, suggesting the indirect effect of oxidant levels [27]. The amount of reactive oxygen species produced in the extracellular environment varies, being lower in the presence of Trx-1 when compared to Trx-$1^{C32/35S}$ [27]. Besides that, the modulation of ADAM17 activity by cell proliferation in cancer is decreased in the presence of the Trx-1 peptide [78]QFFKKGQ [84], indicating that the Trx-1 peptide can be further optimized to be used as a therapeutic approach in cancer [27]. It was also suggested that the cytoplasmic domain shows an important signaling and trafficking function [87]. However, the exact functional role of these interactions is not yet understood. There are some contradictory results regarding the shedding activity in different studies with mutants deprived of the cytoplasmic domain. These controversial results might be related to the use and choice of different cell lines, analyzed substrates, and expression constructs [88, 89].

ADAM17 in development, inflammation, and cancer

The developmental functions of ADAM17 in vertebrates were first studied in the phenotype of the knockout mice by Peschon and collaborators, who associated the importance of this metalloproteinase to normal development, emphasizing the crucial role of protein ectodomain shedding in vivo [38]. In this study, the absence of Zn^{2+}-binding domain inactivates the metalloproteinase activity, lacking the epidermal growth factor receptor (EGFR) or key

EGFR ligands, including TGFα, HB-EGF, or amphiregulin [90]. Perinatal and postnatal imperfections in mice lacking ADAM17 activity were correlated with numerous epithelial abnormalities during mammalian development, such as defects in skin; hair; cornea; and organs like thyroid, parathyroid, stomach, and intestine [38, 43]. Furthermore, it was found that epithelial dysgenesis led to disturbed heart valvulogenesis associated with increased heart size, myocardial trabeculation, mitral valve defects, reduced cell compaction, and hypertrophy of cardiomyocytes [62, 91, 92]. These characteristics were related to the lack of HB-EGF and EGFR proteins, which lead to loss of inhibition on bone morphogenic protein (BMP) signaling, the protein responsible for abnormal valve thickening [91]. Another study also revealed the importance of ADAM17 in energy homeostasis, on which ADAM17-deficient mice exhibit hypermetabolism, without increased appetite, which renders the animals extremely lean [93]. The importance of ADAM17 sheddase activity in bone development was also investigated. Mice lacking ADAM17 in chondrocytes showed a significantly expanded zone of hypertrophic chondrocytes in the growth plate and retarded growth of long bones, suggesting that ADAM17 regulates terminal differentiation of chondrocytes during endochondral ossification by activating the TGFα/EGFR signaling axis [94]. Furthermore, another study showed that forced expression of ADAM17 reduces Runx2 and Alpl expression, indicating that ADAM17 may negatively modulate osteoblast differentiation [95].

One of the first disorders on which ADAM17 enzymatic activity was supposed to be increased was in diseases associated with enhanced circulating or tissue TNFα, taking into account that ADAM17 was first identified as the TNFα cleaving enzyme. The TNFα signaling pathway is overactivated in the inflammatory infiltrates and produces other proinflammatory cytokines that contribute to the loss of cartilage and bone [60]. Actually, increased activity of ADAM17 was found in cartilage [96], and chondrocytes of patients with osteoarthritis [97]; synovial tissue of rheumatoid arthritis patients [98]; monocytes of patients with early systemic sclerosis [99], and patients with bowel disease [100]. In inflammatory diseases like rheumatoid arthritis, ADAM17 mRNA is highly expressed, suggesting that the metalloproteinase is one of the responsible for the overactivation of the TNFα pathway. On the other hand, chronic hypoxia appears to contribute to the upregulation of ADAM17. This low oxygen condition combined with TNFα upregulates the transcription of ADAM17, through the transcription factor hypoxia-inducible factor (HIF-1), in synovial cells [101]. In this way, some researchers are studying the neutralization of the TNFα signaling pathway as an approach to treat some inflammatory diseases, by blocking the interaction of the cytokine with its cognate receptors [102, 103]. However, the correlation between ADAM17 and inflammation is associated with the shedding of other substrates, not only TNFα. An inflamed tissue initiates the recruitment of immune cells to the sites of inflammation by chemoattractants. Leukocytes, the major cellular components of the inflammatory and immune response, can cross blood vessels to attach and roll on the vascular endothelium near the inflammation site. This attachment occurs via transmembrane lectin-like adhesion glycoproteins, called the selectins [104]. Following endothelium adhesion, leukocytes migrate through the basal membrane, which is supported by leukocyte L-selectin cleavage [105]. The cleavage of L-selectin is promoted by ADAM17 [106, 107]. Moreover, other molecules are cleaved by ADAM17, such as CX3CL1, V-CAM, ICAM-1, and JAM-1, which are involved in leukocyte activation [108], mediate leukocyte adhesion to the vascular endothelium [68], participate in leukocyte adhesion to integrins [109], and contribute to leukocyte diapedesis through the endothelial layer [110], respectively. Therefore, besides the ADAM17 enzymatic activity associated with TNFα, ADAM-17 sheddase activity over several factors provides effective recruitment of leukocytes to the inflammation site. Interestingly, there is an indication that ADAM17 activation also governs anti-inflammatory response. The metalloproteinase downregulates the macrophage activation by cleaving colony-stimulating factor-1 (CSF-1) from their surface [111]. Additionally, ADAM17 was recognized as the sheddase for IL-15 receptor α, related to the inhibition of collagen-induced arthritis and cardiac allograph rejection [112]. Another contribution of ADAM17 in anti-inflammatory response is the cleavage of the IL-6 receptor, which promotes a decrease of neutrophil infiltration and stimulates monocyte recruitment, an important process to cease inflammation [113]. In addition, it has been reported that TIMP3 inhibition of ADAM17 controls TNF levels in vivo, avoiding spontaneous inflammation [114]. Therefore, the role of ADAM-17 in pro- or anti-inflammatory signal is, then, complex and depends on the cellular context.

ADAM17 expression is upregulated in different types of cancers, likewise other ADAM family members [30]. The metalloproteinase sheds growth factors implicated in tumor progression and growth, and it also contributes to inflammation often observed in tumors [43]. Ligands for the EGFR receptors are synthesized as transmembrane molecules that are cleaved by ADAM17, among other ADAMs. After binding to EGFR ligands, also known as HER1, HER2, HER3, and HER4, the receptors undergo homo- or hetero-oligomerization and activation [115]. The expression of genes that control cellular proliferation, migration, adhesion, differentiation, and apoptosis is regulated by the activated HER receptors [115]. In this way, the ADAM17 cleavage activates the ligand, which subsequently binds and activates the HER receptors [60], influencing tumor formation and/or progression. The EGFR pathway is overactivated in different types of tumors [116], such as lung [117, 118], breast [119, 120], ovarian [121, 122], and head and neck cancer [30, 123, 124]. Furthermore, the role of ADAM17 in cancer is also related to TGFα [125], IL-6 [126], CX3CL1 [71], and desmoglein-2 [127] expression. TGFα was overexpressed in 61% of nonsmall cell lung cancer, which was not detected in adjacent normal tissues [128]. A significant effort has been made to better understand the EGFR action in promoting cancer, including the development of therapies. Different therapeutic strategies involving the EGFR pathway have been investigated, including the implementation of drugs that act as tyrosine kinase inhibitors, such as Osimertinib [129], and monoclonal antibodies, such as cetuximab and panitumumab [64], that block EGF and TGFα-induced activation of EGFR. Moreover, IL-6 and ADAM17 are responsible for the shedding of the soluble IL-6 receptor, acting on intestinal epithelial cells via IL-6 trans-signaling to induce colon cancer formation [130]. An inhibitor of IL-6 trans-signaling, sgp130Fc, offers a therapeutic strategy for the treatment of colon cancer [131]. Furthermore, in lung adenocarcinoma, the major cause of tumor development is KRAS mutations, which is positively correlated with augmented phospho-ADAM17 levels [132]. The protumorigenic activity of ADAM17 is related to its threonine phosphorylation by p38 MAPK, as well as with the preferential shedding of IL-6R by ADAM17, to release soluble IL-6R that drives IL-6 trans-signaling via the ERK1/2 MAPK pathway [132]. In addition, the chemokine CX3CL1 is directly associated with cancer metastasis, and its secretion is modulated by ADAM17 [71]. In hepatocellular carcinoma, ADAM17 was activated by MAPK in bone marrow endothelial cells and significantly promoted the secretion of CX3CL1, which facilitated the spinal metastasis of hepatocellular carcinoma in a mouse model [71]. Also, as mentioned earlier, desmoglein-2 overexpression is related to malignancies, especially in human skin carcinoma [133]. ADAM17 targets and cleaves desmoglein-2, detected in extracellular vesicle released from squamous cell carcinoma keratinocytes and associated with the tumor microenvironment modulation, a crucial step for cancer progression [134]. In summary, the ADAM17 expression, and its proteolytic activity, is promoted by cytokines and inflammatory stimuli related to immune responses, suggesting that inflammatory signals may stimulate ADAM17-mediated signaling events related to an increased risk of carcinogenesis [24].

Therefore, due to the role of ADAM17 in different diseases, this metalloproteinase has been extensively studied as a potential therapeutic target. The major challenge in developing ADAM17 inhibitors is to selectively inhibit pathological processes while maintaining the normal processes to avoid adverse effects. The next subchapter highlights some approaches used to investigate the potential of ADAM17 as a promising therapeutic strategy to treat the malignancies associated with this metalloproteinase.

ADAM17: A potential therapeutic biomarker

This subchapter aims to address the different approaches that can be used to investigate the role of ADAM17 and its proteolytic reactions in normal physiology and pathological conditions, such as, cancer. ADAM17 has been initially identified as the main sheddase responsible for releasing the soluble form of a variety of cell surface proteins, including growth factors, cytokines, cell adhesion molecules, and receptors (EGFR, TNFα, interleukin-6 receptor, etc.), most of which are signaling molecules that regulate cell proliferation, survival, migration and invasion properties in tumor malignant and inflammatory cells [31]. Considering the relevance of ADAM17 in different pathological processes, including cancer and inflammation, a large number of studies have associated

ADAM17 as a potential therapeutic factor [31, 64, 65]. It is important to recall that ADAM17 has many upstream effectors, such as chemokine-recruited cells, transcription factors, MMP inducers, and activators, as well as several downstream pathways, such as angiogenesis, growth and mitogenesis, synthesis, and migration, that by direct inhibition of ADAM17 can be interrupted as well [135]. Besides, several studies have led to the investigation of molecules that can inhibit ADAM17 activity in vivo, but it has proven not to be specific enough and may also have an inhibitory activity toward other metalloproteinases [61, 136–138]. So, although many efforts have been made to introduce a specific ADAM17-targeted inhibitor, still to date there are no inhibitors available for clinical application and the challenge of encountering inhibitors that are selective to the pathological effects generated by a certain proteinase is one of the main pitfalls in proteolytic activity regulation.

Different studies in the literature have searched for novel ADAM17 inhibitors and inquired about their efficiency (IC50) and selectivity [139–146]. Newly described ADAM17 inhibitors represent research tools to investigate the role of ADAM17 in the progression of various diseases (Table 1). High-throughput screening assays were developed using glycosylated and nonglycosylated substrates and a novel chemotype of ADAM17-selective probes were discovered from the TPIMS (Torrey Pines Institute for Molecular Studies) library [147]. Considering that the cleavage site of TNFα by ADAM17 is only four residues away from a glycosylated residue and glycosylation-enhanced ADAM-17 activities [148], glycosylation may play an important role in modulating ADAM activity as well. In addition, these kinetic studies revealed that ADAM17 inhibition occurred via a nonzinc-binding mechanism. Thus, the modulation of proteolysis via glycosylation may be used for identifying novel and potentially binding compounds. Besides, other inhibitors have been characterized and compared to commercial or broad-spectrum ones, leading to the following conclusions: (i) KP-457 was compared to GM-6001 (panmetalloprotease inhibitor) and p38 MAPK inhibitors, and showed that it enhances the production of functional human-induced pluripotent stem cell (iPSC)-derived platelets, while still inhibiting glycoprotein Ibα (GPIbα) shedding; (ii) addition of INCB3619, a potent inhibitor of ADAM10 and ADAM17, reduces in vitro HER-2/neu and amphiregulin shedding, confirming that it interferes with both HER-2/neu and EGFR ligand ADAM17-dependent cleavage [149]. Combining both inhibitors resulted in a synergistic growth inhibition in MCF-7 and HER-2/neu-transfected MCF-7 human breast cancer cells, suggesting a clinical benefit of combining agents that target the ErbB pathways at multiple points; (iii) PF-5480090 (TMI-002, Pfizer), ADAM17 specific inhibitor, was tested in a panel of breast cancer cell lines and indicates that a combination with chemotherapy or anti-EGFR/HER inhibitors, may contribute as a novel approach for treating breast cancer. (iv) D1(A12), an antibody raised against ADAM17, might also be an effective targeted agent for treating EGFR TKI-resistant HNSCC [150]. Still, although some inhibitors have enrolled and been tested in clinical trials, all have failed so far, mostly because of the poor efficacy or side effects [151]. One of the most recently deposited inhibitors is in phase

Table 1: Selective ADAM inhibitors tested in clinical trials (other metalloproteinases were included when they are also targeted by the inhibitors).

Lead/inhibitors	Targets	Company name and/or references	Specificity/inhibition efficiency	Evidenced activity
Number 17	ADAM17	147	4.2 μM (IC50 for ADAM17). > 100 μM for MMP8, MMP14, and ADAM10	In vitro kinetic inhibition assays with glycosylated substrate reveals ADAM17 inhibition
Number 19	ADAM17	147	4.3 μM (IC50 for ADAM17). > 100 μM for MMP2, MMP8, MMP9, MMP14, and ADAM10	In vitro kinetic inhibition assays with glycosylated substrate reveals ADAM17 inhibition
KP-457	MMP2, MMP8, MMP9, MMP14, ADAM10, ADAM17	136	0.011 μM (IC50 for ADAM17). > 100 μM for MMP1, 0.72 μM for MMP2, 2.2 μM for MMP8, 5.4 μM for MMP9, 2.14 μM for MMP14, and 0.75 μM for ADAM10	KP-457 blocked glycoprotein Ibα (GPIbα) shedding from iPSC platelets, effectively enhancing the production of functional human iPSC-derived platelets at 37°C
INCB3619	ADAM10, ADAM17	Incyte, Wilmington, DE. 149	0.022 and 0.014 μM (IC50 for ADAM19 and ADAM17, respectively). > 0.5 μM for MMP1, 0.045 μM for MMP2, 0.3 μM for MMP9, 0.8 μM for MMP14, 1.03 μM for ADAM33 and > 5 μM for ADAM9	INCB3619, a potent inhibitor of ADAM10 and ADAM17, reduces in vitro HER-2/neu and amphiregulin shedding, confirming that it interferes with both HER-2/neu and EGFR ligand cleavage. Combining INCB3619 with a lapatinib-like dual inhibitor of EGFR and HER-2/neu kinases resulted in synergistic growth inhibition in MCF-7 and HER-2/neu-transfected MCF-7 human breast cancer cells
INCB7839	ADAM10, ADAM17	Incyte, Wilmington, DE. 149	NI	Combining the INCB7839 second-generation sheddase inhibitor with lapatinib prevented the growth of HER-2/neu-positive BT474-SC1 human breast cancer xenografts in vivo
Complex I	ADAM17	139	11 μM (IC50 for ADAM17)	Novel iridium (III) complex 1 as an antitumor necrosis factor agent and the first metal-based inhibitor of TACE enzymatic activity. Complex 1 inhibited TNFα secretion and p38 phosphorylation in human monocytic THP-1 cells
MEDI3622	ADAM17	140	0.0031 μM (IC50 for ADAM17), > 10 μM for MMP12 and ADAM10	ADAM17 inhibitory antibody, MEDI3622, which induces tumor regression or stasis in many EGFR-dependent tumor models. The inhibitory activity of MEDI3622 correlated with EGFR activity both in a series of tumor models across several indications as well as in a focused set of head and neck patient-derived xenograft models

D1(A12)	ADAM17	141	0.0045 μM (IC50 for ADAM17), > 1 μM for ADAM10	A "crossdomain" human antibody-selective TACE antagonist provides a unique alternative to small molecule metalloprotease inhibition
D8	ADAM17	142	0.0012 μM (IC50 for ADAM17)	Developed a "mouse and human crossreactive" specific anti-TACE antibody. First, they reinvestigated the originally selected anti-TACE ectodomain phage-display clones and isolated a lead "mouse-human crossreactive" anti-TACE scFv, clone A9. Matured A9 IgG showed significant cell surface TACE inhibition as a monotherapy or combination therapy with chemotherapeutic agents
4mut	ADAM17	143	0.1 μM (IC50 for ADAM17), > 1 μM for MMP9 and MMP14	A patent has also been issued for this inhibitor (https://patentimages.storage.googleapis.com/80/7c/9e/94be8aeee12f6f/US9764008.pdf)
WAY-022	ADAM17	Wyeth-Aherst, Pearl River, NY. 144	NI	Treatment with mAb528 and WAY-022 enhances growth inhibition of EKI-resistant HCA-7 cells
GW280264X	ADAM10, ADAM17	Glaxo Smith Kline. 145	5.3 μM (IC50 for ADAM10), 541 μM for ADAM17, and no additional ADAM was affected	Presented an improved synthesis of the ADAM10 reference inhibitor GI254023X with a higher overall yield, enhanced detection ability, and increased acid stability, providing easier handling
TMI-2	ADAM17	Pfizer. 146	NI	Therapeutic target for triple-negative breast cancer

NI: not indicated.

Modified from Moss ML, Minond D. Recent advances in ADAM17 research: a promising target for cancer and inflammation. Mediat Inflam https://doi.org/10.1155/2017/9673537; Hirata S, Murata T, Suzuki D, et al. Selective inhibition of ADAM17 efficiently mediates glycoprotein Ibα retention during ex vivo generation of human induced pluripotent stem cell-derived platelets. Stem Cells Trans Med 2017;6(3):720–730. https://doi.org/10.5966/sctm.2016-0104; Leung C-H, Liu L-J, Lu L, et al. A metal-based tumour necrosis factor-alpha converting enzyme inhibitor. Chem Commun (Camb) 2015;51(19):3973–3976. https://doi.org/10.1039/c4cc09251a; Rios-Doria J, Sabol D, Chesebrough J, et al. A monoclonal antibody to ADAM17 inhibits tumor growth by inhibiting EGFR and non-EGFR-mediated pathways. Mol Cancer Ther 2015;14(7):1637–1649. https://doi.org/10.1158/1535-7163.MCT-14-1040; Tape CJ, Willems SH, Dombernowsky SL, et al. Cross-domain inhibition of TACE ectodomain. PNAS 2011;108(14):5578–5583. https://doi.org/10.1073/pnas.1017067108; Kwok HF, Botkjaer KA, Tape CJ, et al. Development of a "mouse and human cross-reactive" affinity-matured exosite inhibitory human antibody specif c to TACE (ADAM17) for cancer immunotherapy. Protein Eng Des Sel 2014;27(6):179–190. https://doi.org/10.1093/protein/gzu010; Wong E, Cohen T, Romi E, et al. Harnessing the natural inhibitory domain to control TNFα converting enzyme (TACE) activity in vivo. Sci Rep 2016;6(1):1–12. https://doi.org/10.1038/srep35598; Merchant NB, Voskresensky I, Rogers CM, et al. TACE/ADAM-17: a component of the epidermal growth factor receptor axis and a promising therapeutic target in colorecta cancer. Clin Cancer Res 2008;14(4):1182–1191. https://doi.org/10.1158/1078-0432.CCR-07-1216; Hoettecke N, Ludwig A, Foro S, et al. Improved synthesis of ADAM 0 inhibitor GI254023X. Neurodegener Dis 2010;7(4):232–238. https://doi.org/10.159/000267865; McGowan PM, Mullooly M, Caiazza F, et al. ADAM-17: a novel therapeutic target for triple negative breast cancer. Ann Oncol 2013;24(2):362–369. https://doi.org/10.1093/annonc/mds279; Minond D, Cudic M, Bionda N, et al. Discovery of novel inhibitors of a disintegrin and metalloprotease 17 (ADAM17) using glycosylated and non-glycosylated substrates. J Biol Chem 2012;287(43):36473–36487. https://doi.org/10.1074/jbc.M112.389114; Witters L, Scherle P, Friedman S, et al. Synergistic inhibition with a dual epidermal growth factor receptor/HER-2/neu tyrosine kinase inhibitor and a disintegrin and metalloprotease inhibitor. Cancer Res 2008;68(17):7083–7089. https://doi.org/10.1158/0008-5472.CAN-08-0739; Duffy M , Mullooly M, O'Donovan N, et al. The ADAMs family of proteases: New biomarkers and therapeutic targets for cancer? Clin Proteomics 2011;8(1):9. https://doi.org/10.1186/1559-0275-8-9.

I/II study using ADAM17 inhibitor (INCB7839) combined with rituximab as consolidated therapy after an autologous hematopoietic cell transplant (HCT) for patients with diffuse large B cell lymphoma (DLBCL) and this study has been completed in 2019 [152].

Failure in clinical trials is caused mainly by the absence of a proper subgroup selection, trials in end-stage disease patients, nonstratification in specific cancer types or stages of the disease, musculoskeletal syndrome, inadequate target validation, and not identifying in vivo MMP substrates or their physiological role and so on [153–155]. Inhibiting ADAM17 catalytic site, for example, has not shown to be a good strategy since the proteinase plays many physiological roles, and its inhibition could lead to adverse effects. That is why searching for novel substrates and aiming to comprehend the signaling pathway that this proteinase is involved in could be by far a more promising strategy.

It is important, though, to understand the role the metalloproteinases, such as ADAM17, plays in the cells under healthy and pathological conditions, and considering that several clinical studies toward ADAM17 inhibition have failed, it may be advisable to combine more than one strategy, such as interfering both directly in ADAM17 and its signaling pathway. In this line, our research group has been working in the field of ADAM17 since 2010, mostly concerned with the different manners of ADAM17 regulation, by searching for protein-protein interactions (PPIs) to better understand ADAM17 and modulate its activity. Therefore, next, we will focus on discussing in detail our findings, which are summarized in Table 2.

ADAM17: Employing different tools to study the mechanism of regulation in health and disease

For that, different strategies have been developed in our group in order to understand ADAM17 activity and modulation. ADAM17 regulation is considered very complex in which for the proteinase to be active in the membrane, different steps must occur starting from protein maturation, posttranslational modification, location in the cell membrane, shedding activity, and intracellular PPIs. All steps can be regulated and affect the final proteinase activity and its substrate levels.

In the past years, we have employed techniques to address the two main pathways: (a) catalytic site-dependent and (b) catalytic site-independent-ADAM17 regulation, including the regulation at full-length ADAM17 expression; the substrate level; at the maturation level; by inside-outside cross talk (Fig. 7). All the different approaches will be addressed in more detail later.

ADAM17 catalytic site-dependent regulation

ADAM17 regulation at the substrate level

One of the main functions of ADAM17 in pathogenesis is caused by its shedding activity on a diversity of membrane-anchored factors, and the release of the soluble protein forms has intrigued the research community for years, due to the specificity and regulation of this

Table 2: Summary of studies from our group evaluating ADAM17 modulation through different biochemical and functional strategies.

Biological function	Adjusted *P* value (Benjamin-Hochberg)	Mouse gene names (potential mADAM17 substrates identified overexpressed or exclusively in mEF WT secretome compared to mEF KO ADAM17) [49]	Number of molecules	Analogous genes to known hADAM17 shedding substrates
Cell adhesion	1.31E − 09	Zyx, Vcam1, Rhoa, Can, Postn, Vcl, Cdh11, Lgals3bp, Col8a1, Ephb4, Col6a1, Ncam1, Mcam, Thbs1, Itgb1, Omd Col12a1, Aimp1, Fbln5, Spp1, ADAM17, Gpnmb Pcdh18, Islr, Lpp, Itga3, B4galt1, Megf10, Ptprk, Cdh2, Col5a1, Tnc, Mfge8 Col6a2, Ccn4, Edil3, Adam9 Nid2, Pcdh19, Spon2	40	Ncam1 [31, 64], Vcam1 [68], Lgals3bp [156], Ephb4 [157], Mcam [158], Cdh2 [159]
Angiogenesis	7.64E − 06	Ephb3, Col4a1, Ncl, Hspg2, Nrp1, Vegfc, Hmox1, Ccl2, Pdgfa, Rtn4, Vegfa, Col8a1, Angpt1, Ephb4, Mfge8, Col4a2, Serpine1, Mcam, Ecm1, Hs6st1, Aimp1, Mmp19, Mapk14	23	Nrp1 [160], Hmox1 [31], Ccl2 [66], Vegfa [161], Ephb4 [157], Serpine1 [162],Mcam [158]
Cell-cell adhesion	4.38E − 05	Pdlim1, Tmpo, Twf2, Prdx6, Glod4, Cxcl12, Hcfc1, Puf60, Fscn1, Atic, Lasp1, Plec, Eno1, Vasn, Capg, Septin2, Slc3a2, Tagln2, Rars1	27	Cxcl12 [163], Vasn [164]
Cell migration	1.46E − 04	Tgfb1, Ephb3, Rhoa, Nrp1, Fat1, Arpc5, Tgfb2, Vegfa, Ptprk, Cdh2, Col5a1, Fscn1 Thbs1, Itgb1, Adam9, Cthrc1, Cd248, Rab1A	17	Nrp1 [160], Vegfa [161], Cdh2 [159]
Positive regulation of cell division	3.28E − 04	Tgfb1, Ybx1, Pdgfc, Vegfc, Pdgfa, Tgfb2, Vegfa	7	Vegfa [161], Tgfb1 [164], Ybx1 [165]
Inflammatory response	3.60E − 04	Tgfb1, Park7, Chst1, Cxcl12, Ccl2, Lxn, Hmgb2, Tnfrsf11b, Ndst1, Ccl7, Axl, Hmgb1, Thbs1, Il1rap, Csf1, Ecm1, C3, Aimp1, Spp1	19	Tgfb1 [164], Cxcl12 [163], Ccl2 [66], Axl [166], Hmgb1 [167], Csfl1 [168]
Positive regulation of cell proliferation	4.82E − 04	Pdgfa, Tgfb2, Vegfa, Gas1, Birc6, Thbs1, Mapk1, Itgb1, Fgf21, Prkaca, Npm1, Adam17, Cd248, Tgfb1, Efemp1, Vegfc, Cxcl12, Cd81, Tnc, Clu, Rps9, Ybx1, Lgals3, Csf1, Pdgfc	25	Vegfa [161], Fgf21 [169], Tgfb1 [164], Cxcl12 [163], Clu [156], Ybx1 [165]

Continued

Table 2: Summary of studies from our group evaluating ADAM17 modulation through different biochemical and functional strategies—cont'd

Biological function	Adjusted *P* value (Benjamin-Hochberg)	Mouse gene names (potential mADAM17 substrates identified overexpressed or exclusively in mEF WT secretome compared to mEF KO ADAM17) [49]	Number of molecules	Analogous genes to known hADAM17 shedding substrates
Extracellular matrix organization	4.75E − 06	Col4a1, Adamtsl2, Hspg2, Postn, Olfml2b, Pxdn, B4galt1, Tnfrsf11b, Tgfb2, Col4a2, Lgals3, Nid2, Fbln5	13	-
Protein folding	1.08E − 08	Fkbp10, Cct7, Hsp90aa1, Dnajb11, Fkbp4, Fkbp9, St13, Ppia, Ppwd1, Cdc37, Grpel1, Ppib, Hsp90b1, Hspa4l, Fkbp3, Hspa8, Cct5	17	-
Collagen fibril organization	8.44E − 08	Lox, Col1a1, Fmod, Col1a2, Tgfb2, Col5a2, Col3a1, Serpinh1, Col5a1, Col11a1	10	Lox [170]
Carbohydrate metabolic process	3.51E − 07	Galk1, Man2a1, Chst1 B3gat3, Manba, B4galt1 Taldo1, Gusb, Fuca2, Man2b2, Neu1, Fuca1, Galc, Idua, Ggta1, Ganab, Aimp1, Slc3a2, Mdh2	9	-
Transforming growth factor beta receptor signaling pathway	4.26E − 07	Zyx, Tgfb1, Usp15, Ltbp3, Itgb1, Col1a2, Adam9, Ccl2, Tgfb2, Col3a1, Ptprk, Ltbp1	12	Tgfb1 [164], Ccl2 [66]
Metabolic process	5.44E − 07	Gns, Acy1, Cad, Gstm2, Tpi1, Pgam1, Acat2, Atic, Man2b2, Aldh2 Fuca1, Tkt, Galc, Pkm, Uap1, Arsk, Samhd1, Galk1, Man2a1, Prdx6, Manba, Gusb, Ndst1, Neu1, Fuca2, Aco2, Idua, Ganab, Rrm1	29	-
Positive regulation of cell migration	1.35E − 06	Tgfb1, Gpnmb, Rhoa, Cxcl12, Pdgfa, Tgfb2, Vegfa, Rdx, Col1a1, Mcam, Hmgb1, Thbs1, Mapk1, Mmp3, Itgb1, Csf1, Pdgfc, ADAM17	18	Tgfb1 [164], Cxcl12 [163], Vegfa [161], Mcam [158]
Chaperone-mediated protein folding	3.54E − 06	Fkbp10, Chordc1, Fkbp4, Fkbp9, Ppib, Hspa8, Fkbp3, Clu	8	Clu [156]
Positive regulation of blood vessel endothelial cell migration	2.66E − 05	Tgfb1, Thbs1, Prl2c2, Vegfc, Vegfa, Angpt1	6	Tgfb1 [164], Vegfa [161]
Wound healing	2.69E − 05	Lox, Tgfb, Dcn, Serpine1, Col1a1, Tgfb2, B4galt1, Col3a1, Tnc, Sdc4, Sparc	11	Lox [170], Tgfb [164], Serpine1 [162]

Positive regulation of angiogenesis	4.71E − 05	Ddah1, Erap1, Serpine1, Thbs1, Lgals3, Prl2c2, ECM1, Vegfc, Hmox1, C3, Anxa3, Vegfa	12	Vegfa [161], Serpine1 [162]
Skeletal system development	8.22E − 05	Tgfb1, Col1a1, Can, Ltbp3, Lgals3, Fbn1, Fam20c, Col1a2, Tgfb2, Col5a2, Col3a1	11	Tgfb1 [164]
Positive regulation of gene expression	1.09E − 04	Tgfb1, Rps3, Msn, Park7, Ezr, Ldlr, Itga3, Hnrnpu, Hmgb2, Tgfb2, Hcfc1, Tnc, Vegfa, Rdx, Serpine1, Serpinb9, Csf1, C1qtnf1, Vim, Lmna, Hspa8, Mapk14	22	Ldlr [31], Tgfb1 [164], Vegfa [161], Serpine1 [162]
Osteoblast differentiation	1.30E − 04	Col6a1, Igfbp3, Gpnmb, Col1a1, Can, Mrc2, Hnrnpc, Tpm4, Hnrnpu, Tnc, Spp1	11	–
Organ regeneration	1.58E − 04	Atic, Tgfb1, Axl, Cad, Pkm, Cxcl12, Ccl2, Anxa3	8	Tgfb1 [164], Cxcl12 [163], Ccl2 [66]
Response to hypoxia	2.08E − 04	Tgfb1, Pdlim1, Vcam1, Rhoa, Postn, Vegfc, Hmox1, Cxcl12, Ccl2, Tgfb2, Vegfa, Adsl, Hsp90b1, ADAM17	14	Vcam [31, 64], Tgfb1 [164], Cxcl12 [163], Ccl2 [66], Vegfa [161]
Positive regulation of cell-substrate adhesion	2.23E − 04	Fbln2, Thbs1, Itgb1, Edil3, Itga3, Col8a1, Spp1	7	–
Glycosaminoglycan biosynthetic process	2.57E − 04	Hspg2, Ext1, Xylt1, B3gat3, Gpc1, Sdc4	6	Gpc1 [26]
Response to mechanical stimulus	3.34E − 04	Dcn, Rhoa, Thbs1, Postn, Cxcl12, Ccl2, Col3a1, Tnc	8	Cxcl12 [163], Ccl2 [66]
Cell migration involved in sprouting angiogenesis	3.61E − 04	Robo1, Nrp1, Itgb1, Vegfa, Ephb4	5	Vegfa [161], Nrp1 [160], Ephb4 [157]
Negative regulation of neuron projection development	3.68E − 04	Rhoa, Lrp1, Fkbp4, Efemp1, Dpysl3, Vim, Rtn4, ADAM17	8	Lrp1 [171]
Positive regulation of protein autophosphorylation	4.56E − 04	Gpnmb, Pdgfc, Vegfc, Pdgfa, Vegfa	5	Vegfa [161]
Heparan sulfate proteoglycan biosynthetic process	4.56E − 04	Ext1, Xylt1, B3gat3, Hs6st1, Ndst1	5	–
Collagen-activated tyrosine kinase receptor signaling pathway	4.69E − 04	Col4a1, Col4a2, Col4a5, Col1a1	4	–

Continued

Table 2: Summary of studies from our group evaluating ADAM17 modulation through different biochemical and functional strategies—cont'd

Biological function	Adjusted *P* value (Benjamin-Hochberg)	Mouse gene names (potential mADAM17 substrates identified overexpressed or exclusively in mEF WT secretome compared to mEF KO ADAM17) [49]	Number of molecules	Analogous genes to known hADAM17 shedding substrates
Protein heterotrimerization	5.68E − 04	Col6a1, Col6a2, Col1a1, C1qtnf1, Col1a2	5	–
Regulation of actin cytoskeleton organization	5.86E − 04	Efna5, Crk, Rhoa, Lrp1, Twf2, Tgfb2, Fscn1	7	Lrp1 [171]

Modified from Aragão AZB, Nogueira MLC, Granato DC, et al. Identification of novel interaction between ADAM17 (a disintegrin and metalloprotease 17) and thioredoxin-1. J Biol Chem 2012;287(51):43071–43082. https://doi.org/10.1074/jbc.M112.364513; Kawahara R, Granato DC, Yokoo S, et al. Mass spectrometry-based proteomics revealed Glypican-1 as a novel ADAM17 substrate. J Proteomics 2017;151:53–65. https://doi.org/10.1016/j.jprot.2016.08.017; Simabuco FM, Kawahara R, Yokoo S, et al. ADAM17 mediates OSCC development in an orthotopic murine model. Mol Cancer 2014;13(1):24. https://doi.org/10.1186/1476-4598-13-24; Zunke F, Rose-John S. The shedding protease ADAM17: physiology and pathophysiology. Biochim Biophys Acta (BBA) Mol Cell Res 2017;1864(11, Pt B):2059–2070. https://doi.org/10.1016/j.bbamcr.2017.07.001; McIlwain DR, Lang PA, Maretzky T, et al. iRhom2 regulation of TACE controls TNF-mediated protection against Listeria and responses to LPS. Science 2012;335(6065):229–232. https://doi.org/10.1126/science.1214448; Saad MI, Rose-John S, Jenkins BJ. ADAM17: an emerging therapeutic target for lung cancer. Cancers (Basel) 2019;11(9). https://doi.org/10.3390/cancers11091218; de Queiroz TM, Lakkappa N, Lazartigues E. ADAM17-mediated shedding of inflammatory cytokines in hypertension. Front Pharmacol 2020;11. https://doi.org/10.3389/fphar.2020.01154; Garton KJ, Gough PJ, Philalay J, et al. Stimulated shedding of vascular cell adhesion molecule 1 (VCAM-1) is mediated by tumor necrosis factor-α-converting enzyme (ADAM 17). J Biol Chem 2003;278(39):37459–37464. https://doi.org/10.1074/jbc.M305877200; Omoteyama K, Sato T, Sato M, et al. Identification of novel substrates of a disintegrin and metalloprotease 17 by specific labeling of surface proteins. Heliyon 2020;6(12):e05804. https://doi.org/10.1016/j.heliyon.2020.e05804; Morancho B, Martínez-Barriocanal Á, Villanueva J, et al. Role of ADAM17 in the non-cell autonomous effects of oncogene-induced senescence. Breast Cancer Res 2015;17:106. https://doi.org/10.1186/s13058-015-0619-7; Nollet M. Etude de l'implication de CD146/CD146 soluble dans l'angiogenèse et le développement tumoral : génération de nouveaux anticorps à visée thérapeutique. Published online December 22, 2017. Accessed January 30, 2021. http://www.theses.fr/2017AIXM0641; Pruessmeyer J, Ludwig A. The good, the bad and the ugly substrates for ADAM10 and ADAM17 in brain pathology, inflammation and cancer. Semin Cell Dev Biol 2009;20(2):164–174. https://doi.org/10.1016/j.semcdb.2008.09.005; Romi E, Gokhman I, Wong E, et al. ADAM metalloproteases promote a developmental switch in responsiveness to the axonal repellant Sema3A. Nat Commun 2014;5(1):4058. https://doi.org/10.1038/ncomms5058; Swendeman S, Mendelson K, Weskamp G, et al. VEGF-A stimulates ADAM17-dependent shedding of VEGFR2 and crosstalk between VEGFR2 and ERK signaling. Circ Res 2008;103(9):916–918. https://doi.org/10.1161/CIRCRESAHA.108.184416; Bernot D, Stalin J, Stocker P, et al. Plasminogen activator inhibitor 1 is an intracellular inhibitor of furin proprotein convertase. J Cell Sci 2011;124(8):1224–1230. https://doi.org/10.1242/jcs.079889; Mustafi R, Dougherty U, Mustafi D, et al. ADAM17 is a tumor promoter and therapeutic target in western diet-associated colon cancer. Clin Cancer Res 2017;23(2):549–561. https://doi.org/10.1158/1078-0432.CCR-15-3140; Malapeira J, Esselens C, Bech-Serra JJ, et al. ADAM17 (TACE) regulates TGFβ signaling through the cleavage of vasorin. Oncogene 2011;30(16):1912–1922. https://doi.org/10.1038/onc.2010.565; Gopal SK, Greening DW, Mathias RA, et al. YBX1/YB-1 induces partial EMT and tumourigenicity through secretion of angiogenic factors into the extracellular microenvironment. Oncotarget 2015;6(15):13718–13730. https://doi.org/10.18632/oncotarget.3764; Orme JJ, Du Y, Vanarsa K, et al. Heightened cleavage of Axl receptor tyrosine kinase by ADAM metalloproteases may contribute to disease pathogenesis in SLE. Clin Immunol 2016;169:58–68. https://doi.org/10.1016/j.clim.2016.05.011; Waller K, James C, de Jong A, et al. ADAM17-mediated reduction in CD14 ++CD16 + monocytes ex vivo and reduction in intermediate monocytes with immune paresis in acute pancreatitis and acute alcoholic hepatitis. Front Immunol 2019;10. https://doi.org/10.3389/fimmu.2019.01902; Tang J, Frey JM, Wilson CL, et al. Neutrophil and macrophage cell surface colony-stimulating factor 1 shed by ADAM17 drives mouse macrophage proliferation in acute and chronic inflammation. Mol Cell Biol 2018;38(17). https://doi.org/10.1128/MCB.00103-18; Badenes M, Amin A, González-García I, et al. Deletion of iRhom2 protects against diet-induced obesity by increasing thermogenesis. Mol Metab 2020;31:67–84. https://doi.org/10.1016/j.molmet.2019.10.006; Zhao XQ, Zhang MW, Wang F, et al. CRP enhances soluble LOX-1 release from macrophages by activating TNF-α converting enzyme. J Lipid Res 2011;52(5):923–933. https://doi.org/10.1194/jlr.M015156; Liu Q, Zhang J, Tran H, et al. LRP1 shedding in human brain: roles of ADAM10 and ADAM17. Mol Neurodegener 2009;4(1):17. https://doi.org/10.1186/1750-1326-4-17.

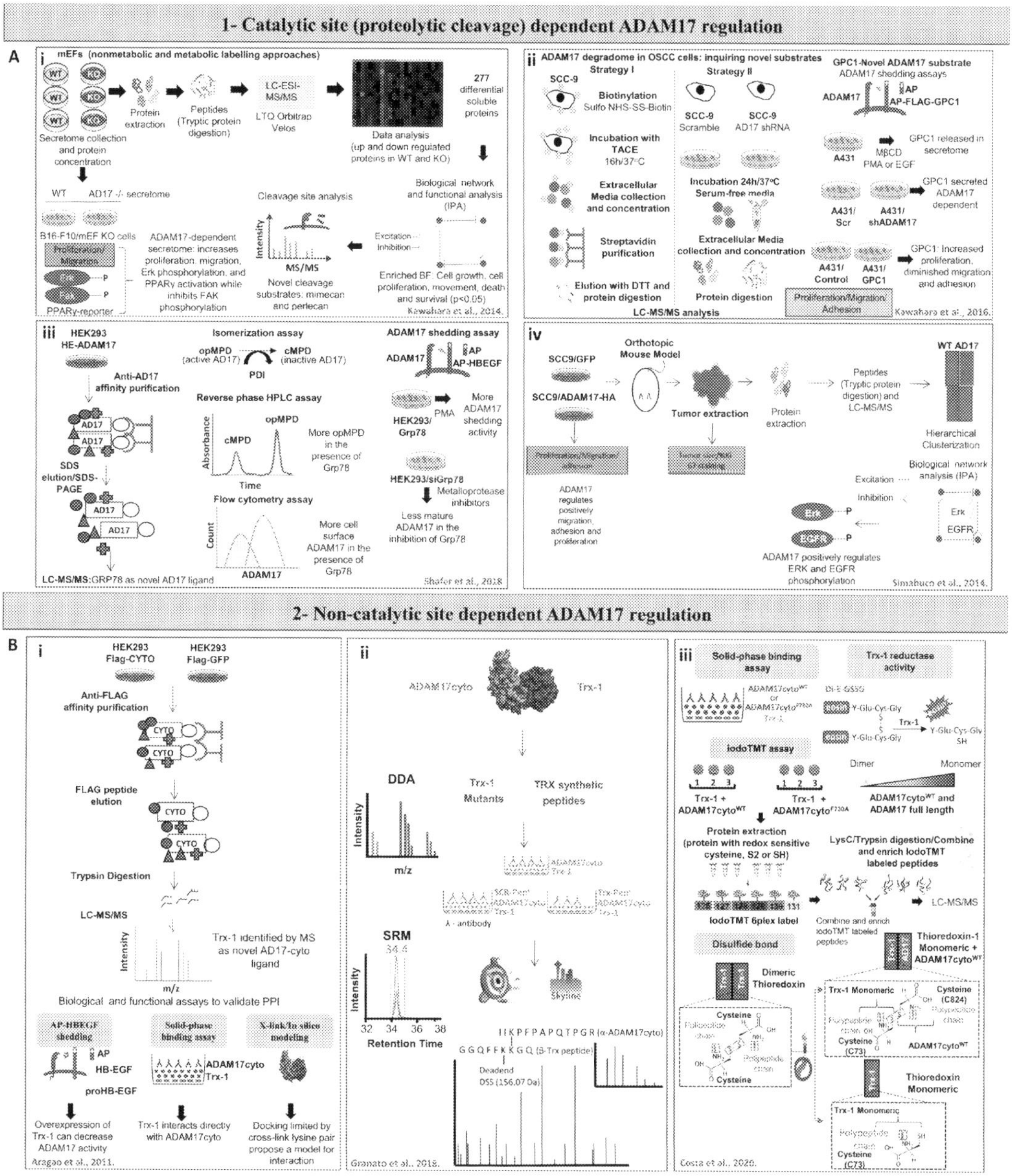

Fig. 7

Strategies used to access ADAM17 regulation and its potential role as a therapeutic target. (A) Catalytic site (proteolytic cleavage)-dependent ADAM17 regulation. (i) Fine-tuning of cell signaling rendered by the soluble molecules mediated by ADAM17-dependent cleavage. Mass spectrometry-based proteomics using nonmetabolic and metabolic labeling approaches has been employed to explore the secretome composition of wild-type (WT) and ADAM17$^{-/-}$ knockout (KO) mouse embryonic fibroblast (mEF) cells. Bioinformatic analyses indicated the differential regulation of 277 soluble proteins in the ADAM17-dependent secretome as well as novel direct ADAM17 cleavage substrates (e.g., mimecan and perlecan, both proteoglycans). ADAM17-dependent secretome has also shown to play an important role in promoting cell proliferation and migration,

Fig. 7, cont'd

and also promoted an opposite regulation of ERK and FAK pathways, as well as PPARγ downstream activation pathways. (ii) Function of glypican-1 (GPC1) modulated by ADAM17 proteolysis signaling. Mass spectrometry-based proteomics approaches revealed GPC1 as a new substrate for ADAM17, and its shedding was confirmed to be metalloproteinase-dependent, induced by a pleiotropic agent, PMA, and physiologic ligand, EGF. In addition, immunoblotting analysis of GPC1 in the extracellular media from control and ADAM17shRNA pointed to the direct involvement of ADAM17 in the cleavage of GPC1. Moreover, pathways related mainly to cellular movement, adhesion, and proliferation were events also modulated by the upregulation of full-length and cleavage GPC1 product. (iii) 17 conformation AMGrp78 influences disulfide-dependent ADAM17 conformation. Here, we show that the chaperone 78-kDa glucose-regulated protein (GRP78) protects the MPD against PDI-dependent disulfide-bond isomerization by binding to this domain and, thereby, preventing ADAM17 inhibition. (iv) Understanding the role of ADAM17 in oral cancer development. The overexpression of ADAM17 in SCC9 cells and orthotopic mouse model were realized to evaluate the proteinase role in oral cancer. ADAM17 overexpression leads to an increase in cellular processes such as adhesion, migration, and proliferation that can be translated in a more complex model by an increase in tumor size, in Ki-67 staining, and in the modulation of the tissue proteome, inferring to changes in Erk and EGFR signaling pathways. (B) Noncatalytic site-dependent ADAM17 regulation. (i) Trx-1 was identified as a novel interaction partner of ADAM17cyto capable to negatively modulate ADAM17 shedding activity when overexpressed in HEK293/AP-HB-EGF cells. (ii) Based on the ADAM17cyto-Trx-1 interaction model obtained in the previous study by in silico modeling, a combination of site-directed mutagenesis and synthetic peptide design was performed to inquire the importance of Trx-1 active site and the interaction interface for the maintenance of Trx-1 role in ADAM17 modulation. (iii) The model indicates that ADAM17cyto regulated Trx-1 enzymatic activity by favoring the monomeric state of the thiol-isomerase. One of the mechanisms that contribute to explain this dimer-monomer modulation is the binding with thioredoxin by a disulfide bond between cysteines C824 of ADAM17cyto and C73 of Trx-1, known as the dimerization site of Trx-1. The mechanism was resolved by employing two mass spectrometry techniques, discovery-based (DDA) and targeted-based (SRM). DDA, data-dependent acquisition; SRM, selective reaction monitoring. *Adapted from Kawahara R, Lima RN, Domingues RR, et al. Deciphering the role of the ADAM17-dependent secretome in cell signaling. J Proteome Res 2014;13(4):2080–2093. https://doi.org/10.1021/pr401224u; Kawahara R, Granato DC, Yokoo S, et al. Mass spectrometry-based proteomics revealed glypican-1 as a novel ADAM17 substrate. J Proteomics 2017;151:53–65. https://doi.org/10.1016/j.jprot.2016.08.017; Schäfer M, Granato DC, Krossa S, et al. GRP78 protects a disintegrin and metalloprotease 17 against protein-disulfide isomerase A6 catalyzed inactivation. FEBS Lett 2017;591(21):3567–3587. https://doi.org/10.1002/1873-3468.12858; Simabuco FM, Kawahara R, Yokoo S, et al. ADAM17 mediates OSCC development in an orthotopic murine model. Mol Cancer 2014;13(1):24. https://doi.org/10.1186/1476-4598-13-24; Aragão AZB, Nogueira MLC, Granato DC, et al. Identification of novel interaction between ADAM17 (a disintegrin and metalloprotease 17) and thioredoxin-1. J Biol Chem 2012;287(51):43071–43082. https://doi.org/10.1074/jbc.M112.364513; Granato DC, e Costa RAP, Kawahara R, et al. Thioredoxin-1 negatively modulates ADAM17 activity through direct binding and indirect reductive activity. Antioxid Redox Signal 2018;29(8):717–734. https://doi.org/10.1089/ars.2017.7297; e Costa RAP, Granato DC, Trino LD, et al. ADAM17 cytoplasmic domain modulates Thioredoxin-1 conformation and activity. Redox Biol 2020;37:101735. https://doi.org/10.1016/j.redox.2020.101735.*

process [61]. To address this topic, we have employed mass spectrometry-based proteomics approaches for the characterization of novel ADAM17 proteolytic substrates in the study of Kawahara et al., 2014 [49] (Fig. 7A-i). First, the function of the ADAM17-dependent secretome was characterized by functional assays that allowed us to query the effect of ADAM17-dependent secretome composition on the modulation of migration and proliferation events. For that, we evaluated migration and proliferation capabilities of wild-type (ADAM17 WT) and ADAM17 −/− knockout (ADAM17 KO) mouse embryonic fibroblast (mEF) cells that showed a reduction in migration and proliferation of ADAM17 KO compared to ADAM17 WT, in the presence of 1% FBS stimulus. In addition, we exchanged the conditioned media to confirm whether the differential secretome (WT vs KO) would be able to modulate these events. We showed that the ADAM17-dependent secretome had an increase in the capability to stimulate migration in B16-F10 melanoma murine cell line compared to the secretome of KO mEF cells. We also demonstrated that the WT secretome enhanced proliferation in KO knockout cells, while the KO secretome inhibited WT proliferation, demonstrating that ADAM17-dependent secretome plays an important role in promoting cell proliferation and migration (Fig. 7A-i).

Later, the composition of the ADAM17-dependent secretome was characterized on a proteome-wide scale, in which, the repertoire of proteins involved in this cross talk was investigated by mass-spectrometry based proteomics using nonmetabolic (label-free) and metabolic labeling (stable isotope labeling by amino acids, SILAC) approaches to explore the secretome composition of wild-type and ADAM17 −/− knockout mouse embryonic fibroblast (mEF) cells (Fig. 7A-i). Bioinformatic analyses indicated the differential regulation of 277 soluble proteins in the ADAM17-dependent secretome by Label-free methodology, while SILAC experiments showed 369 proteins were regulated and quantified with at least 1.5-fold difference. Furthermore, we found that the ADAM17-dependent secretome promoted an opposite regulation of ERK and FAK pathways as well as PPARγ downstream activation (Fig. 7A-i). These findings, finally, demonstrated the fine-tuning of cell signaling rendered by the soluble molecules mediated by ADAM17, as well as highlights novel direct ADAM17 cleavage substrates, such as mimecan and perlecan, both proteoglycan proteins (Fig. 7A-i). To categorize the novel and known substrates identified in this work regarding the biological process in which they belong, an enrichment analysis has been performed with the WT-upregulated and exclusive soluble-identified proteins (Table 2). It is interesting to take into account that ADAM17 could also regulate other proteinases and proteinase inhibitor expression as well. In this context, we have identified that the substrate repertoire of ADAM17 contains upregulated proteinase inhibitors (cystatin-C and cathepsin) and downregulated metalloproteinases.

ADAM17 regulation by novel uncovered substrates in OSCC

There is still a need for deeper investigation into the role of metalloproteinases in cancer, especially at substrate levels and the effect of their cleavage. Therefore, in another study published by our group [25], by using different proteomic techniques, we uncovered novel substrates that can be modulated by ADAM17 in oral squamous cell carcinoma cell lines (Fig. 7A-ii). To explore the repertoire of ADAM17 substrates in oral squamous cell carcinoma cell line, we performed two mass spectrometry-based proteomics strategies. In the first strategy, we used cell surface biotinylation of an HNSCC cell line, the squamous cell carcinoma cell line derived from the tongue (SCC-9), followed by incubation with recombinant ADAM17. In the second strategy, we performed a label-free analysis of the secretome of ADAM17-scrambled and knockdown SCC-9 cell lines. Both approaches showed glypican-1 (GPC1) as a novel substrate candidate for ADAM17. We also confirmed its shedding to be metalloproteinase-dependent, by demonstrating that the cleavage is induced by pleiotropic agent (phorbol 12-myristate 13-acetate, PMA) and physiologic ligand (epidermal growth factor, EGF), while it is inhibited by a broad-spectrum inhibitor of metalloproteinases, marimastat, in a cell-based assay. Moreover, we showed that the depletion of cholesterol increased PMA-induced GPC1 shedding, possibly by modulating GPC1 localization to the cell surface. The involvement of ADAM17 in GPC1 shedding was also validated in the secretome of SCC-9 and A431 cell lines by immunoblotting analysis of GPC1 in the extracellular media from control and ADAM17shRNA pointing to direct involvement of ADAM17 in the cleavage of GPC1.

To evaluate the functional role of soluble GPC1, its interacting partners were identified using co-immunoprecipitation (IP) followed by MS. Using mass spectrometry-based interactome analysis and bioinformatics analyses, we showed an involvement of the GPC1 complex in overrepresented processes such as proliferation, adhesion, and cell movement, which were events validated to be modulated by the upregulation of full-length GPC1 and recombinant N-terminal GPC1 fragment generated at a predicted ADAM17 cleavage site. Taken together, this study underscored GPC1 as a novel substrate of ADAM17 and revealed that the function of GPC1 in adhesion, proliferation, and migration of carcinoma cells may be modulated by proteolysis signaling.

Therefore, this study opens new avenues regarding the proteolysis-mediated function of GPC1 by ADAM17 contributing to one of the major limitations in inhibition of metalloproteinases as a therapeutic approach, which is the limited knowledge of the degradome of individual proteinases as well as the cellular function of the cleaved substrates. Also, this study together with the previous one highlights the role of ADAM17 shedding in substrates from the proteoglycan family, which should be better explored in the future. Also, GPC1 could be a substrate of other metalloproteinases, and the specificity toward ADAM17 cleavage should be further investigated.

ADAM17 regulation at the maturation level

It is known that ADAM17 exists in two conformations that differ in their disulfide connection in the membrane-proximal domain (MPD) [29]. Protein-disulfide isomerases (PDIs) on the cell surface convert the open MPD into a rigid closed form, which corresponds to inactive ADAM17. ADAM17 is expressed in its open activable form in the endoplasmic reticulum (ER) and consequently must be protected against ER-resident PDI activity. To study the players and the mechanism involved in the protection of ADAM17, we investigated the role that the chaperone 78-kDa glucose-regulated protein (GRP78) plays in protecting the MPD against PDI-dependent disulfide-bond isomerization by binding to this domain and, thereby, preventing ADAM17 inhibition [28] (Fig. 7A-iii). Perhaps combining different targets in cancer such as GRP78 and glypican could increase the scope of ADAM17 modulation.

ADAM17 role as a potential therapeutic target

Among cancer types, ADAM17 is involved in the development of HNSCC, which is the sixth most common cancer worldwide [172]. In addition to ADAM17 overexpression in HNSCC [173], its activity has been associated with advanced clinical stages of HNSCC as well as tumor recurrence [174]. Our group has shown that ADAM17 is implicated in oral cancer development [30], such as increasing tumor size and proliferation in an orthotopic murine tumor model, and Lu et al. has demonstrated the importance of ADAM17 in HNSCC cell proliferation and migration [175].

To study the direct role of ADAM17 in OSCC, an orthotopic mouse model combined with mass spectrometry-based proteomics was used in the search for novel potential therapeutic targets associated with ADAM17 expression and activity [30] (Fig. 7A-iv). OSCC is among the diseases that do not have clinical markers available for practical routine. EGFR is the unique known therapeutic target approved by FDA in 2006, and agents targeting EGFR, such as cetuximab, have emerged as a potential adjuvant targeted therapy for OSCC, responsible for 10%–15% of the response rates to monotherapy. The study of novel potential therapeutic targets that can be modulated by specific antibodies or by targeted molecules such as synthetic peptides may help in contributing to novel approaches to treat OSCC patients as well as other types of diseases.

The major mechanism by which ADAM17 contributes to tumor progression is related to the cleavage of ligands of growth factor receptors of EGFR, a widely studied oncogene in head and neck tumors and a therapeutic target in OSCC [176]. In order to access the role of ADAM17 in OSCC, the effect of overexpressing ADAM17 in cell migration, viability, adhesion, and proliferation was comprehensively evaluated in vitro [29]. In addition, the tumor size, tumor proliferative activity, tumor collagenase activity, and MS-based proteomics of tumor tissues have been evaluated by injecting tumorigenic squamous carcinoma cells (SCC-9) overexpressing ADAM17 in immunodeficient mice. The proteomic analysis has identified a total of 2194 proteins in control and tumor tissues. Among these, 110 proteins have been downregulated and 90 have been upregulated in ADAM17 overexpressed tumor tissues. Biological network analysis has uncovered that the overexpression of ADAM17 regulates Erk pathway in OSCC and further indicates proteins regulated by the overexpression of ADAM17 in the respective pathway. MS-based proteomics of those tumors revealed upregulation of several Erk regulatory proteins, which are associated with the Erk phosphorylation. Although ADAM17 is essential for the activation of EGFR signaling, therapeutic strategies using ADAM17 inhibitors have failed clinically [64, 151]. These results can open novel avenues for understanding the role of ADAM17 and its downstream signaling components in oral cancer development, and drug development.

These results are also supported by the evidence of higher viability, migration, adhesion, and proliferation in SCC-9 or A431 cells in vitro along with the increase of tumor size and proliferative activity and higher tissue collagenase activity as an outcome of ADAM17 overexpression. These findings contributed to understanding the role of ADAM17 in oral cancer development and as an emerging potential therapeutic target in this disease. In addition, our study also provided the basis for the development of novel and refined OSCC-targeting approaches.

ADAM17 noncatalytic site-dependent regulation

ADAM17 regulation by inside-outside cross talk promoted by the cytoplasmic domain

Mass spectrometry-based interactome analysis was used as an approach for ADAM17 activity modulation. How a membrane-anchored proteinase with an extracellular catalytic domain can be activated by inside-out regulation is not completely understood and has been an opening question so far. For that, we characterized thioredoxin-1 (Trx-1) as a partner of the ADAM17 cytoplasmic domain that could be involved in the regulation of ADAM17 activity. We induced the overexpression of the ADAM17 cytoplasmic domain in HEK293 cells, and ligands able to bind this domain were identified by MS, following protein immunoprecipitation. Trx-1 was validated as a ligand of the ADAM17 cytoplasmic domain and full-length ADAM17 recombinant proteins by immunoblotting, immunolocalization, and solid-phase binding assay (Fig. 7B-i). In addition, using nuclear magnetic resonance, it was shown in vitro that the titration of the ADAM17 cytoplasmic domain promotes changes in the conformation of Trx-1. The MS analysis of the cross-linked complexes showed cross-linking between the two proteins by lysine residues (K^{728} of ADAM17cyto and K^{82} of Trx-1). To further evaluate the functional role of Trx-1, we used a heparin-binding EGF shedding cell model and observed that the overexpression of Trx-1 in HEK293 cells could decrease the activity of ADAM17, activated by either phorbol 12-myristate 13-acetate (PMA) or EGF. This study identified Trx-1 as a novel interaction partner of the ADAM17 cytoplasmic domain and suggested that Trx-1 was a potential candidate that could be involved in ADAM17 activity regulation [25].

Direct and indirect effect of thiol-isomerase on ADAM17 activity

Furthermore, combining discovery and targeted proteomic approaches, we uncovered that Trx-1 negatively regulates ADAM17 by direct and indirect effects [27] (Fig. 7B-ii).

We performed cell-based assays with synthetic peptides and site-directed mutagenesis, and we demonstrated that the interaction interface of Trx-1 and ADAM17 is important for the negative regulation of ADAM17 activity. However, both Trx-1^{K72A} and catalytic site mutant Trx-$1^{C32/35S}$ rescued ADAM17 activity, although the interaction with Trx-$1^{C32/35S}$ was unaffected, suggesting an indirect effect of Trx-1. We confirmed that the Trx-$1^{C32/35S}$ mutant showed diminished reductive capacity, explaining this indirect effect on increasing ADAM17 activity through oxidant levels. Interestingly, Trx-1^{K72A} mutant showed similar oxidant levels to Trx-$1^{C32/35S}$, even though its catalytic site was preserved. This unexpected Trx-1^{K72A} behavior was due to more dimer formation and, consequently, the reduction of its Trx-1 reductase activity, evaluated through dimer verification, by gel filtration and mass spectrometry analysis. We further demonstrated that the general reactive oxygen species inhibitor, nacetylcysteine (NAC), maintained the regulation of ADAM17 dependent on Trx-1 reductase activity levels, whereas the electron transport chain modulator, rotenone, abolished the Trx-1 effect on ADAM17 activity. However, we showed for the first time that the mechanism of Trx-1-dependent ADAM17 regulation can be both direct and indirect and brings new insights into the cross talk between isomerases and mammalian metalloproteinases.

ADAM17 effect on thiol isomerase activity and protein dimeric conformation

We later investigated if the opposite modulation, ADAM17cyto upon Trx-1, could occur as well (Fig. 7B-iii). We demonstrated that site-directed mutagenesis in ADAM17 (ADAM17$cyto^{F730A}$) also disrupts the interacting interface with Trx-1, resulting in a decrease of Trx-1 reductive capacity and activity [28]. One of the mechanisms that explain this effect might be that ADAM17cyto favors Trx-1 monomerization state—the active state—by forming a disulfide bond between Cys^{824} at the C-terminal of ADAM17cyto with the Cys^{73} of Trx-1, which is involved in the dimerization site. Moreover, we observed that ADAM17 overexpressing or knockdown in cells favors or not the monomeric state of Trx-1, respectively. As a result, there is a decrease of oxidant levels and ADAM17 sheddase activity, and an increase in the reduced cysteine-containing peptides in intracellular proteins in ADAM17cyto overexpressing cells. In summary, we propose that ADAM17 is able to modulate Trx-1 conformation affecting its activity and intracellular redox state, bringing up a novel possibility for positive regulation of thiol isomerase activity in the cell by mammalian metalloproteinases. These results together with the others contribute to the understanding of the ADAM17 physiological and pathological role (Fig. 8).

Perspectives and conclusions

In this chapter, we have provided an insight into the role of the proteinases in health and disease. Proteinases are functionally implicated in different processes of cancer progression as well as they are involved in physiological pathways. In the case of cancers, these

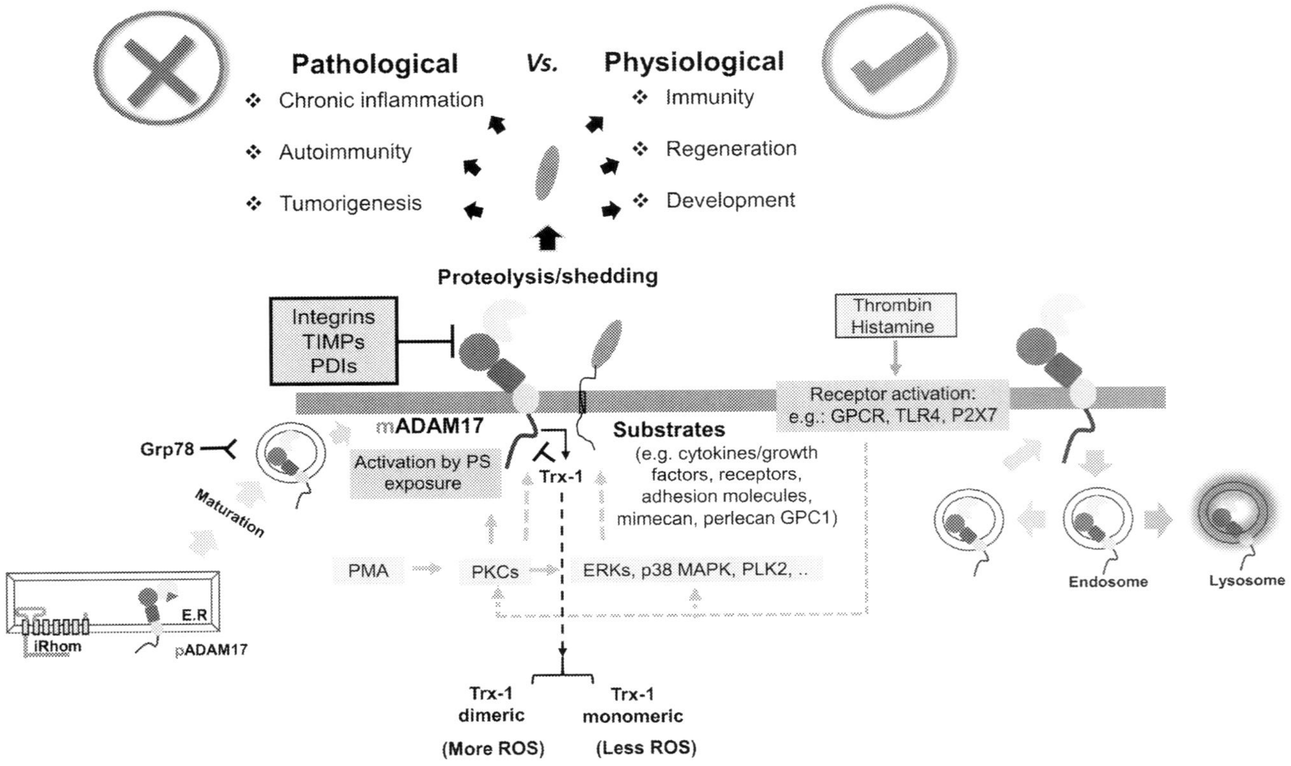

Fig. 8

What is the physiological and pathological role of ADAM17? One of the fundamental questions in ADAM17 modulation is figuring out what players decide when ADAM17 cleaves an ectodomain. There are several effectors in the cell, both extracellular and intracellular. ADAM17 exists as an immature proform (pADAM17) and as a mature proteinase (mADAM17) in cells. After ADAM17 is translated into the ER, it travels iRhom-dependent to the Golgi apparatus, where the maturation takes place. One of the targets involved in this process has been characterized by studies from our group as Grp78, important both for ADAM17 maturation and its maintenance in the open active form, contrary to the effects of PDIs that are responsible for ADAM17 isomerization to an inactive closed state. Mature ADAM17 localized to the membrane acts mainly by promoting shedding in a range of different substrates that upon proteolysis may act on different physiological or pathological responses. Among the different well-known substrates, glypican-1 was characterized in our group as a proteoglycan that plays a role in the adhesion, migration, and proliferation of oral cancer cells. The role of ADAM17 in oral cancer was well characterized also by our group showing that its expression regulates the expression of other proteins that are mostly involved in the signaling pathway of Erk and EGFR both known targets in ADAM17 shedding modulation by inside-out. Also, considering other types of inside-out modulations, we have characterized thioredoxin-1 as a novel ligand to the cytoplasmic domain of ADAM17 and shown that this interaction inhibits ADAM17 shedding while also activates Trx-1 activity by promoting the enzyme's monomerization. Besides, it has been shown that most mature ADAM17 at the cell surface is internalized and recycled by lysosomal degradation. This interconnected mechanism is yet to be deciphered and is a quite good example of the balance between proteolysis and inhibitors that are important to regulate the final proteolytic activity. *Adapted from Düsterhöft S, Babendreyer A, Giese AA, et al. Status update on iRhom and ADAM17: it's still complicated. Biochim Biophys Acta Mol Cell Res 2019;1866(10):1567–1583. https://doi.org/10.1016/j.bbamcr.2019.06.017.*

proteinases were initially considered to stimulate the spread of cancer cells to tissues and organs beyond, contributing to the transpassing of tissue barriers, that occur for example during cancer metastasis. However, proteolytic activity in cancers is a complicated dynamic process involving normal and tumor cells, as well as different cellular processes such as immune response, proliferation, and apoptosis. Besides, the activity of proteinases is mainly regulated by their interactions with their endogenous inhibitors and substrates, which can be depicted as a balance, constantly shifting when one of these members is affected. Our group has investigated in a broad spectrum the activity of proteinases in the saliva of OSCC patients, by determining the peptidome levels, predicting the active proteinases, characterizing the proteolytic sites, and, finally, using the composition and abundance of a set of proteolytic-derived peptides to better classify patients with and without lymph nodal metastasis (pN0 and pN+, respectively).

In addition, in the past years, there have been huge attempts to elucidate the mechanism of action of individual proteinases as well as their biology and structure. The understanding of one proteinase's role in the development of cancer, its substrates or ligands, or the pathways that modulate its activity, is fundamental when considering the proteinase as a therapeutic target. Our group has worked extensively in deciphering the role of ADAM17, one of the main mammalian metalloproteinases, that due to its role in different pathologies, including oral cancer, has been associated as a potential therapeutic factor. Different inhibitors have been developed to inhibit its proteolytic activity; however, very few have succeeded and none are in use for clinical practice. The major challenge in developing inhibitors is to selectively inhibit pathological processes while maintaining the normal processes to avoid adverse effects. Several studies suggest the importance of developing approaches that do not inhibit the catalytic activity of the enzyme but rather attempt to interfere in other allosteric sites, such as protein-protein interactions (PPIs).

The strategies we have employed here aim to develop allosteric inhibitors altering proteinase activity. The discovery of the interactome of proteinases, such as ADAM17, considering their substrates and interacting partners (e.g., glypican, Trx-1, and Grp78), and the consequences of these processes in tumor initiation, progression, and metastasis will guide the development of novel therapeutic strategies against cancer and constitute a promising strategy to move deeper in our understanding of proteolytic dynamics under health and disease contexts.

Acknowledgments

We thank Dra. Carolina M. Carnielli and Ms. Fabio Patroni for assistance in the bioinformatic analysis.

Funding was provided by FAPESP to L.D.T. (2018/12194-8), DCG (2018/15355-0 and 2019/18751-9), L.X.N. (2018/11958-4), A.F.P.L. (2010/19278-0; 2016/07846-0; 2018/18496-6), and CNPq to A.F.P.L. (305851/2017-9).

References

[1] López-Otín C, Bond JS. Proteases: multifunctional enzymes in life and disease. J Biol Chem 2008;283(45):30433–7. https://doi.org/10.1074/jbc.R800035200.

[2] Eatemadi A, Aiyelabegan HT, Negahdari B, Mazlomi MA, Daraee H, Daraee N, Eatemadi R, Sadroddiny E. Role of protease and protease inhibitors in cancer pathogenesis and treatment. Biomed Pharmacother 2017;86:221–31. https://doi.org/10.1016/j.biopha.2016.12.021.

[3] Feng Y, Li Q, Chen J, Yi P, Xu X, Fan Y, Cui B, Yu Y, Li X, Du Y, Chen Q, Zhang L, Jiang J, Zhou X, Zhang P. Salivary protease spectrum biomarkers of oral cancer. Int J Oral Sci 2019;11(1):1–11. https://doi.org/10.1038/s41368-018-0032-z.

[4] Carnielli CM, CCS M, De Rossi T, Granato DC, Rivera C, Domingues RR, Pauletti BA, Yokoo S, Heberle H, Busso-Lopes AF, Cervigne NK, Sawazaki-Calone I, Meirelles GV, Marchi FA, Telles GP, Minghim R, ACP R, Brandão TB, de Castro G, González-Arriagada WA, Gomes A, Penteado F, Santos-Silva AR, Lopes MA, Rodrigues PC, Sundquist E, Salo T, da Silva SD, Alaoui-Jamali MA, Graner E, Fox JW, Coletta RD, Paes Leme AF. Combining discovery and targeted proteomics reveals a prognostic signature in oral cancer. Nat Commun 2018;9(1):3598. https://doi.org/10.1038/s41467-018-05696-2.

[5] Vigneswaran N, Wu J, Nagaraj N, James R, Zeeuwen P, Zacharias W. Silencing of cystatin M in metastatic oral cancer cell line MDA-686Ln by siRNA increases cysteine proteinases and legumain activities, cell proliferation and in vitro invasion. Life Sci 2006;78(8):898–907. https://doi.org/10.1016/j.lfs.2005.05.096.

[6] Strojan P, Oblak I, Svetic B, Šmid L, Kos J. Cysteine proteinase inhibitor cystatin C in squamous cell carcinoma of the head and neck: relation to prognosis. Br J Cancer 2004;90(10):1961–8. https://doi.org/10.1038/sj.bjc.6601830.

[7] Kurahara S, Shinohara M, Ikebe T, Nakamura S, Beppu M, Hiraki A, Takeuchi H, Shirasuna K. Expression of MMPS, MT-MMP, and TIMPs in squamous cell carcinoma of the oral cavity: correlations with tumor invasion and metastasis. Head Neck 1999;21(7):627–38. https://doi.org/10.1002/(sici)1097-0347(199910)21:7<627::aid-hed7>3.0.co;2-2.

[8] Leusink FK, Koudounarakis E, Frank MH, Koole R, van Diest PJ, Willems SM. Cathepsin K associates with lymph node metastasis and poor prognosis in oral squamous cell carcinoma. BMC Cancer 2018;18(1):385. https://doi.org/10.1186/s12885-018-4315-8.

[9] Vigneswaran N, Zhao W, Dassanayake A, Muller S, Miller DM, Zacharias W. Variable expression of cathepsin B and D correlates with highly invasive and metastatic phenotype of oral cancer. Hum Pathol 2000;31(8):931–7. https://doi.org/10.1053/hupa.2000.9035.

[10] Pérez-Silva JG, Español Y, Velasco G, Quesada V. The Degradome database: expanding roles of mammalian proteases in life and disease. Nucleic Acids Res 2016;44(Database issue):D351–5. https://doi.org/10.1093/nar/gkv1201.

[11] Kappelhoff R, Puente XS, Wilson CH, Seth A, López-Otín C, Overall CM. Overview of transcriptomic analysis of all human proteases, non-proteolytic homologs and inhibitors: Organ, tissue and ovarian cancer cell line expression profiling of the human protease degradome by the CLIP-CHIP™ DNA microarray. Biochim Biophys Acta (BBA) Mol Cell Res 2017;1864(11, Pt B):2210–9. https://doi.org/10.1016/j.bbamcr.2017.08.004.

[12] Vitorino R, Ferreira R, Caseiro A, Amado F. Chapter 9—Salivary peptidomics targeting clinical applications. Comprehen Anal Chem 2014;64:223–45. https://doi.org/10.1016/B978-0-444-62650-9.00009-9.

[13] Neves LX, Granato DC, Busso-Lopes AF, Carnielli CM, Patroni FMdS, De Rossi T, Oliveira AK, ACP R, Brandão TB, Rodrigues AN, Lacerda PA, Uno M, Cervigne NK, Santos-Silva AR, Kowalski LP, Lopes MA, AFP L. Peptidomics-driven strategy reveals peptides and predicted proteases associated with oral cancer prognosis. Mol Cell Proteomics 2020;11. https://doi.org/10.1074/mcp.RA120.002227.

[14] Winck FV, Prado Ribeiro AC, Ramos Domingues R, Ling LY, Riaño-Pachón DM, Rivera C, Brandão TB, Gouvea AF, Santos-Silva AR, Coletta RD, Paes Leme AF. Insights into immune responses in oral cancer through proteomic analysis of saliva and salivary extracellular vesicles. Sci Rep 2015;5:16305. https://doi.org/10.1038/srep16305.

[15] Xu Z, Wu C, Xie F, Slysz GW, Tolic N, Monroe ME, Petyuk VA, Payne SH, Fujimoto GM, Moore RJ, Fillmore TL, Schepmoes AA, Levine DA, Townsend RR, Davies SR, Li S, Ellis M, Boja E, Rivers R, Rodriguez H, Rodland KD, Liu T, Smith RD. Comprehensive quantitative analysis of ovarian and breast cancer tumor peptidomes. J Proteome Res 2015;14(1):422–33. https://doi.org/10.1021/pr500840w.

[16] Krochmal M, van Kessel KEM, Zwarthoff EC, Belczacka I, Pejchinovski M, Vlahou A, Mischak H, Frantzi M. Urinary peptide panel for prognostic assessment of bladder cancer relapse. Sci Rep 2019;9(1):1–10. https://doi.org/10.1038/s41598-019-44129-y.

[17] Amado FML, Ferreira RP, Vitorino R. One decade of salivary proteomics: current approaches and outstanding challenges. Clin Biochem 2013;46(6):506–17. https://doi.org/10.1016/j.clinbiochem.2012.10.024.

[18] Amado F, Lobo MJC, Domingues P, Duarte JA, Vitorino R. Salivary peptidomics. Expert Rev Proteomics 2010;7(5):709–21. https://doi.org/10.1586/epr.10.48.

[19] Zhang W-L, Wang S-S, Wang H-F, Tang Y-J, Tang Y-L, Liang X-H. Who is who in oral cancer? Exp Cell Res 2019;384(2):111634. https://doi.org/10.1016/j.yexcr.2019.111634.

[20] Kamarajan P, Ateia I, Shin JM, Fenno JC, Le C, Zhan L, Chang A, Darveau R, Kapila YL. Periodontal pathogens promote cancer aggressivity via TLR/MyD88 triggered activation of integrin/FAK signaling that is therapeutically reversible by a probiotic bacteriocin. PLoS Pathog 2020;16(10). https://doi.org/10.1371/journal.ppat.1008881.

[21] Mager DL, Haffajee AD, Devlin PM, Norris CM, Posner MR, Goodson JM. The salivary microbiota as a diagnostic indicator of oral cancer: a descriptive, non-randomized study of cancer-free and oral squamous cell carcinoma subjects. J Transl Med 2005;3:27. https://doi.org/10.1186/1479-5876-3-27.

[22] Pushalkar S, Ji X, Li Y, Estilo C, Yegnanarayana R, Singh B, Li X, Saxena D. Comparison of oral microbiota in tumor and non-tumor tissues of patients with oral squamous cell carcinoma. BMC Microbiol 2012;12:144. https://doi.org/10.1186/1471-2180-12-144.

[23] Laronha H, Caldeira J. Structure and function of human matrix metalloproteinases. Cell 2020;9(5). https://doi.org/10.3390/cells9051076.

[24] Stautz D, Dombernowsky SL, Kveiborg M. ADAMs in cancer. In: Matrix proteases in health and disease. John Wiley & Sons, Ltd; 2012. p. 299–324. https://doi.org/10.1002/9783527649327.ch12.

[25] Aragão AZB, Nogueira MLC, Granato DC, Simabuco FM, Honorato RV, Hoffman Z, Yokoo S, Laurindo FRM, Squina FM, Zeri ACM, Oliveira PSL, Sherman NE, Paes Leme AF. Identification of novel interaction between ADAM17 (a disintegrin and metalloprotease 17) and thioredoxin-1. J Biol Chem 2012;287(51):43071–82. https://doi.org/10.1074/jbc.M112.364513.

[26] Kawahara R, Granato DC, Yokoo S, Domingues RR, Trindade DM, Paes Leme AF. Mass spectrometry-based proteomics revealed glypican-1 as a novel ADAM17 substrate. J Proteomics 2017;151:53–65. https://doi.org/10.1016/j.jprot.2016.08.017.

[27] Granato DC, e Costa RAP, Kawahara R, Yokoo S, Aragão AZ, Domingues RR, Pauletti BA, Honorato RV, Fattori J, ACM F, PSL O, Consonni SR, Fernandes D, Laurindo F, Hansen HP, Paes Leme AF. Thioredoxin-1 negatively modulates ADAM17 activity through direct binding and indirect reductive activity. Antioxid Redox Signal 2018;29(8):717–34. https://doi.org/10.1089/ars.2017.7297.

[28] e Costa RAP, Granato DC, Trino LD, Yokoo S, Carnielli CM, Kawahara R, Domingues RR, Pauletti BA, Neves LX, Santana AG, Paulo JA, AZB A, Heleno Batista FA, Migliorini Figueira AC, FRM L, Fernandes D, Hansen HP, Squina F, Gygi SP, Paes Leme AF. ADAM17 cytoplasmic domain modulates Thioredoxin-1 conformation and activity. Redox Biol 2020;37. https://doi.org/10.1016/j.redox.2020.101735, 101735.

[29] Schäfer M, Granato DC, Krossa S, Bartels A-K, Yokoo S, Düsterhöft S, Koudelka T, Scheidig AJ, Tholey A, Paes Leme AF, Grötzinger J, Lorenzen I. GRP78 protects a disintegrin and metalloprotease 17 against protein-disulfide isomerase A6 catalyzed inactivation. FEBS Lett 2017;591(21):3567–87. https://doi.org/10.1002/1873-3468.12858.

[30] Simabuco FM, Kawahara R, Yokoo S, Granato DC, Miguel L, Agostini M, Aragão AZ, Domingues RR, Flores IL, Macedo CC, Coletta RD, Graner E, Paes Leme AF. ADAM17 mediates OSCC development in an orthotopic murine model. Mol Cancer 2014;13(1):24. https://doi.org/10.1186/1476-4598-13-24.

[31] Zunke F, Rose-John S. The shedding protease ADAM17: physiology and pathophysiology. Biochim Biophys Acta (BBA) Mol Cell Res 2017;1864(11, Pt B):2059–70. https://doi.org/10.1016/j.bbamcr.2017.07.001.

[32] Kriegler M, Perez C, DeFay K, Albert I, Lu SD. A novel form of TNF/cachectin is a cell surface cytotoxic transmembrane protein: Ramifications for the complex physiology of TNF. Cell 1988;53(1):45–53. https://doi.org/10.1016/0092-8674(88)90486-2.

[33] Zelová H, Hošek J. TNF-α signalling and inflammation: interactions between old acquaintances. Inflam Res Off J Eur Hist Res Soc 2013;62(7):641–51. https://doi.org/10.1007/s00011-013-0633-0.

[34] Mohler KM, Sleath PR, Fitzner JN, Cerretti DP, Alderson M, Kerwar SS, Torranee DS, Otten-Evans C, Greenstreet T, Weerawarna K, Kronheim SR, Petersen M, Gerhart M, Kozlosky CJ, March CJ, Black RA. Protection against a lethal dose of endotoxin by an inhibitor of tumour necrosis factor processing. Nature 1994;370(6486):218–20. https://doi.org/10.1038/370218a0.

[35] Black RA, Rauch CT, Kozlosky CJ, Peschon JJ, Slack JL, Wolfson MF, Castner BJ, Stocking KL, Reddy P, Srinivasan S, Nelson N, Boiani N, Schooley KA, Gerhart M, Davis R, Fitzner JN, Johnson RS, Paxton RJ, March CJ, Cerretti DP. A metalloproteinase disintegrin that releases tumour-necrosis factor-α from cells. Nature 1997;385(6618):729–33. https://doi.org/10.1038/385729a0.

[36] Moss ML, Jin SL, Milla ME, Bickett DM, Burkhart W, Carter HL, Chen WJ, Clay WC, Didsbury JR, Hassler D, Hoffman CR, Kost TA, Lambert MH, Leesnitzer MA, McCauley P, McGeehan G, Mitchell J, Moyer M, Pahel G, Rocque W, Overton LK, Schoenen F, Seaton T, Su JL, Becherer JD. Cloning of a disintegrin metalloproteinase that processes precursor tumour-necrosis factor-alpha. Nature 1997;385(6618):733–6. https://doi.org/10.1038/385733a0.

[37] Maskos K, Fernandez-Catalan C, Huber R, Bourenkov GP, Bartunik H, Ellestad GA, Reddy P, Wolfson MF, Rauch CT, Castner BJ, Davis R, Clarke HR, Petersen M, Fitzner JN, Cerretti DP, March CJ, Paxton RJ, Black RA, Bode W. Crystal structure of the catalytic domain of human tumor necrosis factor-alpha-converting enzyme. Proc Natl Acad Sci U S A 1998;95(7):3408–12. https://doi.org/10.1073/pnas.95.7.3408.

[38] Peschon JJ, Slack JL, Reddy P, Stocking KL, Sunnarborg SW, Lee DC, Russell WE, Castner BJ, Johnson RS, Fitzner JN, Boyce RW, Nelson N, Kozlosky CJ, Wolfson MF, Rauch CT, Cerretti DP, Paxton RJ, March CJ, Black RA. An essential role for ectodomain shedding in mammalian development. Science (New York, NY) 1998;282(5392):1281–4. https://doi.org/10.1126/science.282.5392.1281.

[39] Lal M, Caplan M. Regulated intramembrane proteolysis: signaling pathways and biological functions. Phys Ther 2011;26(1):34–44. https://doi.org/10.1152/physiol.00028.2010.

[40] Giebeler N, Zigrino P. A disintegrin and metalloprotease (ADAM): historical overview of their functions. Toxins (Basel) 2016;8(4). https://doi.org/10.3390/toxins8040122.

[41] Brou C, Logeat F, Gupta N, Bessia C, LeBail O, Doedens JR, Cumano A, Roux P, Black RA, Israël A. A novel proteolytic cleavage involved in Notch signaling: the role of the disintegrin-metalloprotease TACE. Mol Cell 2000;5(2):207–16. https://doi.org/10.1016/s1097-2765(00)80417-7.

[42] Weinmaster G. Notch signal transduction: a real Rip and more. Curr Opin Genet Dev 2000;10(4):363–9. https://doi.org/10.1016/S0959-437X(00)00097-6.

[43] Gooz M. ADAM-17: the enzyme that does it all. Crit Rev Biochem Mol Biol 2010;45(2):146–69. https://doi.org/10.3109/10409231003628015.

[44] McIlwain DR, Lang PA, Maretzky T, Hamada K, Ohishi K, Maney SK, Berger T, Murthy A, Duncan G, Xu HC, Lang KS, Häussinger D, Wakeham A, Itie-Youten A, Khokha R, Ohashi PS, Blobel CP, Mak TW. iRhom2 regulation of TACE controls TNF-mediated protection against Listeria and responses to LPS. Science 2012;335(6065):229–32. https://doi.org/10.1126/science.1214448.

[45] Adrain C, Zettl M, Christova Y, Taylor N, Freeman M. Tumor necrosis factor signaling requires iRhom2 to promote trafficking and activation of TACE. Science 2012;335(6065):225–8. https://doi.org/10.1126/science.1214400.

[46] Düsterhöft S, Babendreyer A, Giese AA, Flasshove C, Ludwig A. Status update on iRhom and ADAM17: it's still complicated. Biochim Biophys Acta Mol Cell Res 2019;1866(10):1567–83. https://doi.org/10.1016/j.bbamcr.2019.06.017.

[47] Düsterhöft S, Höbel K, Oldefest M, Lokau J, Waetzig GH, Chalaris A, Garbers C, Scheller J, Rose-John S, Lorenzen I, Grötzinger J. A disintegrin and metalloprotease 17 dynamic interaction sequence, the sweet tooth for the human interleukin 6 receptor. J Biol Chem 2014;289(23):16336–48. https://doi.org/10.1074/jbc.M114.557322.

[48] Düsterhöft S, Lokau J, Garbers C. The metalloprotease ADAM17 in inflammation and cancer. Pathol Res Pract 2019;215(6):152410. https://doi.org/10.1016/j.prp.2019.04.002.

[49] Kawahara R, Lima RN, Domingues RR, Pauletti BA, Meirelles GV, Assis M, Figueira ACM, Paes Leme AF. Deciphering the role of the ADAM17-dependent secretome in cell signaling. J Proteome Res 2014;13(4):2080–93. https://doi.org/10.1021/pr401224u.

[50] Palau V, Riera M, Soler MJ. ADAM17 inhibition may exert a protective effect on COVID-19. Nephrol Dial Transplant Off Publ Eur Dial Transplant Assoc Eur Renal Assoc 2020;35(6):1071–2. https://doi.org/10.1093/ndt/gfaa093.

[51] Gemmati D, Bramanti B, Serino ML, Secchiero P, Zauli G, Tisato V. COVID-19 and individual genetic susceptibility/receptivity: role of ACE1/ACE2 genes, immunity, inflammation and coagulation. Might the double X-chromosome in females be protective against SARS-CoV-2 compared to the single X-chromosome in males? Int J Mol Sci 2020;21(10). https://doi.org/10.3390/ijms21103474.

[52] Rizzo P, Vieceli Dalla Sega F, Fortini F, Marracino L, Rapezzi C, Ferrari R. COVID-19 in the heart and the lungs: could we "Notch" the inflammatory storm? Basic Res Cardiol 2020;115(3):31. https://doi.org/10.1007/s00395-020-0791-5.

[53] Gheblawi M, Wang K, Viveiros A, Nguyen Q, Zhong J-C, Turner AJ, Raizada MK, Grant MB, Oudit GY. Angiotensin-converting enzyme 2: SARS-CoV-2 receptor and regulator of the renin-angiotensin system: celebrating the 20th anniversary of the discovery of ACE2. Circ Res 2020;126(10):1456–74. https://doi.org/10.1161/CIRCRESAHA.120.317015.

[54] Xu J, Xu X, Jiang L, Dua K, Hansbro PM, Liu G. SARS-CoV-2 induces transcriptional signatures in human lung epithelial cells that promote lung fibrosis. Respir Res 2020;21(1):182. https://doi.org/10.1186/s12931-020-01445-6.

[55] Zamorano Cuervo N, Grandvaux N. ACE2: Evidence of role as entry receptor for SARS-CoV-2 and implications in comorbidities. van de Veerdonk FL, van der Meer JW, eds. *eLife*. 2020;9:e61390. doi: https://doi.org/10.7554/eLife.61390.

[56] Yang J, Petitjean SJL, Koehler M, Zhang Q, Dumitru AC, Chen W, Derclaye S, Vincent SP, Soumillion P, Alsteens D. Molecular interaction and inhibition of SARS-CoV-2 binding to the ACE2 receptor. Nat Commun 2020;11(1):4541. https://doi.org/10.1038/s41467-020-18319-6.

[57] Haga S, Yamamoto N, Nakai-Murakami C, Osawa Y, Tokunaga K, Sata T, Yamamoto N, Sasazuki T, Ishizaka Y. Modulation of TNF-α-converting enzyme by the spike protein of SARS-CoV and ACE2 induces TNF-α production and facilitates viral entry. PNAS 2008;105(22):7809–14. https://doi.org/10.1073/pnas.0711241105.

[58] Xiao L, Sakagami H, Miwa N. ACE2: the key molecule for understanding the pathophysiology of severe and critical conditions of COVID-19: demon or angel? Viruses 2020;12(5). https://doi.org/10.3390/v12050491.

[59] Scheller J, Chalaris A, Garbers C, Rose-John S. ADAM17: a molecular switch to control inflammation and tissue regeneration. Trends Immunol 2011;32(8):380–7. https://doi.org/10.1016/j.it.2011.05.005.

[60] Arribas J, Esselens C. ADAM17 as a therapeutic target in multiple diseases. Curr Pharm Des 2009;15(20):2319–35. https://doi.org/10.2174/138161209788682398.

[61] Grötzinger J, Lorenzen I, Düsterhöft S. Molecular insights into the multilayered regulation of ADAM17: the role of the extracellular region. Biochim Biophys Acta (BBA) Mol Cell Res 2017;1864(11, Pt B):2088–95. https://doi.org/10.1016/j.bbamcr.2017.05.024.

[62] Blobel CP. ADAMs: key components in EGFR signalling and development. Nat Rev Mol Cell Biol 2005;6(1):32–43. https://doi.org/10.1038/nrm1548.

[63] Murphy G, Murthy A, Khokha R. Clipping, shedding and RIPping keep immunity on cue. Trends Immunol 2008;29(2):75–82. https://doi.org/10.1016/j.it.2007.10.009.

[64] Saad MI, Rose-John S, Jenkins BJ. ADAM17: an emerging therapeutic target for lung cancer. Cancers (Basel) 2019;11(9). https://doi.org/10.3390/cancers11091218.

[65] Moss ML, Minond D. Recent advances in ADAM17 research: a promising target for cancer and inflammation. Mediators Inflamm 2017;2017. https://doi.org/10.1155/2017/9673537, 9673537.

[66] de Queiroz TM, Lakkappa N, Lazartigues E. ADAM17-mediated shedding of inflammatory cytokines in hypertension. Front Pharmacol 2020;11. https://doi.org/10.3389/fphar.2020.01154.

[67] Tsakadze NL, Sithu SD, Sen U, English WR, Murphy G, D'Souza SE. Tumor necrosis factor-α-converting enzyme (TACE/ADAM-17) mediates the ectodomain cleavage of intercellular adhesion molecule-1 (ICAM-1). J Biol Chem 2006;281(6):3157–64. https://doi.org/10.1074/jbc.M510797200.

[68] Garton KJ, Gough PJ, Philalay J, Wille PT, Blobel CP, Whitehead RH, Dempsey PJ, Raines EW. Stimulated shedding of vascular cell adhesion molecule 1 (VCAM-1) is mediated by tumor necrosis factor-α-converting enzyme (ADAM 17). J Biol Chem 2003;278(39):37459–64. https://doi.org/10.1074/jbc.M305877200.

[69] Li Y, Brazzell J, Herrera A, Walcheck B. ADAM17 deficiency by mature neutrophils has differential effects on L-selectin shedding. Blood 2006;108(7):2275–9. https://doi.org/10.1182/blood-2006-02-005827.

[70] Tsou CL, Haskell CA, Charo IF. Tumor necrosis factor-alpha-converting enzyme mediates the inducible cleavage of fractalkine. J Biol Chem 2001;276(48):44622–6. https://doi.org/10.1074/jbc.M107327200.

[71] Sun C, Hu A, Wang S, Tian B, Jiang L, Liang Y, Wang H, Dong J. ADAM17-regulated CX3CL1 expression produced by bone marrow endothelial cells promotes spinal metastasis from hepatocellular carcinoma. Int J Oncol 2020;57(1):249–63. https://doi.org/10.3892/ijo.2020.5045.

[72] Garton KJ, Gough PJ, Blobel CP, Murphy G, Greaves DR, Dempsey PJ, Raines EW. TACE (ADAM17) mediates the cleavage and shedding of Fractalkine (CX3CL1). J Biol Chem 2001;8. https://doi.org/10.1074/jbc.M106434200.

[73] Mishra HK, Ma J, Walcheck B. Ectodomain shedding by ADAM17: its role in neutrophil recruitment and the impairment of this process during sepsis. Front Cell Infect Microbiol 2017;7. https://doi.org/10.3389/fcimb.2017.00138.

[74] Stawikowska R, Cudic M, Giulianotti M, Houghten RA, Fields GB, Minond D. Activity of ADAM17 (a disintegrin and metalloprotease 17) is regulated by its noncatalytic domains and secondary structure of its substrates. J Biol Chem 2013;288(31):22871–9. https://doi.org/10.1074/jbc.M113.462267.

[75] Mezyk R, Bzowska M, Bereta J. Structure and functions of tumor necrosis factor-alpha converting enzyme. Acta Biochim Pol 2003;50(3):625–45. doi: 035003625.

[76] Kishimoto TK, Kahn J, Migaki G, Mainolfi E, Shirley F, Ingraham R, Rothlein R. Regulation of L-selectin expression by membrane proximal proteolysis. Agents Actions Suppl 1995;47:121–34. https://doi.org/10.1007/978-3-0348-7343-7_11.

[77] Machado-Pineda Y, Cardeñes B, Reyes R, López-Martín S, Toribio V, Sánchez-Organero P, Suarez H, Grötzinger J, Lorenzen I, Yáñez-Mó M, Cabañas C. CD9 controls integrin α5β1-mediated cell adhesion by modulating its association with the metalloproteinase ADAM17. Front Immunol 2018;9. https://doi.org/10.3389/fimmu.2018.02474.

[78] Gooz P, Dang Y, Higashiyama S, Twal WO, Haycraft CJ, Gooz M. A disintegrin and metalloenzyme (ADAM) 17 activation is regulated by α5β1 integrin in kidney mesangial cells. PLoS One 2012;7(3). https://doi.org/10.1371/journal.pone.0033350.

[79] Reddy P, Slack JL, Davis R, Cerretti DP, Kozlosky CJ, Blanton RA, Shows D, Peschon JJ, Black RA. Functional analysis of the domain structure of tumor necrosis factor-alpha converting enzyme. J Biol Chem 2000;275(19):14608–14. https://doi.org/10.1074/jbc.275.19.14608.

[80] Milla ME, Leesnitzer MA, Moss ML, Clay WC, Carter HL, Miller AB, Su JL, Lambert MH, Willard DH, Sheeley DM, Kost TA, Burkhart W, Moyer M, Blackburn RK, Pahel GL, Mitchell JL, Hoffman CR, Becherer JD. Specific sequence elements are required for the expression of functional tumor necrosis factor-alpha-converting enzyme (TACE). J Biol Chem 1999;274(43):30563–70. https://doi.org/10.1074/jbc.274.43.30563.

[81] Schwab M. Encyclopedia of cancer. 3rd ed. Springer; 2011. https://b-ok.lat/book/2714378/0bea33. [Accessed 8 September 2020].

[82] Düsterhöft S, Michalek M, Kordowski F, Oldefest M, Sommer A, Röseler J, Reiss K, Grötzinger J, Lorenzen I. Extracellular juxtamembrane segment of ADAM17 interacts with membranes and is essential for its shedding activity. Biochemistry 2015;54(38):5791–801. https://doi.org/10.1021/acs.biochem.5b00497.

[83] Sommer A, Kordowski F, Büch J, Maretzky T, Evers A, Andrä J, Düsterhöft S, Michalek M, Lorenzen I, Somasundaram P, Tholey A, Sönnichsen FD, Kunzelmann K, Heinbockel L, Nehls C, Gutsmann T, Grötzinger J, Bhakdi S, Reiss K. Phosphatidylserine exposure is required for ADAM17 sheddase function. Nat Commun 2016;7:11523. https://doi.org/10.1038/ncomms11523.

[84] Le Gall SM, Maretzky T, Issuree PDA, Niu X-D, Reiss K, Saftig P, Khokha R, Lundell D, Blobel CP. ADAM17 is regulated by a rapid and reversible mechanism that controls access to its catalytic site. J Cell Sci 2010;123(22):3913–22. https://doi.org/10.1242/jcs.069997.

[85] Li X, Maretzky T, Perez-Aguilar JM, Monette S, Weskamp G, Le Gall S, Beutler B, Weinstein H, Blobel CP. Structural modeling defines transmembrane residues in ADAM17 that are crucial for Rhbdf2–ADAM17-dependent proteolysis. J Cell Sci 2017;130(5):868–78. https://doi.org/10.1242/jcs.196436.

[86] Xu P, Liu J, Sakaki-Yumoto M, Derynck R. TACE activation by MAPK-mediated regulation of cell surface dimerization and TIMP3 association. Sci Signal 2012;5(222):ra34. https://doi.org/10.1126/scisignal.2002689.

[87] Soond SM, Everson B, Riches DWH, Murphy G. ERK-mediated phosphorylation of Thr735 in TNFalpha-converting enzyme and its potential role in TACE protein trafficking. J Cell Sci 2005;118(Pt 11):2371–80. https://doi.org/10.1242/jcs.02357.

[88] Pavlenko E, Cabron A-S, Arnold P, Dobert JP, Rose-John S, Zunke F. Functional characterization of colon cancer-associated mutations in ADAM17: modifications in the pro-domain interfere with trafficking and maturation. Int J Mol Sci 2019;20(9). https://doi.org/10.3390/ijms20092198.

[89] Cabron A-S, El Azzouzi K, Boss M, Arnold P, Schwarz J, Rosas M, Dobert JP, Pavlenko E, Schumacher N, Renné T, Taylor PR, Linder S, Rose-John S, Zunke F. Structural and functional analyses of the shedding protease ADAM17 in HoxB8-immortalized macrophages and dendritic-like cells. J Immunol 2018;201(10):3106–18. https://doi.org/10.4049/jimmunol.1701556.

[90] Murphy G. Chapter 254—ADAM17, tumor necrosis factor α-convertase. In: Rawlings ND, Salvesen G, editors. Handbook of Proteolytic Enzymes. 3rd ed. Academic Press; 2013. p. 1126–30. https://doi.org/10.1016/B978-0-12-382219-2.00254-4.

[91] Jackson LF, Qiu TH, Sunnarborg SW, Chang A, Zhang C, Patterson C, Lee DC. Defective valvulogenesis in HB-EGF and TACE-null mice is associated with aberrant BMP signaling. EMBO J 2003;22(11):2704–16. https://doi.org/10.1093/emboj/cdg264.

[92] Shi W, Chen H, Sun J, Buckley S, Zhao J, Anderson KD, Williams RG, Warburton D. TACE is required for fetal murine cardiac development and modeling. Dev Biol 2003;261(2):371–80. https://doi.org/10.1016/s0012-1606(03)00315-4.

[93] Gelling RW, Yan W, Al-Noori S, Pardini A, Morton GJ, Ogimoto K, Schwartz MW, Dempsey PJ. Deficiency of TNFα converting enzyme (TACE/ADAM17) causes a lean, hypermetabolic phenotype in mice. Endocrinology 2008;149(12):6053–64. https://doi.org/10.1210/en.2008-0775.

[94] Hall KC, Hill D, Otero M, Plumb DA, Froemel D, Dragomir CL, Maretzky T, Boskey A, Crawford HC, Selleri L, Goldring MB, Blobel CP. ADAM17 controls endochondral ossification by regulating terminal differentiation of chondrocytes. Mol Cell Biol 2013;33(16):3077–90. https://doi.org/10.1128/MCB.00291-13.

[95] Araya HF, Sepulveda H, Lizama CO, Vega OA, Jerez S, Briceño PF, Thaler R, Riester SM, Antonelli M, Salazar-Onfray F, Rodríguez JP, Moreno RD, Montecino M, Charbonneau M, Dubois CM, Stein GS, van Wijnen AJ, Galindo MA. Expression of the ectodomain-releasing protease ADAM17 is directly regulated by the osteosarcoma and bone-related transcription factor RUNX2. J Cell Biochem 2018;119(10):8204–19. https://doi.org/10.1002/jcb.26832.

[96] Patel IR, Attur MG, Patel RN, Stuchin SA, Abagyan RA, Abramson SB, Amin AR. TNF-alpha convertase enzyme from human arthritis-affected cartilage: isolation of cDNA by differential display, expression of the active enzyme, and regulation of TNF-alpha. J Immunol 1998;160(9):4570–9.

[97] Amin AR. Regulation of tumor necrosis factor-alpha and tumor necrosis factor converting enzyme in human osteoarthritis. Osteoarthr Cartil 1999;7(4):392–4. https://doi.org/10.1053/joca.1998.0221.
[98] Ohta S, Harigai M, Tanaka M, Kawaguchi Y, Sugiura T, Takagi K, Fukasawa C, Hara M, Kamatani N. Tumor necrosis factor-alpha (TNF-alpha) converting enzyme contributes to production of TNF-alpha in synovial tissues from patients with rheumatoid arthritis. J Rheumatol 2001;28(8):1756–63.
[99] Bohgaki T, Amasaki Y, Nishimura N, Bohgaki M, Yamashita Y, Nishio M, Sawada K-I, Jodo S, Atsumi T, Koike T. Up regulated expression of tumour necrosis factor {alpha} converting enzyme in peripheral monocytes of patients with early systemic sclerosis. Ann Rheum Dis 2005;64(8):1165–73. https://doi.org/10.1136/ard.2004.030338.
[100] Brynskov J, Foegh P, Pedersen G, Ellervik C, Kirkegaard T, Bingham A, Saermark T. Tumour necrosis factor α converting enzyme (TACE) activity in the colonic mucosa of patients with inflammatory bowel disease. Gut 2002;51(1):37–43.
[101] Charbonneau M, Harper K, Grondin F, Pelmus M, McDonald PP, Dubois CM. Hypoxia-inducible factor mediates hypoxic and tumor necrosis factor alpha-induced increases in tumor necrosis factor-alpha converting enzyme/ADAM17 expression by synovial cells. J Biol Chem 2007;282(46):33714–24. https://doi.org/10.1074/jbc.M704041200.
[102] Dejager L, Dendoncker K, Eggermont M, Souffriau J, Van Hauwermeiren F, Willart M, Van Wonterghem E, Naessens T, Ballegeer M, Vandevyver S, Hammad H, Lambrecht B, De Bosscher K, Grooten J, Libert C. Neutralizing TNFα restores glucocorticoid sensitivity in a mouse model of neutrophilic airway inflammation. Mucosal Immunol 2015;8(6):1212–25. https://doi.org/10.1038/mi.2015.12.
[103] Moreland LW, Baumgartner SW, Schiff MH, Tindall EA, Fleischmann RM, Weaver AL, Ettlinger RE, Cohen S, Koopman WJ, Mohler K, Widmer MB, Blosch CM. Treatment of rheumatoid arthritis with a recombinant human tumor necrosis factor receptor (p75)-Fc fusion protein. N Engl J Med 1997;337(3):141–7. https://doi.org/10.1056/NEJM199707173370301.
[104] Granger DN, Senchenkova E. Leukocyte–Endothelial Cell Adhesion. Morgan & Claypool Life Sciences, https://www.ncbi.nlm.nih.gov/books/NBK53380/; 2010. [Accessed 17 January 2021].
[105] von Andrian UH, Chambers JD, McEvoy LM, Bargatze RF, Arfors KE, Butcher EC. Two-step model of leukocyte-endothelial cell interaction in inflammation: distinct roles for LECAM-1 and the leukocyte beta 2 integrins in vivo. Proc Natl Acad Sci U S A 1991;88(17):7538–42.
[106] Mohammed RN, Wehenkel SC, Galkina EV, Yates E-K, Preece G, Newman A, Watson HA, Ohme J, Bridgeman JS, Durairaj RRP, Moon OR, Ladell K, Miners KL, Dolton G, Troeberg L, Kashiwagi M, Murphy G, Nagase H, Price DA, Matthews RJ, Knäuper V, Ager A. ADAM17-dependent proteolysis of L-selectin promotes early clonal expansion of cytotoxic T cells. Sci Rep 2019;9(1):5487. https://doi.org/10.1038/s41598-019-41811-z.
[107] Faveeuw C, Preece G, Ager A. Transendothelial migration of lymphocytes across high endothelial venules into lymph nodes is affected by metalloproteinases. Blood 2001;98(3):688–95. https://doi.org/10.1182/blood.v98.3.688.
[108] Garton KJ, Gough PJ, Blobel CP, Murphy G, Greaves DR, Dempsey PJ, Raines EW. Tumor necrosis factor-alpha-converting enzyme (ADAM17) mediates the cleavage and shedding of fractalkine (CX3CL1). J Biol Chem 2001;276(41):37993–8001. https://doi.org/10.1074/jbc.M106434200.
[109] Tsakadze NL, Sithu SD, Sen U, English WR, Murphy G, D'Souza SE. Tumor necrosis factor-alpha-converting enzyme (TACE/ADAM-17) mediates the ectodomain cleavage of intercellular adhesion molecule-1 (ICAM-1). J Biol Chem 2006;281(6):3157–64. https://doi.org/10.1074/jbc.M510797200.
[110] Koenen RR, Pruessmeyer J, Soehnlein O, Fraemohs L, Zernecke A, Schwarz N, Reiss K, Sarabi A, Lindbom L, Hackeng TM, Weber C, Ludwig A. Regulated release and functional modulation of junctional adhesion molecule A by disintegrin metalloproteinases. Blood 2009;113(19):4799–809. https://doi.org/10.1182/blood-2008-04-152330.
[111] Rovida E, Paccagnini A, Rosso MD, Peschon J, Sbarba PD. TNF-α-converting enzyme cleaves the macrophage colony-stimulating factor receptor in macrophages undergoing activation. J Immunol 2001;166(3):1583–9. https://doi.org/10.4049/jimmunol.166.3.1583.

[112] Budagian V, Bulanova E, Orinska Z, Ludwig A, Rose-John S, Saftig P, Borden EC, Bulfone-Paus S. Natural soluble interleukin-15Rα is generated by cleavage that involves the tumor necrosis factor-α-converting enzyme (TACE/ADAM17). J Biol Chem 2011;286(11):9894. https://doi.org/10.1074/jbc.A110.404125.
[113] Marin V, Montero-Julian F, Grès S, Bongrand P, Farnarier C, Kaplanski G. Chemotactic agents induce IL-6Ralpha shedding from polymorphonuclear cells: involvement of a metalloproteinase of the TNF-alpha-converting enzyme (TACE) type. Eur J Immunol 2002;32(10):2965–70. https://doi.org/10.1002/1521-4141(2002010)32:10<2965::AID-IMMU2965>3.0.CO;2-V.
[114] Mohammed FF, Smookler DS, Taylor SEM, Fingleton B, Kassiri Z, Sanchez OH, English JL, Matrisian LM, Au B, Yeh W-C, Khokha R. Abnormal TNF activity in Timp3-/- mice leads to chronic hepatic inflammation and failure of liver regeneration. Nat Genet 2004;36(9):969–77. https://doi.org/10.1038/ng1413.
[115] Yarden Y, Sliwkowski MX. Untangling the ErbB signalling network. Nat Rev Mol Cell Biol 2001;2(2):127–37. https://doi.org/10.1038/35052073.
[116] Nicholson RI, Gee JMW, Harper ME. EGFR and cancer prognosis. Eur J Cancer 2001;37:9–15. https://doi.org/10.1016/S0959-8049(01)00231-3.
[117] Bethune G, Bethune D, Ridgway N, Xu Z. Epidermal growth factor receptor (EGFR) in lung cancer: an overview and update. J Thorac Dis 2010;2(1):48–51.
[118] Mitsudomi T, Kosaka T, Yatabe Y. Biological and clinical implications of EGFR mutations in lung cancer. Int J Clin Oncol 2006;11(3):190–8. https://doi.org/10.1007/s10147-006-0583-4.
[119] Lo H-W, Hsu S-C, Hung M-C. EGFR signaling pathway in breast cancers: from traditional signal transduction to direct nuclear translocalization. Breast Cancer Res Treat 2006;95(3):211–8. https://doi.org/10.1007/s10549-005-9011-0.
[120] Foley J, Nickerson NK, Nam S, Allen KT, Gilmore JL, Nephew KP, Riese DJ. EGFR signaling in breast cancer: Bad to the bone. Semin Cell Dev Biol 2010;21(9):951–60. https://doi.org/10.1016/j.semcdb.2010.08.009.
[121] Parashar D, Nair B, Geethadevi A, George J, Nair A, Tsaih S-W, Kadamberi IP, Nair GKG, Lu Y, Ramchandran R, Uyar DS, Rader JS, Ram PT, Mills GB, Pradeep S, Chaluvally-Raghavan P. Peritoneal spread of ovarian cancer harbors therapeutic vulnerabilities regulated by FOXM1 and EGFR/ERBB2 signaling. Cancer Res 2020;80(24):5554–68. https://doi.org/10.1158/0008-5472.CAN-19-3717.
[122] Bull Phelps SL, Schorge JO, Peyton MJ, Shigematsu H, Xiang L-L, Miller DS, Lea JS. Implications of EGFR inhibition in ovarian cancer cell proliferation. Gynecol Oncol 2008;109(3):411–7. https://doi.org/10.1016/j.ygyno.2008.02.030.
[123] Kalyankrishna S, Grandis JR. Epidermal growth factor receptor biology in head and neck cancer. J Clin Oncol 2006;24(17):2666–72. https://doi.org/10.1200/JCO.2005.04.8306.
[124] Feng B, Shen Y, Hostench XP, Bieg M, Plath M, Ishaque N, Eils R, Freier K, Weichert W, Zaoui K, Hess J. Integrative analysis of multi-omics data identified EGFR and PTGS2 as key nodes in a gene regulatory network related to immune phenotypes in head and neck cancer. Clin Cancer Res 2020;26(14):3616–28. https://doi.org/10.1158/1078-0432.CCR-19-3997.
[125] McGowan PM, Ryan BM, Hill ADK, McDermott E, O'Higgins N, Duffy MJ. ADAM-17 expression in breast cancer correlates with variables of tumor progression. Clin Cancer Res 2007;13(8):2335–43. https://doi.org/10.1158/1078-0432.CCR-06-2092.
[126] Brooks GD, McLeod L, Alhayyani S, Miller A, Russell PA, Ferlin W, Rose-John S, Ruwanpura S, Jenkins BJ. IL6 trans-signaling promotes KRAS-driven lung carcinogenesis. Cancer Res 2016;76(4):866–76. https://doi.org/10.1158/0008-5472.CAN-15-2388.
[127] Bech-Serra JJ, Santiago-Josefat B, Esselens C, Saftig P, Baselga J, Arribas J, Canals F. Proteomic identification of desmoglein 2 and activated leukocyte cell adhesion molecule as substrates of ADAM17 and ADAM10 by difference gel electrophoresis. Mol Cell Biol 2006;26(13):5086–95. https://doi.org/10.1128/MCB.02380-05.
[128] Rusch V, Baselga J, Cordon-Cardo C, Orazem J, Zaman M, Hoda S, McIntosh J, Kurie J, Dmitrovsky E. Differential expression of the epidermal growth factor receptor and its ligands in primary non-small cell lung cancers and adjacent benign lung. Cancer Res 1993;53(10 Suppl):2379–85.

[129] Sequist LV, Han J-Y, Ahn M-J, Cho BC, Yu H, Kim S-W, Yang JC-H, Lee JS, Su W-C, Kowalski D, Orlov S, Cantarini M, Verheijen RB, Mellemgaard A, Ottesen L, Frewer P, Ou X, Oxnard G. Osimertinib plus savolitinib in patients with EGFR mutation-positive, MET-amplified, non-small-cell lung cancer after progression on EGFR tyrosine kinase inhibitors: interim results from a multicentre, open-label, phase 1b study. Lancet Oncol 2020;21(3):373–86. https://doi.org/10.1016/S1470-2045(19)30785-5.

[130] Schumacher N, Rose-John S. ADAM17 activity and IL-6 trans-signaling in inflammation and cancer. Cancers (Basel) 2019;11(11). https://doi.org/10.3390/cancers11111736.

[131] Hurst SM, Wilkinson TS, McLoughlin RM, Jones S, Horiuchi S, Yamamoto N, Rose-John S, Fuller GM, Topley N, Jones SA. Il-6 and its soluble receptor orchestrate a temporal switch in the pattern of leukocyte recruitment seen during acute inflammation. Immunity 2001;14(6):705–14. https://doi.org/10.1016/s1074-7613(01)00151-0.

[132] Saad MI, Alhayyani S, McLeod L, Yu L, Alanazi M, Deswaerte V, Tang K, Jarde T, Smith JA, Prodanovic Z, Tate MD, Balic JJ, Watkins DN, Cain JE, Bozinovski S, Algar E, Kohmoto T, Ebi H, Ferlin W, Garbers C, Ruwanpura S, Sagi I, Rose-John S, Jenkins BJ. ADAM17 selectively activates the IL-6 trans-signaling/ERK MAPK axis in KRAS-addicted lung cancer. EMBO Mol Med 2019;11(4). https://doi.org/10.15252/emmm.201809976.

[133] Brennan D, Peltonen S, Dowling A, Medhat W, Green KJ, Wahl JK, Del Galdo F, Mahoney MG. A role for caveolin-1 in desmoglein binding and desmosome dynamics. Oncogene 2012;31(13):1636–48. https://doi.org/10.1038/onc.2011.346.

[134] Overmiller AM, Pierluissi JA, Wermuth PJ, Sauma S, Martinez-Outschoorn U, Tuluc M, Luginbuhl A, Curry J, Harshyne LA, Wahl JK, South AP, Mahoney MG. Desmoglein 2 modulates extracellular vesicle release from squamous cell carcinoma keratinocytes. FASEB J 2017;31(8):3412–24. https://doi.org/10.1096/fj.201601138RR.

[135] White JM. ADAMs: modulators of cell-cell and cell-matrix interactions. Curr Opin Cell Biol 2003;15(5):598–606. https://doi.org/10.1016/j.ceb.2003.08.001.

[136] Hirata S, Murata T, Suzuki D, Nakamura S, Jono-Ohnishi R, Hirose H, Sawaguchi A, Nishimura S, Sugimoto N, Eto K. Selective inhibition of ADAM17 efficiently mediates glycoprotein ibα retention during ex vivo generation of human induced pluripotent stem cell-derived platelets. Stem Cells Transl Med 2017;6(3):720–30. https://doi.org/10.5966/sctm.2016-0104.

[137] Seifert A, Düsterhöft S, Wozniak J, Koo CZ, Tomlinson MG, Nuti E, Rossello A, Cuffaro D, Yildiz D, Ludwig A. The metalloproteinase ADAM10 requires its activity to sustain surface expression. Cell Mol Life Sci 2020. https://doi.org/10.1007/s00018-020-03507-w.

[138] Richards FM, Tape CJ, Jodrell DI, Murphy G. Anti-tumour effects of a specific anti-ADAM17 antibody in an ovarian cancer model in vivo. PLoS One 2012;7(7):1–10. https://doi.org/10.1371/journal.pone.0040597.

[139] Leung C-H, Liu L-J, Lu L, He B, Kwong DWJ, Wong C-Y, Ma D-L. A metal-based tumour necrosis factor-alpha converting enzyme inhibitor. Chem Commun (Camb) 2015;51(19):3973–6. https://doi.org/10.1039/c4cc09251a.

[140] Rios-Doria J, Sabol D, Chesebrough J, Stewart D, Xu L, Tammali R, Cheng L, Du Q, Schifferli K, Rothstein R, Leow CC, Heidbrink-Thompson J, Jin X, Gao C, Friedman J, Wilkinson B, Damschroder M, Pierce AJ, Hollingsworth RE, Tice DA, Michelotti EF. A monoclonal antibody to ADAM17 inhibits tumor growth by inhibiting EGFR and non-EGFR-mediated pathways. Mol Cancer Ther 2015;14(7):1637–49. https://doi.org/10.1158/1535-7163.MCT-14-1040.

[141] Tape CJ, Willems SH, Dombernowsky SL, Stanley PL, Fogarasi M, Ouwehand W, McCafferty J, Murphy G. Cross-domain inhibition of TACE ectodomain. PNAS 2011;108(14):5578–83. https://doi.org/10.1073/pnas.1017067108.

[142] Kwok HF, Botkjaer KA, Tape CJ, Huang Y, McCafferty J, Murphy G. Development of a "mouse and human cross-reactive" affinity-matured exosite inhibitory human antibody specific to TACE (ADAM17) for cancer immunotherapy. Protein Eng Des Sel 2014;27(6):179–90. https://doi.org/10.1093/protein/gzu010.

[143] Wong E, Cohen T, Romi E, Levin M, Peleg Y, Arad U, Yaron A, Milla ME, Sagi I. Harnessing the natural inhibitory domain to control TNFα converting enzyme (TACE) activity in vivo. Sci Rep 2016;6(1):1–12. https://doi.org/10.1038/srep35598.

[144] Merchant NB, Voskresensky I, Rogers CM, Lafleur B, Dempsey PJ, Graves-Deal R, Revetta F, Foutch AC, Rothenberg ML, Washington MK, Coffey RJ. TACE/ADAM-17: a component of the epidermal growth factor receptor axis and a promising therapeutic target in colorectal cancer. Clin Cancer Res 2008;14(4):1182–91. https://doi.org/10.1158/1078-0432.CCR-07-1216.
[145] Hoettecke N, Ludwig A, Foro S, Schmidt B. Improved synthesis of ADAM10 inhibitor GI254023X. Neurodegener Dis 2010;7(4):232–8. https://doi.org/10.1159/000267865.
[146] McGowan PM, Mullooly M, Caiazza F, Sukor S, Madden SF, Maguire AA, Pierce A, McDermott EW, Crown J, O'Donovan N, Duffy MJ. ADAM-17: a novel therapeutic target for triple negative breast cancer. Ann Oncol 2013;24(2):362–9. https://doi.org/10.1093/annonc/mds279.
[147] Minond D, Cudic M, Bionda N, Giulianotti M, Maida L, Houghten RA, Fields GB. Discovery of novel inhibitors of a disintegrin and metalloprotease 17 (ADAM17) using glycosylated and non-glycosylated substrates. J Biol Chem 2012;287(43):36473–87. https://doi.org/10.1074/jbc.M112.389114.
[148] Takakura-Yamamoto R, Yamamoto S, Fukuda S, Kurimoto M. O-glycosylated species of natural human tumor-necrosis factor-alpha. Eur J Biochem 1996;235(1-2):431–7. https://doi.org/10.1111/j.1432-1033.1996.00431.x.
[149] Witters L, Scherle P, Friedman S, Fridman J, Caulder E, Newton R, Lipton A. Synergistic inhibition with a dual epidermal growth factor receptor/HER-2/neu tyrosine kinase inhibitor and a disintegrin and metalloprotease inhibitor. Cancer Res 2008;68(17):7083–9. https://doi.org/10.1158/0008-5472.CAN-08-0739.
[150] Huang Y, Benaich N, Tape C, Kwok HF, Murphy G. Targeting the sheddase activity of ADAM17 by an anti-ADAM17 antibody D1(A12) inhibits head and neck squamous cell carcinoma cell proliferation and motility via blockage of bradykinin induced HERs transactivation. Int J Biol Sci 2014;10(7):702–14. https://doi.org/10.7150/ijbs.9326.
[151] Ieguchi K, Maru Y. Savior or not: ADAM17 inhibitors overcome radiotherapy-resistance in non-small cell lung cancer. J Thorac Dis 2016;8(8):E813–5. https://doi.org/10.21037/jtd.2016.07.56.
[152] Masonic Cancer Center, University of Minnesota. Study of the ADAM17 inhibitor INCB7839 combined with rituximab after autologous hematopoietic cell transplantation (HCT) for patients with diffuse large B cell non-Hodgkin lymphoma (DLBCL). clinicaltrials.gov; 2020. Accessed January 27, 2021. https://clinicaltrials.gov/ct2/show/NCT02141451.
[153] Pavlaki M, Zucker S. Matrix metalloproteinase inhibitors (MMPIs): the beginning of phase I or the termination of phase III clinical trials. Cancer Metastasis Rev 2003;22(2-3):177–203. https://doi.org/10.1023/a:1023047431869.
[154] Coussens LM, Fingleton B, Matrisian LM. Matrix metalloproteinase inhibitors and cancer: trials and tribulations. Science 2002;295(5564):2387–92. https://doi.org/10.1126/science.1067100.
[155] Fogel DB. Factors associated with clinical trials that fail and opportunities for improving the likelihood of success: a review. Contemp Clin Trials Commun 2018;11:156–64. https://doi.org/10.1016/j.conctc.2018.08.001.
[156] Omoteyama K, Sato T, Sato M, Tsutiya A, Arito M, Suematsu N, Kurokawa MS, Kato T. Identification of novel substrates of a disintegrin and metalloprotease 17 by specific labeling of surface proteins. Heliyon 2020;6(12). https://doi.org/10.1016/j.heliyon.2020.e05804, e05804.
[157] Morancho B, Martínez-Barriocanal Á, Villanueva J, Arribas J. Role of ADAM17 in the non-cell autonomous effects of oncogene-induced senescence. Breast Cancer Res 2015;17:106. https://doi.org/10.1186/s13058-015-0619-7.
[158] Nollet M. Etude de l'implication de CD146/CD146 soluble dans l'angiogenèse et le développement tumoral : génération de nouveaux anticorps à visée thérapeutique. Published online December 22 http://www.theses.fr/2017AIXM0641; 2017. [Accessed 30 January 2021].
[159] Pruessmeyer J, Ludwig A. The good, the bad and the ugly substrates for ADAM10 and ADAM17 in brain pathology, inflammation and cancer. Semin Cell Dev Biol 2009;20(2):164–74. https://doi.org/10.1016/j.semcdb.2008.09.005.
[160] Romi E, Gokhman I, Wong E, Antonovsky N, Ludwig A, Sagi I, Saftig P, Tessier-Lavigne M, Yaron A. ADAM metalloproteases promote a developmental switch in responsiveness to the axonal repellant Sema3A. Nat Commun 2014;5(1):4058. https://doi.org/10.1038/ncomms5058.

[161] Swendeman S, Mendelson K, Weskamp G, Horiuchi K, Deutsch U, Scherle P, Hooper A, Rafii S, Blobel CP. VEGF-A stimulates ADAM17-dependent shedding of VEGFR2 and crosstalk between VEGFR2 and ERK signaling. Circ Res 2008;103(9):916–8. https://doi.org/10.1161/CIRCRESAHA.108.184416.

[162] Bernot D, Stalin J, Stocker P, Bonardo B, Scroyen I, Alessi M-C, Peiretti F. Plasminogen activator inhibitor 1 is an intracellular inhibitor of furin proprotein convertase. J Cell Sci 2011;124(8):1224–30. https://doi.org/10.1242/jcs.079889.

[163] Mustafi R, Dougherty U, Mustafi D, Ayaloglu-Butun F, Fletcher M, Adhikari S, Sadiq F, Meckel K, Haider HI, Khalil A, Pekow J, Konda V, Joseph L, Hart J, Fichera A, Li YC, Bissonnette M. ADAM17 is a tumor promoter and therapeutic target in western diet-associated colon cancer. Clin Cancer Res 2017;23(2):549–61. https://doi.org/10.1158/1078-0432.CCR-15-3140.

[164] Malapeira J, Esselens C, Bech-Serra JJ, Canals F, Arribas J. ADAM17 (TACE) regulates TGFβ signaling through the cleavage of vasorin. Oncogene 2011;30(16):1912–22. https://doi.org/10.1038/onc.2010.565.

[165] Gopal SK, Greening DW, Mathias RA, Ji H, Rai A, Chen M, Zhu H-J, Simpson RJ. YBX1/YB-1 induces partial EMT and tumourigenicity through secretion of angiogenic factors into the extracellular microenvironment. Oncotarget 2015;6(15):13718–30. https://doi.org/10.18632/oncotarget.3764.

[166] Orme JJ, Du Y, Vanarsa K, Mayeux J, Li L, Mutwally A, Arriens C, Min S, Hutcheson J, Davis LS, Chong BF, Satterthwaite AB, Wu T, Mohan C. Heightened cleavage of Axl receptor tyrosine kinase by ADAM metalloproteases may contribute to disease pathogenesis in SLE. Clin Immunol 2016;169:58–68. https://doi.org/10.1016/j.clim.2016.05.011.

[167] Waller K, James C, de Jong A, Blackmore L, Ma Y, Stagg A, Kelsell D, O'Dwyer M, Hutchins R, Alazawi W. ADAM17-mediated reduction in CD14++CD16+ monocytes ex vivo and reduction in intermediate monocytes with immune paresis in acute pancreatitis and acute alcoholic hepatitis. Front Immunol 2019;10. https://doi.org/10.3389/fimmu.2019.01902.

[168] Tang J, Frey JM, Wilson CL, Moncada-Pazos A, Levet C, Freeman M, Rosenfeld ME, Stanley ER, Raines EW, Bornfeldt KE. Neutrophil and macrophage cell surface colony-stimulating factor 1 shed by ADAM17 drives mouse macrophage proliferation in acute and chronic inflammation. Mol Cell Biol 2018;38(17). https://doi.org/10.1128/MCB.00103-18.

[169] Badenes M, Amin A, González-García I, Félix I, Burbridge E, Cavadas M, Ortega FJ, de Carvalho É, Faísca P, Carobbio S, Seixas E, Pedroso D, Neves-Costa A, Moita LF, Fernández-Real JM, Vidal-Puig A, Domingos A, López M, Adrain C. Deletion of iRhom2 protects against diet-induced obesity by increasing thermogenesis. Mol Metab 2020;31:67–84. https://doi.org/10.1016/j.molmet.2019.10.006.

[170] Zhao XQ, Zhang MW, Wang F, Zhao YX, Li JJ, Wang XP, Bu PL, Yang JM, Liu XL, Zhang MX, Gao F, Zhang C, Zhang Y. CRP enhances soluble LOX-1 release from macrophages by activating TNF-α converting enzyme. J Lipid Res 2011;52(5):923–33. https://doi.org/10.1194/jlr.M015156.

[171] Liu Q, Zhang J, Tran H, Verbeek MM, Reiss K, Estus S, Bu G. LRP1 shedding in human brain: roles of ADAM10 and ADAM17. Mol Neurodegener 2009;4(1):17. https://doi.org/10.1186/1750-1326-4-17.

[172] Chin D, Boyle GM, Porceddu S, Theile DR, Parsons PG, Coman WB. Head and neck cancer: past, present and future. Expert Rev Anticancer Ther 2006;6(7):1111–8. https://doi.org/10.1586/14737140.6.7.1111.

[173] Jimenez L, Jayakar SK, Ow TJ, Segall JE. Mechanisms of invasion in head and neck cancer. Arch Pathol Lab Med 2015;139(11):1334–48. https://doi.org/10.5858/arpa.2014-0498-RA.

[174] Gaździcka J, Gołąbek K, Strzelczyk JK, Ostrowska Z. Epigenetic modifications in head and neck cancer. Biochem Genet 2020;58(2):213–44. https://doi.org/10.1007/s10528-019-09941-1.

[175] Lu Y, Huang W, Chen H, Wei H, Luo A, Xia G, Deng X, Zhang G. MicroRNA-224, negatively regulated by c-jun, inhibits growth and epithelial-to-mesenchymal transition phenotype via targeting ADAM17 in oral squamous cell carcinoma. J Cell Mol Med 2019;23(8):4913–20. https://doi.org/10.1111/jcmm.14107.

[176] Berndt A, Büttner R, Gühne S, Gleinig A, Richter P, Chen Y, Franz M, Liebmann C. Effects of activated fibroblasts on phenotype modulation, EGFR signalling and cell cycle regulation in OSCC cells. Exp Cell Res 2014;322(2):402–14. https://doi.org/10.1016/j.yexcr.2013.12.024.

CHAPTER 10

"Omics" approaches to determine protease degradomes in complex biological matrices

Maithreyan Kuppusamy[a] **and Pitter F. Huesgen**[a,b,c]

[a]Central Institute for Engineering, Electronics and Analytics, Jülich, Germany, [b]Cluster of Excellence Cellular Stress Responses in Aging-Associated Diseases, CECAD, Medical Faculty and University Hospital Cologne, University of Cologne, Cologne, Germany, [c]Institute of Biochemistry, Department of Chemistry, University of Cologne, Cologne, Germany

Introduction

Proteases are master regulators of the proteome: On the one hand, some are processive proteolytic machines that regulate protein abundance and ensure functionality [1, 2]; on the other hand, others act as more delicate protein sculptors, trimming target proteins by a few N- or C-terminal amino acids or processing at selective cleavage sites [3]. Such site-specific processing represents a regulatory protein modification that irreversibly converts substrate proteins into new proteoforms with altered activity, location, and interactions [4]. Notably, site-selective processing can earmark target substrates for degradation, enabling, for example, recognition by the protease-generated protein termini (N-end rule) by the degradative machines like the ubiquitin-proteasomal system [5]. Proteases thus collectively determine the functional state of the proteome and are thereby intimately involved in all cellular processes. Consequently, the misregulation of proteolytic activities is implicated in many severe acute and chronic human diseases [6]. That is why proteases are attractive drug targets [7], while their specific cleavage products may provide specific disease biomarkers [8, 9].

The key to understanding the function of a protease is knowledge of its substrates, also termed its degradome [10]. This is not only of fundamental interest but also required to discriminate differentially activated proteases as targets or antitargets in diseases [11]. However, even for most of the more than 550 proteases encoded in the human genome, only a few substrates are known [12]. Other multicellular organisms likewise contain hundreds of proteases, most with poorly defined or completely unknown functions.

Proteolytic Signaling in Health and Disease. https://doi.org/10.1016/B978-0-323-85696-6.00004-4

Intact proteins can be ionized and directly identified using tandem mass spectrometry (MS/MS) [13]. Also known as "top-down" proteomics, these approaches resolve individual proteoforms and are therefore highly attractive to identify processed proteoforms [14]. Indeed, one recent study identified more than a 1000 different truncated proteoforms derived from 225 precursor proteins in myeloid-derived suppressor cell (MDSC)-derived extracellular vesicles [15]. However, top-down proteomics still faces considerable technical and computational challenges that impede robustness and sensitivity [16], currently restricting the application of this exciting technology primarily to purified proteins or highly simplified proteome fractions. In contrast, "bottom-up" or "shotgun" proteome approaches enable nearly complete coverage of the proteome. In shotgun/bottom-up proteomics, proteins are digested into peptides using enzymes to facilitate mass spectrometric analysis and enable routine identification and quantification of thousands of proteins across multiple conditions [17]. Consequently, shotgun proteomics has become an essential tool to study proteases and their impact on the proteome in an unbiased manner [18]. In this tutorial-style chapter, we briefly discuss the merits and drawbacks of different experimental systems for unbiased protease degradome characterization, followed by a brief introduction to different bottom-up proteomics methods for peptide and protein quantification. We also summarize recent developments in termini-centric "degradomics" methods tailored to protease cleavage site identification in complex samples.

Experimental design for protease degradome profiling

Common to all mass spectrometry (MS) approaches is that no method will achieve complete proteome/peptidome coverage. As a consequence, lack of identification cannot be considered proof of absence. Therefore, protease substrate identification requires relative quantification of peptides and proteins in samples exposed to a varying degree of activity of the protease of interest. Protease activity can be modulated in many different ways, ranging from selective incubation with purified enzyme over cell-based assays and pharmacological interventions to elaborated animal models with tissue-specific alterations of protease activity.

Substrate discovery with recombinant proteases in vitro

The arguably most simple and most direct experimental design is to incubate a purified, active protease of interest with a relevant proteome extracted from cells or tissues under "native-like," nondenaturing conditions (Fig. 1A). Typically, the test proteome is treated with protease inhibitors during extraction to minimize "background" hydrolysis due to other proteases present in the proteome that may confound analysis and may result in complete degradation of the sample during the incubation. One half of the isolated proteome is incubated with the active protease and the other with appropriate control, such as the purified enzyme inactivated either by point mutation or by the addition of a specific inhibitor. If the

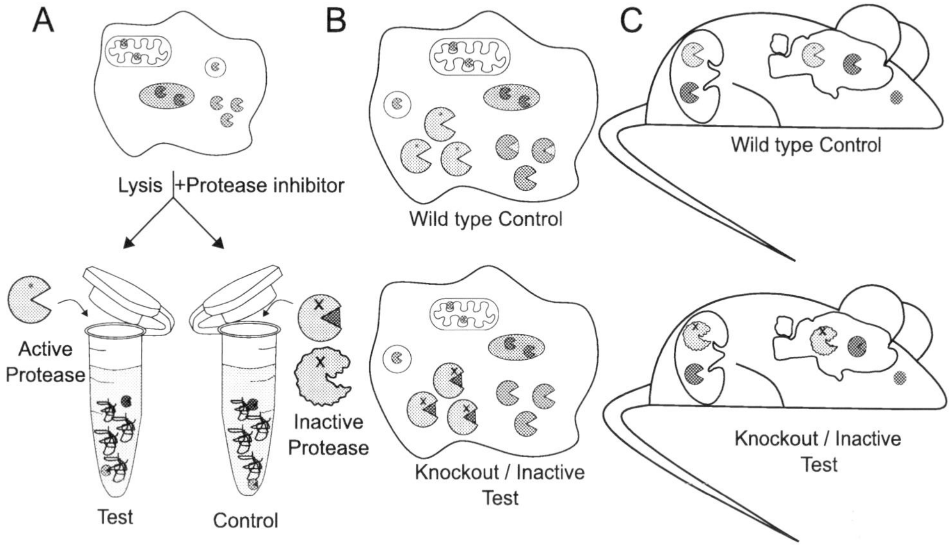

Fig. 1
Experimental design for protease degradome analysis in vivo and in vitro. (A) In an in vitro approach, a purified protease of interest (*indicated orange* (gray in the print version)) is added to whole cell lysates, using a catalytically inactive point-mutant or chemically inhibited form as a control. However, this approach can result in nonrelevant substrate cleavage due to the lack of physiological compartmentation, excess protease, and nonphysiological conditions that can render proteins unstable. (B) Protease activity can be modulated in cell culture by genetic and pharmacological means, as depicted by selective inhibition. The main advantage of this approach is that the protease has negligible chances of interacting with proteins from other compartments leading to confident identification of substrates. However, many indirect effects can confound analysis, for example, unanticipated compensatory activation of other proteases (*indicated pink* (light gray in the print version)). (C) Protease activity can be altered in vivo approach, for example, by knocking out the protease of interest in animal models like mice. Still, this process consumes time and money, and there is a chance that another protease may compensate for the missing activity. (For interpretation of the references to color in this figure legend, the reader is referred to the web version of this article.)

test protease is active under the chosen incubation condition, this type of experiment typically provides numerous candidate substrates and cleavage sites. However, at the same time, this approach is prone to deliver "false-positive" substrates that are not cleaved in a cellular context [19]. For example, in vivo, the protease of interest is usually present in much lower concentrations, substrate protein fold more tightly maintained, additional interacting factors may restrict access to substrate cleavage sites, or active protease and substrates may be separated in time and space, for example, reside in different compartments. These drawbacks can be mitigated by isolating proteomes from the physiologically relevant tissue sources and subcellular compartments, for example, by the incubation of extracellular proteases such as MMP2 and MMP9 with the secreted proteins isolated from the cell-culture conditioned media [20]. Time-resolved analysis of substrate cleavage during incubation with recombinant proteases can reveal kinetically preferred substrates [21, 22]. In summary, this direct in vitro approach, enriched by integrating additional information such as subcellular localization and co-expression, provides a reasonable basis for selecting candidate substrates and validation in a physiologically relevant spatiotemporal context.

Modulation of protease abundance in cell culture

Many proteases cannot be obtained in recombinant form, may require endogenous cofactors, or may depend on proper physiological conditions for their proper function. In these cases, manipulating the proteolytic activity in cell-based systems provides a convenient opportunity to screen for physiologically relevant substrates of the protease of interest. Suited cell-based models provide endogenous factors for protease activity and primarily maintain appropriate subcellular compartmentation (Fig. 1B). Protease activity can be manipulated genetically, either by knockout, knockdown, or overexpression of the protease of interest, or by pharmacological intervention. However, chemical modulators such as protease inhibitors frequently target multiple co-expressed proteases simultaneously. Genetic manipulation requires careful validation that the desired effect is achieved, especially since many proteases are under posttranslational control, resulting in a poor correlation of the overall protease abundance with enzymatic activity. Ideally, the activity of the protease of interest is assayed using specific substrates or activity-based probes [23, 24]. Furthermore, the modulation of the activity of the protease of interest may inadvertently alter the activity of other proteases as a consequence of the multitude of direct and indirect interactions in the protease web [25]. For example, depletion of a protease of interest may directly dampen or activate, respectively, downstream-acting proteases in proteolytic cascades such as the clotting factors or initiator/executioner caspases during apoptosis or altered transcription and translation as a consequence of altered signaling.

Modulation of protease activity in vivo

Proteases are often expressed in a tissue-specific manner [26] and may target different substrates with different functional consequences in different tissues (Fig. 1C). Furthermore, substrate availability and sometimes cleavage may be regulated by physiological stimuli that are not easily replicated in standard cell culture experiments, for example, mechanical stress [27] or shear forces [28]. Therefore, animal models with manipulated protease activity, including tissue- or cell-type specific manipulation of changes in expression, are indispensable tools for protease degradome discovery [19]. Also, biopsy material from individuals suffering from genetically altered protease activity can be valuable source material for protease substrate discovery [29]. However, human specimens are rarely available; generating animal model systems are costly and time-consuming and may not reflect the human disease. In all cases, the observed changes are even more likely to be indirect than in cell culture systems, requiring additional validation to establish direct protease-substrate relationships.

In summary, a large variety of experimental approaches are available to initiate a quest for protease substrates, each with advantages and limitations that have to be considered during data interpretation. In many cases, a combination of in vivo and in vitro approaches may be required to verify that an identified substrate cleavage by a specific protease is both direct and physiologically relevant [22].

Quantitative bottom-up proteomics

As indicated earlier, MS-based proteomics enables largely unbiased identification of thousands of proteins with relative quantification across multiple conditions. It is, therefore, a natural choice to investigate protease degradomes. The basis of these efficient bottom-up proteomics techniques, also known as shotgun proteomics, is the deliberate destruction of proteins using specific endoproteases as tools to obtain peptides with predictable sequences [30]. The digested peptides are then separated by chromatographic separation, ionized, and analyzed by MS/MS, followed by computational identification of the peptide sequences using genome-derived proteome databases/spectral libraries to infer the proteins present in the sample (Fig. 2A).

In a typical bottom-up proteomics experiment, proteomes are extracted from the sample of interest using a lysis buffer containing detergents and chaotropic agents to denature the proteins and lyse membrane-bound organelles. The lysis buffer typically contains a wide range of protease inhibitors to prevent unspecific degradation in the proteome. The extracted proteome is reduced to assist denaturation followed by alkylation at the cysteine residue to prevent uncontrolled modification or reformation of disulfide bridges. The proteome is then purified either by polyacrylamide gel electrophoresis (in gel-based approaches) or by precipitation, buffer exchange, or dilution (in gel-free approaches). The purified proteome is then digested with a specific endoprotease, most frequently by cleavage after the basic residues Arg and Lys using trypsin. The resulting peptides are further purified/desalted and separated by C18 reverse-phase liquid chromatography (C18-RP-LC) under acidic conditions using a nano- or microflow high-performance liquid chromatography (nano –/ micro-HPLC) system. Eluting peptides are directly introduced to a high-resolution MS/MS system by electrospray ionization (ESI). For complex samples, peptides are often fractionated using orthogonal methods such as high-pH-RP-LC or strong cation exchange (SCX) chromatography to obtain additional dimension of separation followed by tandem mass spectrometry measurements in different modes.

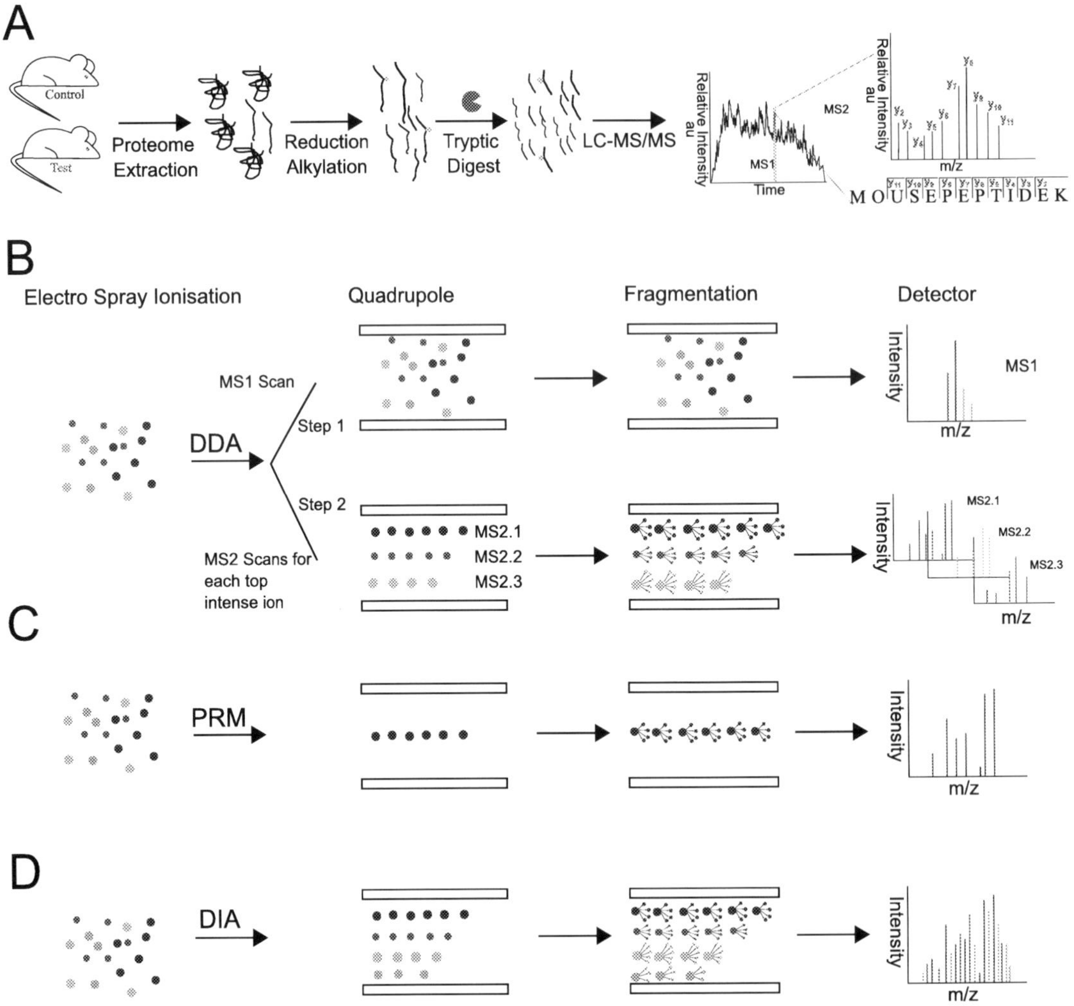

Fig. 2

Quantitative shotgun proteomics. (A) In shotgun proteomics, a typical bottom-up approach, the proteome is extracted using a lysis buffer under denaturing conditions followed by reduction-alkylation. Peptides are generated by digestion using sequence-specific endoproteases like trypsin. They are analyzed by C18 reverse-phase liquid chromatography coupled to a tandem mass spectrometer, and the peptide sequence is identified based on peptide fragment ion spectra. (B) The most common spectrum acquisition method in a bottom-up mass spec is data-dependent acquisition (DDA). First, the intensity and *m/z* of peptide ions eluting from the chromatography are recorded in regular intervals. Then, the mass spec is programmed to select a number of the most intense ion for fragmentation. Peptides are identified based on the peptide mass recorded in the MS1 spectra in conjunction with the MS2 fragment ion spectra. (C) In parallel reaction monitoring (PRM), preselected peptide ions are isolated and fragment spectra are recorded using tailored MS settings. The selected ions are scanned in frequent cycles resulting in full MS2 coverage of peptide peaks eluting from the column. Accurate quantification is achieved by averaging multiple fragmentation chromatograms for each peptide. (D) In data-independent acquisition (DIA), the

Data-dependent acquisition of peptide fragment spectra

Data-dependent acquisition (DDA) mode is the most frequently used approach for peptide identification in shotgun proteomics. In DDA, the mass spectrometer performs a scan of peptide ions (precursor ions) co-eluting at a given time, termed the MS1 spectrum, and then selects the most intense peptides for fragmentation (Fig. 2B). Fragment ion spectra are recorded individually, and fragmented precursor ions are dynamically excluded to avoid repeated fragmentation of intense peaks. Peptide sequences are then computationally determined by matching the acquired fragment ion spectra to spectral libraries or theoretical spectra from genome-derived protein sequence databases predicted using mass spectrometer-dependent variables such as mass accuracy (for both precursor and fragment ions) and experimental variables such as the specificity of the protease used for proteome digestion and expected peptide modifications as constraints.

Typically, MS1 spectra are acquired at the latest every 3 s, so that chromatograms of eluting peptides can be reconstructed from the peptide ion intensity in the MS1 spectra. Label-free quantification methods directly compare independently measured samples using the precursor ion MS1 intensities [31]. A disadvantage to this approach is that the stochastic nature of precursors selected for fragmentation means that not every peptide is identified and quantified in every sample, leading to "missing values," i.e., peptide/proteins quantified in only a subset of the analyzed replicates. This problem can be overcome by transferring identifications to corresponding MS1 peaks within certain mass and retention time tolerances. However, this can lead to erroneous assignments and thereby affect quantification accuracy, particularly if the proteome composition between the analyzed samples changes. Also, quantification accuracy can be affected by changes in LC column performance and the amount of peptide injected.

Differential labeling with stable isotope reagents enables reliable quantification of a peptide across several experimental conditions within the same mass spectrometry run [32]. It is, therefore, an attractive alternative to label-free quantification, especially for complex workflows where significant differential losses can occur. Stable isotope labeling by amino acids in cell culture (SILAC) is one technique where the labels are introduced metabolically using stable isotope-labeled essential amino acids [33]. In contrast, other approaches introduce stable isotope-marked modifications using predominantly cysteine- or

---

Fig. 2, cont'd
mass spec records MS1 spectra in regular intervals. It then cycles through a series of specific *m*/*z* windows, co-fragmenting all co-eluting multiple precursor ions eluting in the *m*/*z* window at the same time. The resulting complex fragmentation spectra are deconvoluted with the help of spectral libraries, using retention time (RT) and *m*/*z* window mass as constraints mass range. DIA thus combines aspects such as unbiased peptide identification and improved MS2-based peptide quantification accuracy like in DDA and PRM assays, respectively.

amine-reactive reagents [34, 35]. Most labeling reagents introduce a distinct, differential mass shift of the modified peptides in the MS1 spectrum, resulting in more complex spectra and the need to consider variable mass shifts during database searches. Isobaric tagging reagents like the isobaric tags for relative and absolute quantification (iTRAQ) and the tandem mass tags (TMTs) overcome this by introducing bipartite chemical modifications that can fragment with the precursor ion in the collision cell [36]. Fragmentation of the isobaric tags results in a neutral loss of a balance group and release of unique reporter ions that are detected in the low mass range. Different distribution of stable carbon and nitrogen isotopes between balance and reporter ion enables multiplexing by up to 16 samples in a single mass spectrometry run and ensures quantification of most identified peptides [37, 38]. However, co-eluting peptides and isotope impurities can impair quantification accuracy but can be improved at the expense of sensitivity using a third fragmentation step MS3 [36, 39].

Improved quantification by targeted mass spectrometry

Targeted methods such as the selected and parallel reaction monitoring (SRM and PRM) are alternative MS/MS data acquisition modes that are a natural extension of data-dependent acquisition [40, 41]. These methods generally require prior knowledge of the target peptide or peptides (precursor ion(s)), such as its mass-to-charge ratio (*m/z*), elution time, and fragment ions (MS2) [42]. This information, if not available in the literature or custom databases [43], is generally extracted from the DDA data using software like Skyline [44], PeptideMass [45], and PeptidePicker [46]. Once target peptides of interest have been identified, the mass spectrometer can be set to select and fragment these peptides (precursor ions) and record the intensity of selected (SRM) or all fragment ions (PRM, Fig. 2C). Optimization of mass spectrometer settings for the selected peptides of interest, such as ionization and fragmentation energies, enables increased sensitivity and accuracy of quantification compared to DDA methods, thereby enabling reliable quantification across large sample cohorts [40]. Depending on the type of mass spectrometers/mass analyzers, additional instrument parameters such as ion fill time also need adjusting to obtain better results [42]. The recorded spectra are then identified and quantified using free software packages like Skyline [44] or commercial alternatives. Relative quantification of these target peptides is done by comparing the peptide intensities to an internal standard. However, absolute quantification can be achieved using stable isotope-labeled synthetic peptides spiked into the sample [42].

Data-independent acquisition of peptide fragment spectra

More recently, a range of untargeted acquisition methods, collectively termed data-independent acquisition (DIA), are rapidly gaining popularity. DIA methods combine the extensive coverage achieved by DDA with the accuracy of targeted methods [47, 48]. DIA is an untargeted method in the sense that MS1 and MS2 acquisition is uncoupled: Rather than selecting specific precursors for fragmentation, the analyzed mass range is divided into broad *m/z* windows (typically 20–30 *m/z*), where all co-eluting precursor ions are fragmented simultaneously (Fig. 2D). This results in highly convoluted MS2 spectra but provides fragment ion information for all co-eluting peaks. Peptide identification is achieved by matching to comprehensive experimental spectral libraries acquired by extensive DDA analysis of the same or a similar sample, using the retention time as an additional criterion. Most recently, sophisticated neural network algorithms have been developed that reliably predict peptide fragmentation patterns and retention times from protein sequences, removing the requirement of DDA analysis [49, 50]. As all fragmentation ions are recorded, peptide quantification can be averaged from several fragment ions similar to the targeted data acquisition methods. Another benefit is that no precursor ion is excluded from the analysis. The same full range of data is acquired for each run, meaning that the data can easily be queried for the presence of additional peptides or modifications that have not been considered in the initial analysis. DIA methods are currently a hot topic of research, and both methods and software are experiencing rapid development [48].

Protease substrate discovery by quantitative shotgun proteomics

Irrespective of the precise quantification mode and data acquisition, proteins present in the sample are inferred from the peptide sequences identified by the acquired tandem mass spectra. Closely related proteins, isoforms, and redundant protein sequences frequently result in ambiguous assignments of peptides to multiple protein sequences in a database. Therefore, Occam's razor principle is applied, and the proteins are restricted to a minimal set of sequences explaining all identified peptides in the dataset [51]. Proteins that cannot be distinguished by the identified peptides are reported as protein groups, and quantification of individual peptides is typically averaged for each protein group. This enables direct identification of protease substrates in systems where ablated protease activity results in accumulation of the substrate or vice versa [52]. Proteolytically processed substrates can also be identified using the shotgun approach. Since bottom-up proteomics is intrinsically peptide-centric, a detailed investigation of the peptides assigned to a protein can also provide information on specific regions affected by proteolytic processing [53]. For example, truncation of a protein can lead to the loss of the peptides that belong to that particular removed domain or altered abundance in specific subcellular compartments, e.g., during ectodomain shedding [54]. Likewise, individual peptides containing the proteolytic processing sites will be less abundant in the sample with increased processing, even when the overall abundance—as determined by the majority of peptides attributed to the protein—may not be significantly altered [53]. Identification of protease substrates and proteolytic cleavages from shotgun proteomics data is further facilitated by peptide-centric computational tools, such as ImproViser [55] and QUARIP [56], that visualize peptide position and quantity relative to the protein sequence and domain structure.

Gel-based protein-level prefractionation before the digestion can further assist in identifying protease degradomes [57]. A particularly elegant approach is PROtein TOpography and Migration Analysis Platform (PROTOMAP) [58, 59], where samples exposed to differential proteolytic activity, either in vivo or in vitro, are separated by a simple SDS-PAGE before tryptic digestion of multiple gel bands [58]. Processed proteins are then identified by "peptographs" that integrate the apparent molecular mass estimated from the SDS-PAGE migration with protein sequence coverage and quantitative information from spectral counting [58] or in combination with SILAC labeling [60].

Selective enrichment and proteome-wide characterization of protein termini

For many regulatory proteases, knowledge of the specific substrate cleavage sites is required to understand their impact on the substrate(s) and thus protease function. However, the neoN and neoC termini generated by site-specific proteolytic cleavage are rarely identified in standard shotgun proteomics experiments. First, protein termini only form a small minority among all peptides generated during a shotgun digest and are therefore also rarely selected for identification in DDA-MS. Secondly, typically only peptide sequences delimited on both sides by the cleavage site of the protease used for the proteome digestion are considered during the matching to computationally predicted spectra. These constraints limit the search space, which minimizes the computational requirements and decreases the chance of random matches. In fact, considering peptides matching the specificity of the digestion protease only on one end in a "semispecific" search increases the search space approximately 10–30-fold, depending on the protease employed [61] with deleterious impact on the number of peptide identifications. Furthermore, proteolytic cleavage often affects only a fraction of the total substrate protein, resulting in a very low abundance of the processed proteoforms. Overall, this results in low identification of proteolytic cleavages even if semispecific database searches are applied. Therefore, specific enrichment is required to facilitate identification, just like the enrichments done for other posttranslationally modified peptides [62]. A wide variety of approaches for selective enrichment of protein N- and C-termini have been developed and are discussed in detail in recent reviews [62–65]. In general, two approaches are followed: In positive selection, proteins termini are selectively tagged with an affinity reagent before the digest and modified terminal peptides are selectively enriched by affinity capture (Fig. 3). In the alternative negative selection workflows, protein termini and amino acid side chains with the same reactive group are modified before digest. The internal peptides with new functional primary amine or carboxyl group generated from the proteolytic proteome digest are then removed by a second reaction or modification that enables selective depletion (Fig. 3).

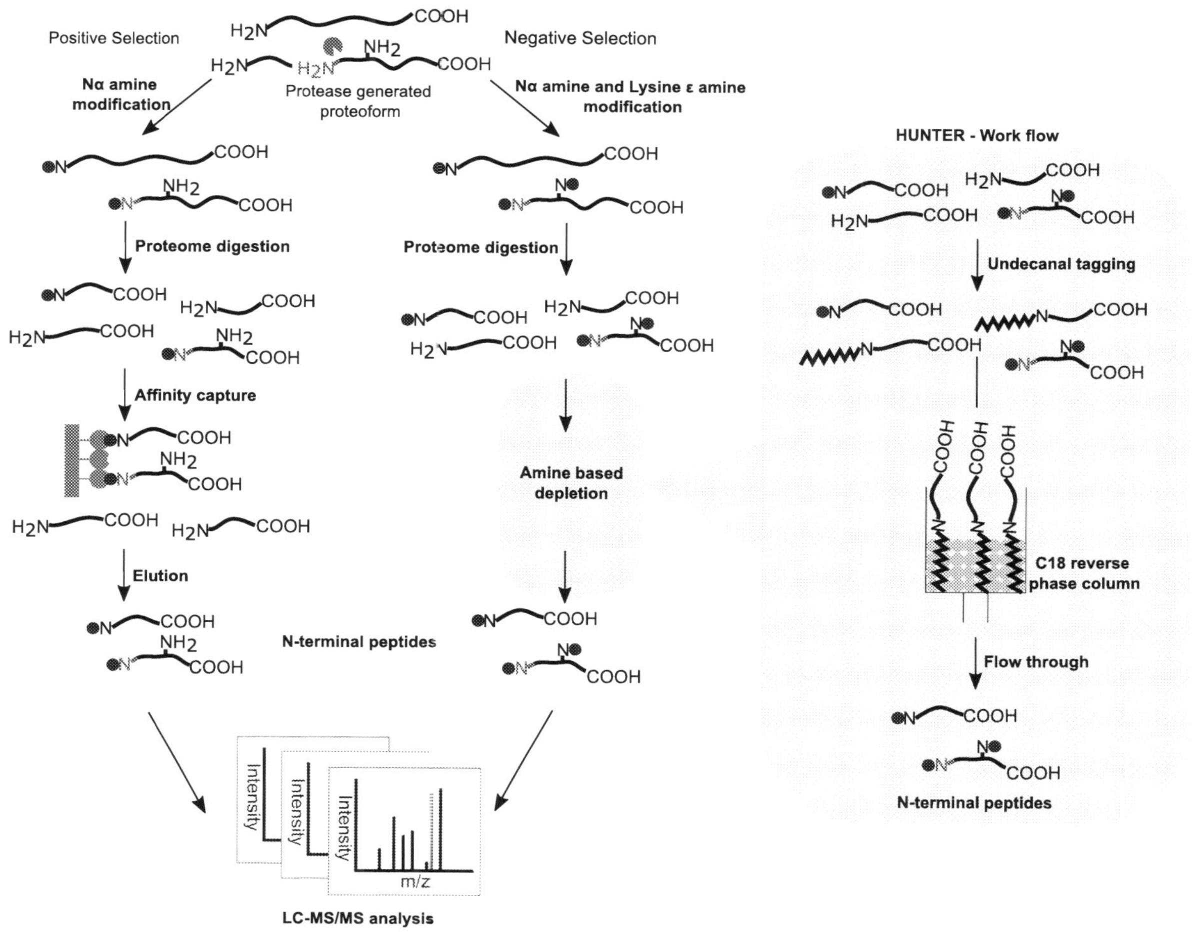

Fig. 3

Enrichment of protein N-termini by positive or negative selection reveals protease cleavage sites. In positive selection techniques, the N-terminal peptides are captured, and other unwanted peptides are washed away. In contrast, negative selection methods capture non-N-terminal peptides, leaving only the N-terminal peptides behind. In both techniques, chemical modification of the N-terminal primary amine is used to alter the properties of the peptides like affinity, charge, and hydrophobicity. In positive selection workflows, the protein N-termini are selectively modified with an affinity group such as biotin, captured by affinity reagents to enrich N-terminal peptides after enzymatic digestion. In negative selection workflows, the both protein N-terminal amines and the lysine side-chain amines are blocked by chemical modification. Then, the proteins are subjected to digestion, and the resulting new generated peptides have a free N-terminal primary amine except for the protein N-termini, which were modified in the first step. These free amines are then exploited for modification. For example, in the HUNTER protocol primary amines generated by proteome digest are modified with undecanal, making the non-N-terminal peptides very hydrophobic. When passed through a reverse-phase column, the hydrophobic undecanal-modified peptides are retained, while N-terminal peptides remain in the flow-through.

Enrichment of protein N-termini

Positive selection of N-terminal peptides involves coupling an affinity tag such as biotin to the α-amines at free protein N-termini, which can be achieved chemically [66, 67] or enzymatically using, for example, the engineered peptide ligase subtiligase [68]. Following the proteome digestion, the tagged peptides arising from the protein N-termini are selectively enriched (Fig. 3). However, obtaining sufficient selectivity for α-amines over ε-amines in chemical reactions is challenging. It requires delicate control of the reaction conditions [66], while enzymatic reactions usually require large amounts of material [63]. Therefore, the alternative negative selection methods are more frequently used for protease substrate discovery. In negative selection methods, all primary amines in a protein are chemically modified, followed by the proteome digest. The digestion releases new primary amines in all internal and C-terminal peptides, which is used to capture these peptides in a second reaction and separate them from the inert, endogenously, or previously modified N-terminal peptides (Fig. 3). A large variety of different methods have been established that utilize the digest-generated peptide α-amines for depletion. These include direct capture by aldehyde-functionalized materials [69–71], affinity depletion after biotinylation [72], or separation after amine-based modification to alter one or multiple properties such as the charge [73, 74] or hydrophobicity [75–77]. A substantial advantage of negative selection techniques for protease substrate discovery is that a large variety of amine-reactive isotope labeling reagents such as formaldehyde, iTRAQ, and TMT enable multiplexing, thereby avoiding differential losses during subsequent sample handling steps [20, 70]. As an example, the recently developed High-efficiency Undecanal-based N-Termini EnRichment (HUNTER) protocol differentially labels primary N-terminal amine and lysine side-chain amines with formaldehyde isotopes before digestion [75]. The peptides generated after proteome digestion with trypsin are further modified with undecanal, increasing their hydrophobicity to allow retention on a C18-RP column while the less hydrophobic formaldehyde-labeled N-terminal peptides elute in the flow-through [75]. A challenge for N-termini enrichment by negative selection is that both steps, the initial blocking reaction and the depletion step, use the same chemistry and therefore require proteome cleanup with concomitant sample losses. In HUNTER, such cleanup steps are performed with paramagnetic SP3 beads [78], enabling enrichment of N-terminal peptides from micrograms starting materials [75]. Likewise, recent improvements also enabled enrichment from small tissue biopsies by combining efficient sample lysis and multiplexing using the TAILS technique [79] and increased sensitivity of charge-based selection by ChaFraTip [80].

Enrichment of protein C-termini

C-termini are similarly enriched using positive and negative selection methods [62, 65]. However, these are not as sensitive or efficient as the N-termini enrichment methods because of the poor reactivity of the C-terminal carboxyl group, which requires activation before modification. In carboxy-terminal amine-based isotope labeling of substrates (C-TAILS) [81, 82], the primary amines are modified by reductive dimethylation using formaldehyde to avoid cross-reactivity in the subsequent modification step of the carboxyl groups. The proteins are then digested using enzymes like trypsin. The peptides generated after digestion containing free amines at the peptide N-termini are again dimethylated, followed by covalent coupling of the digestion-generated C-terminal carboxyl groups to a water-soluble, high molecular weight amine-functionalized polymer. The C-terminal peptides present in the original proteome, modified in the first reaction, remain inert and are then isolated in the flow-through of an ultrafiltration. However, C-terminal peptides rarely contain basic amino acids in tryptic digests, resulting in poor identification because of poor MS properties. This can be ameliorated using LysargiNase, which cleaves before Arg and Lys [83, 84]. Alternatively, C-terminal COFRADIC [85] can be applied to simultaneously enriched for N- and C-terminal peptides. In this technique, the primary amines are acetylated, followed by tryptic digestion. The N- and C-terminal peptides are then selectively enriched based on their charge and hydrophobicity using SCX and RP in sequential orthogonal chromatographic separations. A further butyrylation of the free amines will enable separating the N- and C-terminal peptides from each other via RP chromatography.

Positional proteomics data analysis

Enriched N- and C-terminal peptides are then typically analyzed by LC–MS/MS using DDA. Conventional software for shotgun proteome data analysis can also be used for N-terminome data with small changes in the search settings. Digestion specificity must be set as semispecific, and all amine and carboxyl modifications have to be included in the search settings. In negative selection approaches, additional modifications such as N-terminal acetylation and pyroglutamate formation should also be considered during the search. N-termini acetylation is one of the major endogenous modifications with regulatory functions found in most cytosolic proteins [86]. As the peptides with this modification remain inert in the amine-modification step after digest, they are also co-enriched with the other N-terminal peptides in negative selection methods. Glu and Gln can undergo cyclization to form an N-terminal pyro-Glu in vivo, similarly protecting the α-amine. However, Gln can also cyclize spontaneously in vitro and thus protect peptides with N-terminal Gln after digestion. These peptides are, therefore, usually not considered true N-terminal peptides [87].

Enrichment of protein termini effectively selects the originally synthesized or few protease-generated neotermini as representatives for each protein, which significantly simplifies the proteome [77] and enables deeper proteome coverage in the same measurement time as that of a shotgun approach [70]. However, identified peptides may match multiple protein sequences in the database, hampering the application of established tie-breaker rules [51].

To understand the function of a proteolytic cleavage, it is vital to consider the exact position of the cleavage in the protein in relation to known functional domains and protein maturation sites. To create such positional annotations, dedicated software tools have been developed, such as CLIPPER [88] as an add-on for the TransProteomicPipeline [89] and MANTI [87] for terminome data analyzed by MaxQuant [90].

Overall, these termini-centric positional proteomics techniques have remarkably accelerated protease substrate discovery [3, 91]. Furthermore, targeted analysis of cleavage product and the corresponding tryptic peptide containing the cleavage site can reveal the proportion of substrate cleavage, such as the proportion of active MMP2 in relation to the amount of proMMP2 [92].

However, all termini-centric approaches face two additional challenges. First, not all peptides generated by one particular digestion enzyme, such as trypsin, will be suitable for MS. For example, the peptides may be too short, too long, or unfavorable in amino acid composition. This restricts the number of protein termini that can be observed in a single experiment, but increased N-terminome coverage can be achieved using multiple digestion enzymes [80, 93, 94]. Secondly, quantification in termini-centric approaches relies on a single peptide. This is inevitably more variable than the quantification over several peptides in typical shotgun proteomics approaches. One way to improve accuracy and reproducibility for quantifying these single peptides are targeted data acquisition methods such as SRM, which enabled, for example, determination of cleavage rates for apoptotic caspase substrates [21] and detection of caspase cleavage products as markers of chemotherapy-induced cell death in human plasma [95]. A more recent alternative are DIA methods that similarly increase quantification accuracy for all identified termini by quantifying several MS2 peaks [75]. However, no software can predict spectral libraries for semispecific library-free-termini DIA data analysis. This makes it compulsory to generate dedicated spectral libraries with high terminome coverage, meaning each sample type has to be extensively characterized by high-quality DDA measurements to enable DIA analysis.

Peptidomics

Regulatory peptides constitute a significant class of proteolysis-derived signals with crucial physiological functions [96]. Likewise, proteolysis-derived immune peptides are a key element of adaptive immunity in mammals [97, 98]. Identification of peptides releases by a protease of interest can be achieved by direct mass spectrometry-based analysis, often without the need for further proteolytic cleavage, using principally the same peptide-centric DDA and DIA measurement methods and model systems outlined previously. However, peptidome analysis often requires dedicated sample collection and cleanup. Additional challenges include instability during sample preparation, more heterogeneous behavior in mass spectrometry due to a lack of terminal-basic residues, and data analysis. We refer to excellent reviews [99–101] and protocols [102, 103] for a detailed discussion.

Conclusions

Mass spectrometry-based methods have become the "go-to" methods of choice for unbiased interrogation of protease degradomes in complex proteomes. A large variety of experimental approaches for comparative analysis of protein, peptide, and neo-N and neo-C-termini abundance across multiple genotypes, experimental conditions, and time points have been established. Many popular methods are well-described in detailed step-by-step protocols available (reviewed in [63]). The choice thus largely depends on the experimental system(s) and (suspected) mechanism of action of the protease(s) of interest. For processive and degradative proteases, determination of protein abundance changes may suffice. In contrast, for a large variety of *exo*- and endoproteases, determination of the precise cleavage site is required to infer the consequences of substrate cleavage [4]. Here, termini-centric positional proteomics approaches enable identifying precise cleavage sites on a proteome-wide scale [63]. Recent methodological improvements have increased sensitivity and throughput of unbiased N-termini enrichment by negative selection [75, 79, 80], and new software tools facilitate data analysis [87]. In contrast, enrichment of C-termini is still challenging. While limited to specific digestion enzymes and therefore to a subset of C-terminal peptides, improved methods enabling enrichment based on selective digestion and charge-based selection appear promising due to their ease of use [104].

All termini-centric approaches share the limitation of quantification based on a single peptide. This decreases the reliability of quantification compared to shotgun proteome approaches. Here, targeted [21] and DIA [75] MS methods appear as promising solutions to achieve accurate and reliable quantification across multiple samples. DIA methods additionally offer the opportunity to detect protein termini in complex matrices without selective enrichment [105] and are likely to find increased use.

Acknowledgments

Work in the author's laboratory is supported by grants of the Deutsche Forschungsgemeinschaft (D.F.G., FOR 2743-HU1756/3-1 and SFB 1403-414786233, to P.F.H.).

References

[1] Jackson MP, Hewitt EW. Cellular proteostasis: degradation of misfolded proteins by lysosomes. In: Van Oosten Hawle P, editor. Proteostasis. Essays in biochemistry, 60. London: Portland Press Ltd; 2016. p. 173–80.

[2] Majumder P, Baumeister W. Proteasomes: unfoldase-assisted protein degradation machines. Biol Chem 2019;401(1):183–99.

[3] Klein T, Eckhard U, Dufour A, Solis N, Overall CM. Proteolytic cleavage—mechanisms, function, and "Omic" approaches for a near-ubiquitous posttranslational modification. Chem Rev 2018;118(3):1137–68.

[4] Lange PF, Overall CM. Protein TAILS: when termini tell tales of proteolysis and function. Curr Opin Chem Biol 2013;17(1):73–82.

[5] Varshavsky A. N-degron and C-degron pathways of protein degradation. Proc Natl Acad Sci U S A 2019;116(2):358–66.

[6] Turk V, Turk B, Turk D. Lysosomal cysteine proteases: facts and opportunities. EMBO J 2001;20(17):4629–33.

[7] Drag M, Salvesen GS. Emerging principles in protease-based drug discovery. Nat Rev Drug Discov 2010;9(9):690–701.

[8] Grozdanić M, Vidmar R, Vizovišek M, Fonović M. Degradomics in biomarker discovery. PROTEOMICS Clin Appl 2019;13(6):1800138.

[9] Huesgen PF, Lange PF, Overall CM. Ensembles of protein termini and specific proteolytic signatures as candidate biomarkers of disease. Proteomics Clin Appl 2014;8(5–6):338–50.

[10] Puente XS, Sánchez LM, Overall CM, López-Otín C. Human and mouse proteases: a comparative genomic approach. Nat Rev Genet 2003;4(7):544–58.

[11] Dufour A, Overall CM. Missing the target: matrix metalloproteinase antitargets in inflammation and cancer. Trends Pharmacol Sci 2013;34(4):233–42.

[12] Perez-Silva JG, Espanol Y, Velasco G, Quesada V. The Degradome database: expanding roles of mammalian proteases in life and disease. Nucleic Acids Res 2016;44(D1):D351–5.

[13] Donnelly DP, Rawlins CM, DeHart CJ, Fornelli L, Schachner LF, Lin ZQ, et al. Best practices and benchmarks for intact protein analysis for top-down mass spectrometry. Nat Methods 2019;16(7):587–94.

[14] Tholey A, Becker A. Top-down proteomics for the analysis of proteolytic events—methods, applications and perspectives. Biochim Biophys Acta Mol Cell Res 2017;1864(11 Pt B):2191–9.

[15] Chen DP, Geis-Asteggiante L, Gomes FP, Ostrand-Rosenberg S, Fenselau C. Top-down proteomic characterization of truncated proteoforms. J Proteome Res 2019;18(11):4013–9.

[16] Schaffer LV, Millikin RJ, Miller RM, Anderson LC, Fellers RT, Ge Y, et al. Identification and quantification of proteoforms by mass spectrometry. Proteomics 2019;19(10):15.

[17] Aebersold R, Mann M. Mass-spectrometric exploration of proteome structure and function. Nature 2016;537(7620):347–55.

[18] López-Otín C, Overall CM. Protease degradomics: a new challenge for proteomics. Nat Rev Mol Cell Biol 2002;3(7):509–19.

[19] Overall CM, Blobel CP. In search of partners: linking extracellular proteases to substrates. Nat Rev Mol Cell Biol 2007;8(3):245–57.

[20] Prudova A, Auf Dem Keller U, Butler GS, Overall CM. Multiplex N-terminome analysis of MMP-2 and MMP-9 substrate degradomes by iTRAQ-TAILS quantitative proteomics. Mol Cell Proteomics 2010;9(5):894–911.

[21] Agard NJ, Mahrus S, Trinidad JC, Lynn A, Burlingame AL, Wells JA. Global kinetic analysis of proteolysis via quantitative targeted proteomics. Proc Natl Acad Sci 2012;109(6):1913–8.

[22] Schlage P, Kockmann T, Sabino F, Kizhakkedathu JN, Auf Dem Keller U. Matrix metalloproteinase 10 degradomics in keratinocytes and epidermal tissue identifies bioactive substrates with pleiotropic functions. Mol Cell Proteomics 2015;14(12):3234–46.

[23] Liu Y, Patricelli MP, Cravatt BF. Activity-based protein profiling: the serine hydrolases. Proc Natl Acad Sci 1999;96(26):14694–9.

[24] Cravatt BF, Wright AT, Kozarich JW. Activity-based protein profiling: from enzyme chemistry to proteomic chemistry. Annu Rev Biochem 2008;77:383–414.

[25] Fortelny N, Cox JH, Kappelhoff R, Starr AE, Lange PF, Pavlidis P, et al. Network analyses reveal pervasive functional regulation between proteases in the human protease web. PLoS Biol 2014;12(5), e1001869.

[26] Kappelhoff R, Puente XS, Wilson CH, Seth A, Lopez-Otin C, Overall CM. Overview of transcriptomic analysis of all human proteases, non-proteolytic homologs and inhibitors: organ, tissue and ovarian cancer cell line expression profiling of the human protease degradome by the CLIP-CHIP (TM) DNA microarray. Biochim Biophys Acta Mol Cell Res 2017;1864(11):2210–9.

[27] Reimann L, Wiese H, Leber Y, Schwäble AN, Fricke AL, Rohland A, et al. Myofibrillar Z-discs are a protein phosphorylation hot spot with protein kinase C (PKCα) modulating protein dynamics. Mol Cell Proteomics 2017;16(3):346–67.

[28] Zhang X, Halvorsen K, Zhang CZ, Wong WP, Springer TA. Mechanoenzymatic cleavage of the ultralarge vascular protein von Willebrand factor. Science 2009;324(5932):1330–4.

[29] Klein T, Fung S-Y, Renner F, Blank MA, Dufour A, Kang S, et al. The paracaspase MALT1 cleaves HOIL1 reducing linear ubiquitination by LUBAC to dampen lymphocyte NF-kappa B signalling. Nat Commun 2015;6.

[30] Zhang Y, Fonslow BR, Shan B, Baek MC, Yates 3rd JR. Protein analysis by shotgun/bottom-up proteomics. Chem Rev 2013;113(4):2343–94.

[31] Cox J, Hein MY, Luber CA, Paron I, Nagaraj N, Mann M. Accurate proteome-wide label-free quantification by delayed normalization and maximal peptide ratio extraction, termed MaxLFQ. Mol Cell Proteomics 2014;13(9):2513–26.

[32] Bantscheff M, Lemeer S, Savitski MM, Kuster B. Quantitative mass spectrometry in proteomics: critical review update from 2007 to the present. Anal Bioanal Chem 2012;404(4):939–65.

[33] Deng J, Erdjument-Bromage H, Neubert TA. Quantitative comparison of proteomes using SILAC. Curr Protoc Protein Sci 2019;95(1), e74.

[34] Lassowskat I, Hartl M, Hosp F, Boersema PJ, Mann M, Finkemeier I. Dimethyl-labeling-based quantification of the lysine Acetylome and proteome of plants. Methods Mol Biol 1653;2017:65–81.

[35] Shiio Y, Aebersold R. Quantitative proteome analysis using isotope-coded affinity tags and mass spectrometry. Nat Protoc 2006;1(1):139–45.

[36] Rauniyar N, Yates JR. Isobaric labeling-based relative quantification in shotgun proteomics. J Proteome Res 2014;13(12):5293–309.

[37] Li J, Van Vranken JG, Pontano Vaites L, Schweppe DK, Huttlin EL, Etienne C, et al. TMTpro reagents: a set of isobaric labeling mass tags enables simultaneous proteome-wide measurements across 16 samples. Nat Methods 2020;17(4):399–404.

[38] Thompson A, Wölmer N, Koncarevic S, Selzer S, Böhm G, Legner H, et al. TMTpro: design, synthesis, and initial evaluation of a proline-based isobaric 16-Plex tandem mass tag reagent set. Anal Chem 2019;91(24):15941–50.

[39] McAlister GC, Nusinow DP, Jedrychowski MP, Wühr M, Huttlin EL, Erickson BK, et al. MultiNotch MS3 enables accurate, sensitive, and multiplexed detection of differential expression across cancer cell line proteomes. Anal Chem 2014;86(14):7150–8.

[40] Picotti P, Aebersold R. Selected reaction monitoring-based proteomics: workflows, potential, pitfalls and future directions. Nat Methods 2012;9(6):555–66.

[41] Rauniyar N. Parallel reaction monitoring: a targeted experiment performed using high resolution and high mass accuracy mass spectrometry. Int J Mol Sci 2015;16(12):28566–81.

[42] Bourmaud A, Gallien S, Domon B. Parallel reaction monitoring using quadrupole-orbitrap mass spectrometer: principle and applications. Proteomics 2016;16(15–16):2146–59.

[43] Kusebauch U, Deutsch EW, Campbell DS, Sun Z, Farrah T, Moritz RL. Using PeptideAtlas, SRMAtlas, and PASSEL: comprehensive resources for discovery and targeted proteomics. Curr Protoc Bioinformatics 2014;46:13.25.1–8.

[44] Maclean B, Tomazela DM, Shulman N, Chambers M, Finney GL, Frewen B, et al. Skyline: an open source document editor for creating and analyzing targeted proteomics experiments. Bioinformatics 2010;26(7):966–8.

[45] Wilkins MR, Lindskog I, Gasteiger E, Bairoch A, Sanchez JC, Hochstrasser DF, et al. Detailed peptide characterization using PEPTIDEMASS—a world-wide-web-accessible tool. Electrophoresis 1997;18(3–4):403–8.

[46] Mohammed Y, Domański D, Jackson AM, Smith DS, Deelder AM, Palmblad M, et al. PeptidePicker: a scientific workflow with web interface for selecting appropriate peptides for targeted proteomics experiments. J Proteomics 2014;106:151–61.

[47] Ludwig C, Gillet L, Rosenberger G, Amon S, Collins B, Aebersold R. Data-independent acquisition-based SWATH-MS for quantitative proteomics: a tutorial. Mol Syst Biol 2018;14(8):23.

[48] Zhang FF, Ge WG, Ruan G, Cai X, Guo TN. Data-independent acquisition mass spectrometry-based proteomics and software tools: a glimpse in 2020. Proteomics 2020;20(17–18):12.

[49] Demichev V, Messner CB, Vernardis SI, Lilley KS, Ralser M. DIA-NN: neural networks and interference correction enable deep proteome coverage in high throughput. Nat Methods 2020;17(1):41.

[50] Yang Y, Liu XH, Shen CP, Lin Y, Yang PY, Qiao L. In silico spectral libraries by deep learning facilitate data-independent acquisition proteomics. Nat Commun 2020;11(1):11.

[51] Nesvizhskii AI, Aebersold R. Interpretation of shotgun proteomic data: the protein inference problem. Mol Cell Proteomics 2005;4(10):1419–40.

[52] Tam EM, Morrison CJ, Wu YI, Stack MS, Overall CM. Membrane protease proteomics: isotope-coded affinity tag MS identification of undescribed MT1-matrix metalloproteinase substrates. Proc Natl Acad Sci U S A 2004;101(18):6917–22.

[53] Dean RA, Overall CM. Proteornics discovery of metalloproteinase substrates in the cellular context by iTRAQ(TM) labeling reveals a diverse MMP-2 substrate degradome. Mol Cell Proteomics 2007;6(4):611–23.

[54] Lichtenthaler SF, Lemberg MK, Fluhrer R. Proteolytic ectodomain shedding of membrane proteins in mammals-hardware, concepts, and recent developments. EMBO J 2018;37(15).

[55] Videm P, Gunasekaran D, Schröder B, Mayer B, Biniossek ML, Schilling O. Automated peptide mapping and protein-topographical annotation of proteomics data. BMC Bioinformatics 2014;15(1):207.

[56] Ivankov DN, Bogatyreva NS, Honigschmid P, Dislich B, Hogl S, Kuhn PH, et al. QARIP: a web server for quantitative proteomic analysis of regulated intramembrane proteolysis. Nucleic Acids Res 2013;41(W1):W459–64.

[57] Agard NJ, Wells JA. Methods for the proteomic identification of protease substrates. Curr Opin Chem Biol 2009;13(5–6):503–9.

[58] Dix MM, Simon GM, Cravatt BF. Global mapping of the topography and magnitude of proteolytic events in apoptosis. Cell 2008;134(4):679–91.

[59] Niessen S, Hoover H, Gale AJ. Proteomic analysis of the coagulation reaction in plasma and whole blood using PROTOMAP. Proteomics 2011;11(12):2377–88.

[60] Dix MM, Simon GM, Wang C, Okerberg E, Patricelli MP, Cravatt BF. Functional interplay between caspase cleavage and phosphorylation sculpts the apoptotic proteome. Cell 2012;150(2):426–40.

[61] Willems P, Ndah E, Jonckheere V, Stael S, Sticker A, Martens L, et al. N-terminal proteomics assisted profiling of the unexplored translation initiation landscape in Arabidopsis thaliana. Mol Cell Proteomics 2017;16(6):1064–80.

[62] Niedermaier S, Huesgen PF. Positional proteomics for identification of secreted proteoforms released by site-specific processing of membrane proteins. Biochim Biophys Acta (BBA) Proteins Proteomics 2019;1867(12):140138.

[63] Perrar A, Dissmeyer N, Huesgen PF. New beginnings and new ends: methods for large-scale characterization of protein termini and their use in plant biology. J Exp Bot 2019;70(7):2021–38.

[64] Kaushal P, Lee C. N-terminomics—its past and recent advancements. J Proteomics 2021;233:104089.

[65] Tanco S, Gevaert K, Van Damme P. C-terminomics: targeted analysis of natural and posttranslationally modified protein and peptide C-termini. Proteomics 2015;15(5–6):903–14.

[66] Chen D, Disotuar MM, Xiong X, Wang Y, Chou DH. Selective N-terminal functionalization of native peptides and proteins. Chem Sci 2017;8(4):2717–22.

[67] Timmer JC, Salvesen GS. N-terminomics: a high-content screen for protease substrates and their cleavage sites. In: Gevaert K, Vandekerckhove J, editors. Gel-free proteomics: methods and protocols. Methods in molecular biology, 753; 2011. p. 243–55.

[68] Weeks AM, Wells JA. Subtiligase-catalyzed peptide ligation. Chem Rev 2020;120(6):3127–60.

[69] Shen P-T, Hsu J-L, Chen S-H. Dimethyl isotope-coded affinity selection for the analysis of free and blocked N-termini of proteins using LC–MS/MS. Anal Chem 2007;79(24):9520–30.

[70] Kleifeld O, Doucet A, Keller UAD, Prudova A, Schilling O, Kainthan RK, et al. Isotopic labeling of terminal amines in complex samples identifies protein N-termini and protease cleavage products. Nat Biotechnol 2010;28(3):281–U144.

[71] McDonald L, Beynon RJ. Positional proteomics: preparation of amino-terminal peptides as a strategy for proteome simplification and characterization. Nat Protoc 2006;1(4):1790–8.

[72] McDonald L, Robertson DHL, Hurst JL, Beynon RJ. Positional proteomics: selective recovery and analysis of N-terminal proteolytic peptides. Nat Methods 2005;2(12):955–7.
[73] Venne AS, Voegtle FN, Meisinger C, Sickmann A, Zahedi RP. Novel highly sensitive, specific, and straightforward strategy for comprehensive N-terminal proteomics reveals unknown substrates of the mitochondrial peptidase Icp55. J Proteome Res 2013;12(9):3823–30.
[74] Venne AS, Zahedi RP. The potential of fractional diagonal chromatography strategies for the enrichment of post-translational modifications. EuPA Open Proteom 2014;4:165–70.
[75] Weng SSH, Demir F, Ergin EK, Dirnberger S, Uzozie A, Tuscher D, et al. Sensitive determination of proteolytic proteoforms in limited microscale proteome samples. Mol Cell Proteomics 2019;18(11):2335–47.
[76] Chen L, Shan Y, Weng Y, Sui Z, Zhang X, Liang Z, et al. Hydrophobic tagging-assisted N-termini enrichment for in-depth N-terminome analysis. Anal Chem 2016;88(17):8390–5.
[77] Gevaert K, Goethals M, Martens L, Van Damme J, Staes A, Thomas GR, et al. Exploring proteomes and analyzing protein processing by mass spectrometric identification of sorted N-terminal peptides. Nat Biotechnol 2003;21(5):566–9.
[78] Hughes CS, Moggridge S, Müller T, Sorensen PH, Morin GB, Krijgsveld J. Single-pot, solid-phase-enhanced sample preparation for proteomics experiments. Nat Protoc 2019;14(1):68–85.
[79] Bundgaard L, Savickas S, UAD K. Mapping the N-terminome in tissue biopsies by PCT-TAILS. In: Apte SS, editor. Adamts proteases: methods and protocols. Methods in molecular biology, 2043. Totowa: Humana Press Inc; 2020. p. 285–96.
[80] Shema G, Nguyen MTN, Solari FA, Loroch S, Venne AS, Kollipara L, et al. Simple, scalable, and ultrasensitive tip-based identification of protease substrates. Mol Cell Proteomics 2018;17(4):826–34.
[81] Schilling O, Barre O, Huesgen PF, Overall CM. Proteome-wide analysis of protein carboxy termini: C terminomics. Nat Methods 2010;7(7):508–U33.
[82] Schilling O, Huesgen PF, Barre O, Overall CM. Identification and relative quantification of native and proteolytically generated protein C-termini from complex proteomes: C-terminome analysis. In: Cagney G, Emili A, editors. Network biology: methods and applications. Methods in molecular biology, 781; 2011. p. 59–69.
[83] Zhang Y, Li Q, Huang J, Wu Z, Huang J, Huang L, et al. An approach to incorporate multi-enzyme digestion into C-TAILS for C-terminomics studies. Proteomics 2018;18(1).
[84] Huesgen PF, Lange PF, Rogers LD, Solis N, Eckhard U, Kleifeld O, et al. LysargiNase mirrors trypsin for protein C-terminal and methylation-site identification. Nat Methods 2015;12(1):55–8.
[85] Van Damme P, Staes A, Bronsoms S, Helsens K, Colaert N, Timmerman E, et al. Complementary positional proteomics for screening substrates of endo- and exoproteases. Nat Methods 2010;7(7):512–U39.
[86] Drazic A, Myklebust LM, Ree R, Arnesen T. The world of protein acetylation. Biochim Biophys Acta (BBA) Proteins Proteomics 2016;1864(10):1372–401.
[87] Demir F, Kizhakkedathu JN, Rinschen MM, Huesgen PF. MANTI: automated annotation of protein N-termini for rapid interpretation of N-terminome data sets. Anal Chem 2021;93(13):5596–605.
[88] Uad K, Overall CM. CLIPPER: an add-on to the trans-proteomic pipeline for the automated analysis of TAILS N-terminomics data. Biol Chem 2012;393(12):1477–83.
[89] Deutsch EW, Mendoza L, Shteynberg D, Slagel J, Sun Z, Moritz RL. Trans-proteomic pipeline, a standardized data processing pipeline for large-scale reproducible proteomics informatics. Proteomics Clin Appl 2015;9(7–8):745–54.
[90] Tyanova S, Temu T, Cox J. The MaxQuant computational platform for mass spectrometry-based shotgun proteomics. Nat Protoc 2016;11(12):2301–19.
[91] Luo SY, Araya LE, Julien O. Protease substrate identification using N-terminomics. ACS Chem Biol 2019;14(11):2361–71.
[92] Fahlman RP, Chen W, Overall CM. Absolute proteomic quantification of the activity state of proteases and proteolytic cleavages using proteolytic signature peptides and isobaric tags. J Proteomics 2014;100:79–91.
[93] Lange PF, Huesgen PF, Nguyen K, Overall CM. Annotating N termini for the human proteome project: N termini and Nα-acetylation status differentiate stable cleaved protein species from degradation remnants in the human erythrocyte proteome. J Proteome Res 2014;13(4):2028–44.

[94] Soh WT, Demir F, Dall E, Perrar A, Dahms SO, Kuppusamy M, et al. ExteNDing proteome coverage with legumain as a highly specific digestion protease. Anal Chem 2020;92(4):2961–71.
[95] Wiita AP, Hsu GW, Lu CM, Esensten JH, Wells JA. Circulating proteolytic signatures of chemotherapy-induced cell death in humans discovered by N-terminal labeling. Proc Natl Acad Sci U S A 2014;111(21):7594–9.
[96] Russo AF. Overview of neuropeptides: awakening the senses? Headache 2017;57:37–46.
[97] Ramarathinam SH, Croft NP, Illing PT, Faridi P, Purcell AW. Employing proteomics in the study of antigen presentation: an update. Expert Rev Proteomics 2018;15(8):637–45.
[98] Chen R, Fulton KM, Twine SM, Li JJ. Identification of MHC peptides using mass spectrometry for neoantigen discovery and cancer vaccine development. Mass Spectrom Rev 2021;40(2):110–25.
[99] Dallas DC, Guerrero A, Parker EA, Robinson RC, Gan JN, German JB, et al. Current peptidomics: applications, purification, identification, quantification, and functional analysis. Proteomics 2015;15(5–6):1026–38.
[100] Schrader M. Origins technological development, and applications of peptidomics. In: Schrader M, Fricker L, editors. Peptidomics: methods and strategies. Methods in molecular biology, 1719. Totowa: Humana Press Inc; 2018. p. 3–39.
[101] Maes E, Oeyen E, Boonen K, Schildermans K, Mertens I, Pauwels P, et al. The challenges of peptidomics in complementing proteomics in a clinical context. Mass Spectrom Rev 2019;38(3):253–64.
[102] Azkargorta M, Escobes I, Iloro I, Elortza F. Mass spectrometric identification of endogenous peptides. In: Schrader M, Fricker L, editors. Peptidomics: methods and strategies. Methods in molecular biology, 1719. Totowa: Humana Press Inc; 2018. p. 59–70.
[103] Fricker L. Quantitative peptidomics: general considerations. In: Schrader M, Fricker L, editors. Peptidomics: methods and strategies. Methods in molecular biology, 1719. Totowa: Humana Press Inc; 2018. p. 121–40.
[104] Tsumagari K, Chang CH, Ishihama Y. Exploring the landscape of ectodomain shedding by quantitative protein terminomics. iScience 2021;24(4), 102259.
[105] Canbay V, Auf Dem Keller U. New strategies to identify protease substrates. Curr Opin Chem Biol 2021;60:89–96.

CHAPTER 11

The protease web

Wolfgang Esser-Skala and Nikolaus Fortelny
Computational Systems Biology Group, Department of Biosciences, University of Salzburg, Salzburg, Austria

Introduction

The protease web is a term used to describe large-scale, proteolysis-driven, and regulatory interactions between proteases. Proteases cleave other proteases, resulting in a modified activity of the cleaved, downstream proteases. Such proteolytic interactions are well studied in classical proteolytic cascades, for example, the complement system, coagulation, or apoptosis. However, recently developed high-throughput methods have shed light onto a far greater extent of proteolytic interactions. Furthermore, it was found that proteases cleave inhibitors of other proteases, resulting in a second mechanism of regulation between proteases. Consequently, proteolytic interactions are thought to connect protease groups in a large network—the protease web [1].

Proteolytic interactions in the protease web can greatly affect proteases in both health and disease. Target substrates of a protease can extend well beyond the direct substrates, if the protease cleaves downstream proteases or protease inhibitors. As a result, drug targeting of proteases may affect many more proteins (and biological processes) than is expected from the number of direct substrates.

Given the potential impact of the protease web, it is crucial to study how protease activity can be modulated in this network. Importantly, it is not sufficient to only identify the downstream proteases that are directly affected by a given protease. As demonstrated by classical protease cascades, protease interactions can be indirect, following multiple steps of proteases cleaving one another. The complexity arising from these indirect proteolytic interactions (across cascades and other protease groups) may only be understood if appropriate computational models are available.

Using a computational model of the protease web based on graph theory, we previously demonstrated the large connectivity in the protease web. Moreover, we used this network to predict novel indirect proteolytic pathways. Graph theory is a modeling approach primarily suited to study structural properties of large networks, for example, to define groups or hierarchies of proteases. This approach has been extensively applied to a wide range of biological networks beyond proteases. Nevertheless, graph theory is limited in its ability

Proteolytic Signaling in Health and Disease. https://doi.org/10.1016/B978-0-323-85696-6.00001-9

to model quantitative, dynamic systems, which would require parameters such as enzyme kinetics, protein concentration, or a time dimension. To enable such detailed studies of the protease web, more sophisticated computational models are required.

A key challenge for the computational modeling of biological systems is the availability of appropriate data. While biochemists have characterized individual protease interactions over decades, proteomics approaches have revealed many more substrates in recent years. These data are collected in databases such as MEROPS, which is a central resource of protease-specific data. However, the resulting data are often significantly biased toward a small number of highly studied proteases, while few or no substrates are known for the vast majority. In addition, obtaining kinetic information is much more laborious and is therefore missing for most annotated protease interactions.

Given the limitations of protease-specific data, it is interesting to identify complementary data from other biological systems. In this respect, the prediction of novel proteolytic interactions from data on general, physical protein–protein interactions is a promising approach. In the latter field, multiple types of data have been collected that are, in principle, transferable to the prediction of protease interactions. For example, coexpression, colocalization, or phylogenetic similarity of two genes can all be considered to predict their interaction. Also, the prediction of protease cleavage based on sequence motifs and other information may aid in closing the gaps in our current knowledge.

In summary, the protease web is complex, and a research approach integrating biochemical and computational methods is required to study this system. This chapter describes how classically defined protease groups are connected in the larger protease web and how computational methods enable modeling of these pervasive connections. The first sections outline traditional classification systems of proteases based on biochemical characteristics, as well as classical protease cascades that are the best-studied examples of regulatory proteolytic interactions between proteases. The subsequent sections more generally describe the regulation of proteases by protease cleavage and inhibition, and the effects of these interactions on failures of protease drug development. Moreover, these sections present results from high-throughput profiling techniques, which have led to a conceptional model of the protease web. The next section then introduces mathematical modeling approaches applicable to the proteases web, especially focusing on network biology, the data underlying protease web network models, and applications in the prediction of proteolytic pathways. This chapter concludes with a description of strategies for extending the protease web by discovering novel proteolytic interactions.

Classification systems of proteases

The protease web has key relevance for the study of proteases as it defines the downstream proteases and thus indirect substrates that may be affected when experimentally or medically targeting a protease. This biological complexity is increased by the large number of proteases, which encompass 460 gene members in humans and 525 in mouse [2]. In order to simplify this problem, proteases were originally classified in defined groups, assuming that these groups are functionally distinct. This section will summarize commonly used groupings of proteases.

Proteases (also termed peptidases or proteinases [3, 4]) catalyze the cleavage of peptide bonds in a process called proteolysis. The coarsest separation of proteases is consequently based on their molecular function, i.e., the type of bond cleaved by a protease. Endopeptidases cleave inside a protein chain, generating two protein fragments. In contrast, exopeptidases cleave terminal residues of a protein and are consequently further divided into aminopeptidases and carboxypeptidases that cleave a single residue off the N- and C-terminus, respectively. An alternative classification system is based on the nucleophile of the active site and thus comprises serine, threonine, cysteine, aspartic, glutamic, metallo, and asparagine proteases. The above classifications of proteases are annotated across organisms based on the sequence similarity and biochemical properties of proteases in the MEROPS database [4]. Finally, proteases may be grouped into families and clans based on sequence homology and three-dimensional structure, respectively.

Protease structure importantly affects the protease's amino acid preference in the substrates (sequence specificity) [5]. Proteases with similar specificity tend to cleave similar substrates and sequence specificity, thus suggesting functional similarity. For example, trypsin and similar proteases cleave specifically after lysine or arginine residues, while caspases generally cleave after aspartate residues. Grouping of proteases based on sequence specificity is thus useful and more fine-grained than proteases classes. However, reliable specificity profiles are only available for a minority of proteases [5]. In addition, other factors such as exosites also influence substrate specificity [6], limiting a grouping purely based on sequence specificity.

To functionally isolate protease groups, a more direct grouping is based on their biological functions, the biological processes in which a protease is involved. Assigning biological functions to proteases is generally difficult, as new biological roles are constantly discovered and functional annotations thus are instable [7]. Furthermore, some genes are highly multifunctional, which leads to overlaps and blurring between functionally separate groups [8]. Yet, functional groupings are commonly used to simplify biological complexity. For example, proteases are often separated into intra- and extracellular functions [9]. However, surprising relocalizations of proteases have been observed [10–15], demonstrating the uncertainty inherent in functional annotations.

Biological pathways represent a specific type of functional annotation that is particularly relevant for protease research, where pathways are commonly referred to as "proteolytic cascades." Cascades consist of consecutive cleavage events and thus form directional pathways that start by the activation of an initiator protease and end in the cleavage of proteins that perform a specific biological function. Examples of cascades include the caspases in apoptosis [16–19], the coagulation cascade that forms blood clots in wound healing [20–22], and the complement system in immunity [23–27]. Evidently, proteolytic cascades are signal transduction mechanisms, where one active protease can result in the rapid turnover of a large number of downstream substrates. Due to this capability of strong signal amplification, proteolytic cascades must be tightly regulated by multiple feedback loops, similar to other biological pathways.

The above examples demonstrate the various efforts undertaken to group proteases in order to simplify biological complexity arising from the large number of proteases, and thereby enable focused studies of individual protease groups. The protease web suggests that regulatory interactions occur between these groups. The next section will therefore outline in more depth how proteases regulate the function of other proteases.

Regulation of protease activity by proteases

Proteases are key enzymes in the regulation of protein activity. The majority of proteins undergo proteolysis through protein degradation; in addition, many proteins are proteolytically processed into stable protein fragments during maturation, signal peptide cleavage, or similar processes. These types of regulation are highly relevant for proteases, which often require zymogen activation or signal peptide cleavage to become active. This section summarizes the relevance of protein degradation, maturation, and other cleavage mechanisms for protease interactions in the protease web.

Protein degradation is a pervasive, well-understood process, and failure of protein degradation is associated with a range of diseases [28–30]. Within cells, protein degradation into small peptides is mediated by proteasome, a molecular complex with multiple subunits. A similar complex is the immunoproteasome that is relevant in antigen presentation [31]. Extracellular and mitochondrial proteins are thought to be degraded by cathepsins, proteases found in the lysosome [29, 32]. Degradation processes broadly affect most proteins, are biologically relevant, and are also highly regulated (for example, by ubiquitination). However, they do not constitute specific, regulatory interactions between proteases relevant for the protease web. Notwithstanding, some of the involved proteases do carry out specific roles (for example, cathepsins are involved in multiple intra- and extracellular processes such as antigen processing and bone remodeling), exemplifying the multifunctionality of proteases [33–35].

In contrast to protein degradation, protein processing in maturation generates stable protein species. Processing is a limited and controlled process [9, 36]. Protease processing is highly relevant for proteases, which require cleavage of the inhibitory pro-peptide to be activated [37, 38]. Similarly, signal peptides are proteolytically removed as part of protein secretion, and similar cleaved sequences are known for mitochondria and chloroplast localization

[39–41]. Protein maturation by proteolysis can extend beyond cleavage of signal peptides, for example, in angiotensin maturation, where subsequent protease cleavages generate peptides that control vasoconstriction and blood pressure [42, 43].

Beyond protein maturation, proteases further control protein functions by precise proteolytic processing. Well-studied examples of such processing are available for MMPs: These proteases carry out limited proteolysis of numerous substrates [10, 44, 45], for example, chemokines [46], laminin [47], and fibronectin [48], often altering the functional activity of a mature and functional protein. These examples demonstrate the regulatory potential inherent to proteases, which may posttranslationally modify protein function similar to kinases. Indeed, proteases contribute to cell signaling networks through their interactions with kinases [49, 50] and glycosylation [51].

Regulation of protease activity by protease inhibitor proteins

The activity of proteases must be finely controlled, as protease cleavage is often rapid and generally irreversible. This is in contrast to other posttranslational modifications such as phosphorylation that may be removed by phosphatases, which facilitates fine-tuning of downstream effects. Since aberrant protease activity can have devastating effects, including a rapid turnover of various substrates and spurious initiation of protease cascades, it is controlled by various factors. A key regulatory feature of proteases is their inhibition by dedicated proteins [19, 22, 27]. Such protease inhibitors crucially confine protease activity in time and space. Examples include stefins, which inhibit cysteine cathepsins that erroneously localize to the cytosol [33], or antithrombin, which inhibits coagulation proteases diffusing around the injury site [22].

Protease inhibitors are grouped in clans and families similar to proteases and captured in the MEROPS database [4, 52]. Fewer protease inhibitors than proteases have been identified (159 human inhibitors compared to 460 proteases), and inhibitors have been shown to often inhibit multiple proteases. As a result, inhibitors have often been assigned to classes of proteases, where kallistatins inhibit kallikreins, cystatins inhibit cathepsins, inhibitors of apoptosis inhibit caspases, and tissue inhibitors of metalloproteases (TIMPs) inhibit MMPs [9, 52]. This grouping of inhibitors based on protease classes is driven by biochemical studies, which have revealed the mechanism of action for most inhibitors in great detail. In many cases, an inhibitor reversibly binds a protease; however, irreversible inhibition mechanisms are also known. In suicide inhibition, the inhibitor is initially cleaved by the protease, resulting in a conformational change of the inhibitor and trapping of the protease. This mechanism is especially well studied in serpins, which remain covalently bound to the active site of the protease. Serpins can thereby inhibit serine and cysteine proteases, but not metalloproteases, where a water molecule acts as the nucleophile. Other examples of protease suicide inhibition are alpha-2-macroglobulin (A2M) and pregnancy zone protein [53], which engulf their target proteases upon cleavage, enabling them to inhibit proteases of any class.

Although biochemical mechanisms of protease inhibition are well understood, identifying the physiologically relevant target(s) of a protease inhibitor remains a key challenge. Protease inhibitors do not always inhibit all proteases of a given class, as inhibitory strength varies for individual proteases within one class [54] and is modified by the cellular surroundings. Furthermore, protease inhibitors can target multiple classes, as demonstrated by serpins and A2M. Protease inhibitors thereby contribute to the complexity of protease biology and the interplay of protease classes in the protease web.

Conceptualization of the protease web

Driven by the reductionist, biochemical perspective as outlined above, early protease research has provided an in-depth understanding of the biochemical mechanisms of protease function and inhibition. However, when attempting to translate these findings to medical applications in organisms, protease research required extending this approach to account for the complexity inherent to biological systems. This section describes examples of complex protease biology, which eventually led to a conceptualization of the protease web.

A striking example of the complexity of protease systems biology is drug development. Proteases are implicated in various pathologies and were considered promising drug targets [43, 49, 55, 56]. Inhibitor drugs are clinically employed to treat, for example, coagulopathies, hypertension, and infections by viruses such as HIV. However, proteases fulfill many more biological roles and thus are potential therapeutic targets for diverse diseases (especially given their well-studied biochemical mechanisms), including COVID-19 [57]. Yet, only few protease-targeting drugs have been successfully developed [43]. MMPs, which were originally thought to facilitate cancer invasion and metastasis by degrading the extracellular matrix, proved to be particularly difficult targets for drug development. Over 70 companies attempted to develop MMP inhibitors, but failed [11, 44, 58]. Clinical causes of these failures were manifold, including the design of clinical trials, patient selection, drug dosage, and drug timing. Importantly, difficulties in drug development stimulated further research, in particular, identification of additional substrates of these proteases. Indeed, substrate identification experiments revealed additional immunological roles of MMPs, explaining some of the observed side effects encountered in drug development [58].

The importance of identifying the target substrates of a protease (the "substrate repertoire" [36]) has led to the development of proteomics-based methods to specifically identify protein termini, so-called terminomics [59], which are described in detail in Chapter 11. Terminomics have provided a strong boost to protease research, enabling an unbiased, high-throughput identification of protease substrates across the full proteome. Early applications of this approach were performed in vitro, adding an activated protease to a secretome or proteome followed by the identification of protein termini that arise in a protease-treated sample compared to an untreated control. Later applications mirrored this approach in vivo, comparing proteomes extracted from samples with protease knock-out to wild-type samples. The resulting, large datasets have crucially revealed the complexity of protease biology.

In particular, terminomics profiling methods have revealed the vast extent of protease cleavage. Terminomics revealed that the majority of identified N-termini map within mature proteins, with 61% in murine skin [60], 64% in red blood cells [61], 77% in platelets [62], and 78% in dental pulp [63]. These N-terminal truncations potentially shape the activity and localization of proteins but require follow-up experiments to ascertain their biological relevance. N-terminal truncations can result from protease-unrelated processes, such as alternative translation [64] and alternative splicing [65], but meta-analyses have demonstrated the high propensity of protease cleavage [66].

By identifying the vast extent of protease cleavage, terminomics have also revealed the great extent of interactions between proteases. Protease interactions within defined classes and cascades were well established as described above, with isolated examples of cross-talk known. Terminomics have expanded our knowledge of cross-talk, identifying numerous examples of protease cleaving other proteases. In addition to direct interactions between proteases, protease can also cross-talk by cleaving protease inhibitors. In particular, examples include the cleavage of serpins by MMPs, which can deactivate the inhibitor and thus increase the activity of downstream proteases [1, 60, 67, 68].

Taken together, the vast amount of proteolytic processing and the interactions between proteases beyond classes and cascades suggest the existence of a large network. This network consists of proteolytic interactions, i.e., protease cleavage and inhibition, and was termed the "protease web" [69]. While providing a novel perspective on protease biology, the complexity arising from the protease web hinders its study with conventional, purely biochemical methods. Initial attempts to manually assemble subnetworks of the protease web resulted in initial insights [9, 10, 70, 71], but a systematic assessment of the extent of this network was only possible through computational modeling.

Network biology and the protease web

Network biology is a research field aimed at understanding the complexity of biological networks, using computational network models that enable systematic characterization of their properties. Network biology is based upon systems biology datasets, which often result in complex problems akin to the protease web, such as protein–protein interaction networks, coexpression networks, gene-regulatory networks, and metabolic networks.

Introduction to network biology

The prevailing approach to computationally model and study biological networks (in silico) is graph theory (see Fig. 1 for an explanation of basic graph-theoretical concepts). A graph is

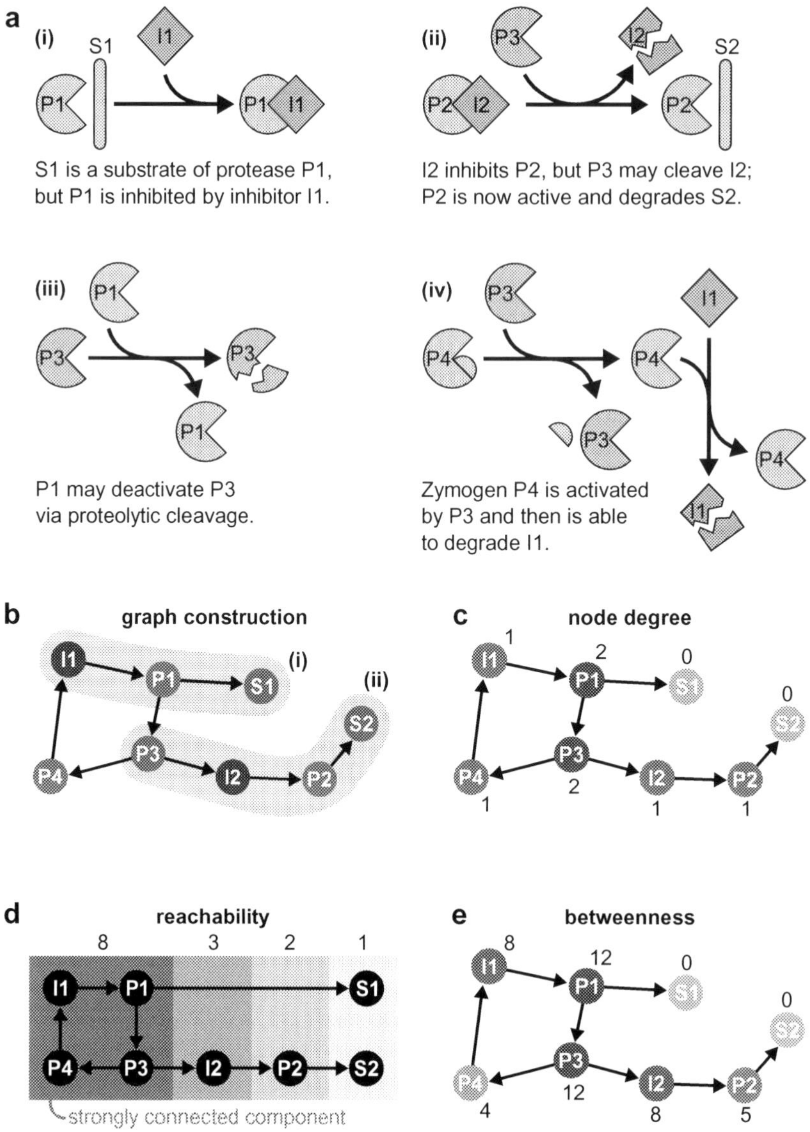

Fig. 1

Graph modeling and analysis of the protease web. (A) Verbal descriptions and corresponding schematics of exemplary proteolytic interactions comprising four proteases P1 to P4, two inhibitors I1 and I2, and two substrates S1 and S2. (B) Representation of these interactions as a graph consisting of eight nodes (*colored circles*) and eight directed edges (*black arrows*). *Gray areas* highlight parts of the graph that correspond to the indicated schematics in (A). (C–E) Illustration of graph-theoretical measures commonly used to describe biological networks: (C) The node degree counts the number of direct neighbors of each node. (D) Reachability corresponds to the number of nodes that can be reached from a given node along an arbitrary number of edges. The four nodes with a reachability of eight (i.e., I1, P1, P3, and P4) form a strongly connected component, since each of these nodes is reachable from each other. (E) Betweenness measures the number of shortest paths in which a node appears and thus represents a measure of centrality.

a mathematical representation of a network that consists of nodes and edges. Nodes are the objects in the network, and edges are their relationships (links or connections). For example, in protein–protein interaction networks [72], nodes generally represent proteins and edges represent interactions that were identified by experiments such as yeast two-hybrid screening or co-immunoprecipitation. Importantly, graph models are flexible. For example, metabolic networks [73] can be modeled such that nodes represent chemical compounds and edges represent metabolic reactions, or such that nodes represent both chemical compounds and reactions, while edges connect compounds to the reactions they are involved in. Further modeling decisions are available, for example, whether to include edge weights (to represent interaction strength) or to use directed edges (if interactions are directional).

Once a network is defined and constructed, computational algorithms can extract properties from the network in order to provide insights into its structure, thereby revealing its topology and organizational principles. For example, a common structural property is the node degree, the number of neighbors of each node (Fig. 1C). Analysis of the node degree distribution in biological networks has suggested a "scale-free" architecture of biological networks [74], where a few nodes (proteins) have many connections (interacting proteins) and many nodes have few connections. The highly connected nodes are termed "hubs" and can be identified as major regulatory proteins. Another result of structural network analyses is the small world effect [73, 75], suggesting that pairs of nodes are generally highly connected, enabling extensive cross-talk. Network analysis also identified network motifs [76, 77], small structural patterns such as feed-forward loops in transcription factor networks. While structural network properties have sparked intriguing hypotheses, they have also been criticized [78, 79]. Bias in the data [80] and modeling decisions can critically affect results, which thus need to be interpreted cautiously.

Network models of the protease web reveal structural properties

A graph model of the protease web can be constructed similarly to the described protein interaction networks, with proteins modeled as nodes and interactions modeled as edges (Fig. 1A and B). However, in contrast to general protein interactions, proteolytic interactions are inherently directional, i.e., a protease cleaves its substrate, and not vice versa. Protein interactions, where two proteins bind to form a complex, affect both proteins equally and are thus generally modeled with undirected edges, notwithstanding the existence of other directional protein interactions such as phosphorylation or ubiquitination. The first graph model of the protease web [1] thus included protease and protease inhibitors as nodes, and proteolytic interactions (protease cleavage and inhibition) as directed edges in the network (Fig. 2). The model was built on biochemical knowledge collected in databases (discussed in detail below) and enabled first systematic insights into this network.

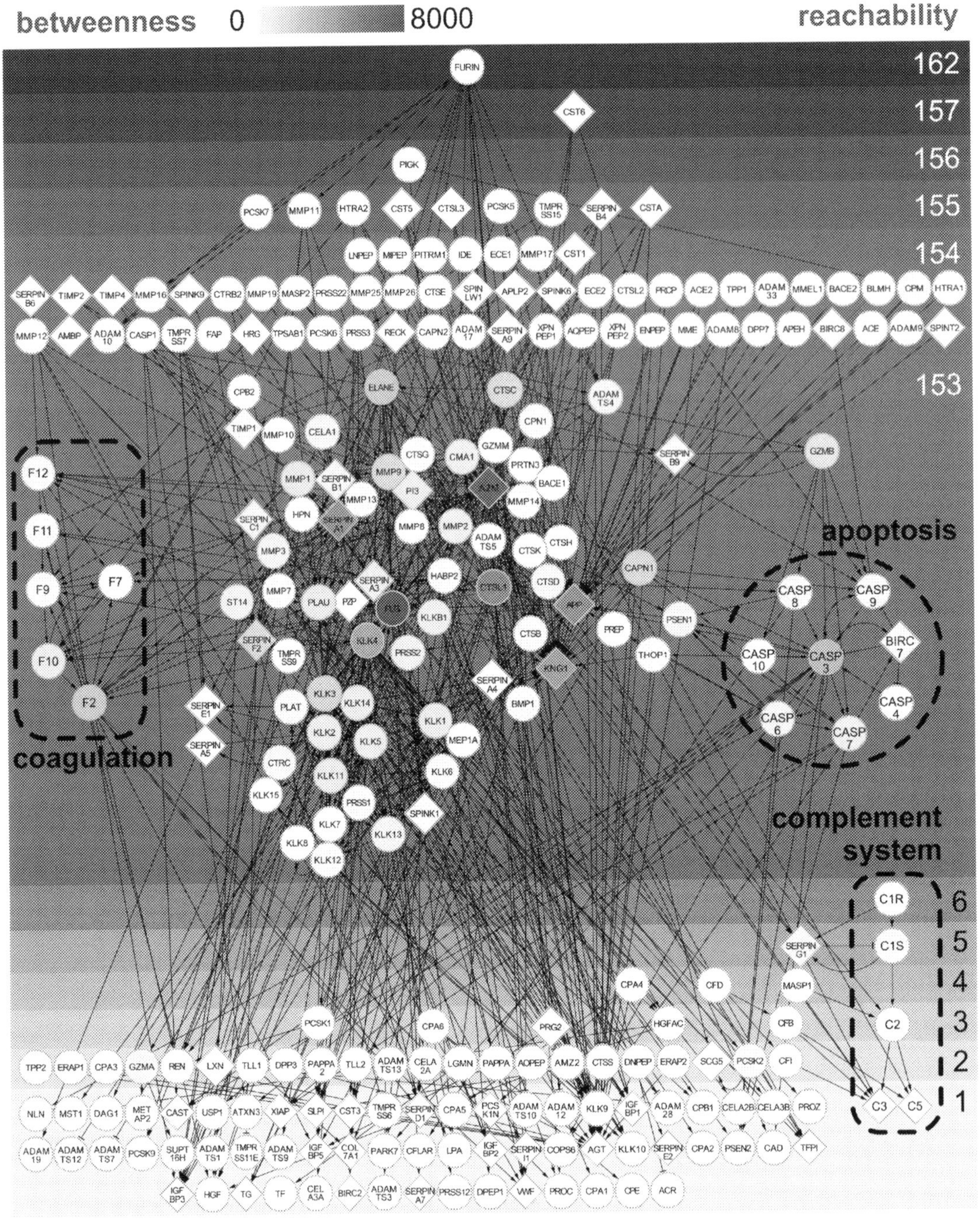

Fig. 2

The largest connected component of the human protease web, comprising 255 proteases (*circles*) and inhibitors (*diamonds*) labeled by their UniProt gene names. Each node is characterized by betweenness (*node color, shades of blue* (dark gray in the print version)) and reachability (*background color, shades of green* (gray in the print version)). *Dashed shapes* highlight three classical proteolytic cascades. This figure is a derivative of fig. 3 in Fortelny et al. [1], used under a Creative Commons Attribution 4.0 International License (https://creativecommons.org/licenses/by/4.0/). (For interpretation of the references to color in this figure legend, the reader is referred to the web version of this article.)

Topological analysis of the protease web model [1] highlighted the strong connectivity between different protease groups, classes, and cascades. This was demonstrated using mainly two types of analyses. First, component analysis revealed that the main groups of proteases are connected in a single component. (In the context of graph theory, a component is a set of nodes in which all nodes are connected to each other but not connected to any node in the remaining graph.) In the protease web, one large component was identified in addition to individual disconnected nodes. Within this component, major classes and cascades of proteases were connected by multiple interactions.

Second, reachability analysis confirmed the strong connectivity within the protease web [1]. The reachability metric is calculated for each node in the network and quantifies the number of other nodes than can be "reached" through direct and indirect connections (i.e., along an arbitrary number of edges; Fig. 1D). In the protease web, if protease A reaches protease B, then there is a series of proteolytic interactions that connect A to B. Reachability thus has an important interpretation for biology, as it reflects the regulatory potential of a protease. All reachable proteases represent downstream proteases, whose activity is potentially altered if the source protease is modified. Reachability thus informs on the regulatory hierarchy of protease interactions. The protease web indeed appeared hierarchically structured, where some nodes had a high reachability and other had low reachability. Interestingly, reachability analysis also revealed a large cluster of proteases and inhibitors that was not hierarchically structured but strongly connected such that every protease could "reach" (and thus potentially regulate) every other protease. This so-called strongly connected component encompassed 87 proteases and inhibitors, representing the bulk of well-studied classes and cascades of proteases. Importantly, this lack of hierarchy is in stark contrast to the general conception of proteases acting in cascades, as cascades are inherently structured hierarchically. Moreover, the lack of hierarchy suggests that ample feedback loops exist and thus provides an important perspective on protease regulation. Interestingly, topological analysis also revealed that protease web connectivity is to a large extent mediated by protease inhibitors, where the removal of inhibitors from the protease web model greatly reduced connectivity. Inhibitors are known key regulators of proteases and cascades, and their central role in the protease web model suggests that protease cleavage of inhibitors is pivotal in the cross-talk between proteases. In summary, the results from topological analysis of the protease web model suggest a highly connected network between proteases, with potential for complex and indirect regulation.

The strong connectivity of the protease web suggests that perturbing one protease can have extensive effects on the full network, altering the activity of numerous downstream proteases throughout the system. However, this is likely not the case in any specific biological system, where several physiological processes are expected to limit the connectivity in the protease web, such as the temporal and spatial separation of proteases. For example, proteases can be spatially confined to vesicles, intracellular compartments, or the extracellular space, and further limited by their specific expression in a subset of cell types and tissues. Proteases and their inhibitors may thereby only encounter their respective targets under very specific conditions. To computationally simulate tissue-specific networks, the full network can be pruned by removing all nodes (proteins) not expressed in a given tissue [1]. Interestingly, this analysis has demonstrated that many tissue-specific protease web models are also highly connected. Importantly, proteases do not only interact within a single tissue but also across tissues. For example, many proteases found in blood serum are expressed in the liver [81]. Beyond tissues, subcellular localization is expected to restrict the connectivity in the protease web, in particular the separation into intra- and extracellular proteases. Although annotations of subcellular localization exist for many proteases, these data are highly biased and prone to exceptions [10], hindering an in-depth analysis of cellular localization effects on protease connectivity. Taken together, network models of the protease web have demonstrated an unexpected degree of connectivity within this network, which appears to persist within the tissues but is likely fine-tuned by further regulatory processes and spatiotemporal separation of proteases. Protease web models thus reflect the regulatory "potential" of a protease, which can be exploited to gain insights into specific biological processes and systems.

Prediction of novel proteolytic pathways in the protease web

In addition to revealing system-level, structural properties of biological networks, network models can also provide insights into specific biological processes. For example, predictions can be made for specific nodes by "node labeling," where a node is labeled based on the labels of its neighbors, for example, in the prediction of gene functions based on the "guilt-by-association" principle [82, 83]. A second approach that is highly relevant to the protease web is pathway prediction [84]. Pathways are consecutive molecular interactions, where an input signal is transduced to an output signal, with protease cascades representing classically identified pathways of proteases. Pathway prediction is an important application for biological research, because in contrast to structural network properties, pathways can be biochemically validated by perturbing pathway elements, and because signal transduction through pathways is generally considered as a key molecular mechanism governing biological systems.

Pathway prediction identifies the series of nodes and edges (a path) that needs to be traversed through the network to reach from a given "start" to a given "end" node, where the number of traversed edges is the length of the path. In the protease web, pathfinding can be applied to identify all paths from a given protease to a specific target protein. If a path of length one exists, then the target protein is a direct substrate of the protease. If paths of length greater than one exist, then the target protein is an indirect substrate and the nodes between the protease and the substrates represent intermediary proteases or inhibitors. Importantly, multiple direct and indirect paths of different lengths can exist between two nodes. To identify paths in networks, dedicated graph algorithms have been developed [85]. The "shortest path" is of particular interest, since only the shortest path(s) between two nodes may be computed efficiently, i.e., without the requirement to enumerate all possible paths within a network. This is especially relevant for strongly connected components (such as the protease web), where a large number of potential paths can be identified. Moreover, the determination of all shortest paths allows further characterization of a network, for example, by calculating node betweenness (Fig. 1E).

The first application of pathfinding in the protease web helped in the understanding of Mmp8-dependent cleavage of the chemokine Cxcl5 [1]. Loss of Cxcl5 cleavage was observed in a murine in vivo airpouch model after Mmp8 knockout [86], but a direct cleavage of Cxcl5 by Mmp8 could be ruled out, as Mmp8 did not cleavage Cxcl5 at the site observed in vivo when tested in vitro. Pathfinding in the protease web model revealed a proteolytic pathway starting with cleavage and inactivation of the protease inhibitor Slpi by Mmp8. Slpi deactivation liberates downstream proteases such as neutrophil elastase (Elane), which was ultimately shown to cleave Cxcl5 at the position observed in vivo. By the definition of a start point (Mmp8) and end point (Elane), it was thus possible to identify paths through the network that reflect biological pathways.

Protease pathway prediction is particularly relevant in the analysis of in vivo terminomics data, for example, when comparing cells with a protease knock-out to the wild type, where indirect connections can make it difficult to assign proteases to the cleaved proteins identified in vivo. By defining the perturbed protease as the start point and the observed cleaved proteins as end points, pathfinding can predict and suggest a series of indirect interactions that explain how the protease could have led to the observed cleavage sites. Such analyses may be performed by the freely available PathFINDer software [87].

Data and databases of protease interactions

Availability of relevant data is key to an accurate and relevant model of the protease web. In protease research, the MEROPS database (https://www.ebi.ac.uk/merops/) represents the main database covering information on protease targets, families, classes, and inhibition [4]. MEROPS is focused on annotating proteases based on genomic data of various organisms and also stores annotated protease-substrate and protease-inhibitor pairs. Other databases of protease cleavage data include CutDB [88], DegraBase [89], CASBAH [90], and TOPPR [91], each focusing on a specific research question within protease biology. Generally, data are cross-referenced within MEROPS, which is thus the most comprehensive resource for the systematic study of the protease web. The TopFIND knowledgebase (https://topfind.clip.msl.ubc.ca) complements MEROPS by capturing data on N- and C-termini (where the responsible protease may not be known) [92]. TopFIND includes data from MEROPS and additionally provides software for the analysis of protein termini and the protease web, e.g., pathfinding via the PathFINDer tool [87].

A key aspect of data storage is annotation with metadata, which in biochemical research generally refers to the experimental conditions relevant for a specific finding. For example, metadata for cleavage of a given protease substrate may include the biochemical method used or the biological system of the experiment (in vivo or in vitro). In TopFIND [92], metadata are annotated as "evidences" that are linked to each data point in the database. This enables a detailed filtering of data for a specific biological problem. Indeed, experimental details are essential consideration points when aggregating large data into network analyses, which can otherwise be highly biased. For example, a large number of protease-substrate pairs have been annotated based on detailed low-throughput studies, which are generally considered to be of high quality and reliability. More recent high-throughput studies provide much greater coverage (many protease-substrate pairs), but possibly include a greater number of false positives. Similarly, kinetic parameters of protease cleavage (such as Michaelis–Menten constants) would be highly relevant for constructing realistic network models: Since they represent the "strength" of the links in such models, they could be incorporated as edge weights. Yet kinetic parameters are rarely available for most cleavage data. Finally, the physiological relevance of protease cleavages has to be judged based on experimental metadata: In vitro experiments generally provide greater detail of biochemical parameters, but their relevance in the specific biological systems is less clear.

In summary, knowledge of protease biology is well organized and documented in dedicated databases enabling a rapid retrieval of proteolytic interactions for protease web analysis. These databases also include metadata to annotate the complexity of biochemical parameters relevant for the study of proteases. Nevertheless, the comprehensive study of the protease web is difficult due to the lack of knowledge of proteolytic interactions. Indeed, it was shown that for many of the observed N-termini in TopFIND, no responsible protease is known [66]. Protease web data is thus expected to be largely incomplete, and more comprehensive knowledge will improve understanding of the connectivity in this network. Different biochemical and computational approaches exist to identify proteolytic interactions, which are outlined in the following section.

Identification of protease interactions

Biochemical methods used to identify cleaved proteins and thus protease substrates have a long-standing history. Basic methods such as gel electrophoresis and Western blots are widely used but do not identify the precise cleavage site and cannot distinguish smaller truncations. Edman degradation [93] enables a precise identification of the N-terminal amino acids of a peptide, but is limited to short peptides of about 30 amino acids. Mass spectrometry is a highly sensitive and parallelizable method and has enabled the high-throughput quantification of proteins in proteomics analyses. In particular, it has led to the development of methodologies to identify protease cleavage sites, which focus on the enrichment of N- or C-terminal peptides [59] and are therefore referred to as terminomics (see Chapter 11).

For the purpose of identifying protease substrates, terminomics techniques are generally applied by comparing samples (e.g., proteomes) where a protease is present and active to samples where the same protease is deactivated or absent.

Computational prediction (for instance, of protein–protein interactions) complements and guides biochemical experiments in identifying protease substrates [94, 95]. Applied to the identification of protease cleavage, in silico predictions are commonly based on substrate specificity of the protease and often include other features such as secondary structure, native disorder, and/or solvent accessibility of the substrate [96]. Originally based on simpler algorithms such as regular expressions [97], more recent methods combine sets of features using machine learning [98]. In silico cleavage predictors are often focused on specific proteases such as caspases and granzyme B [99–101] or calpains [102]. Importantly, in silico prediction algorithms have reported high prediction performance but their application in biology and biochemical validation is rare, and computational tools based on protease specificity may thus be overly optimistic. In addition, protease cleavage predictions require robust specificity profiles, which depend on the number of known protease substrates and are thus not available for a large number of proteases [5]. Specialized proteomics approaches have been developed to obtain high-quality substrate specificity profiles of proteases [103, 104], but only a small number of proteases have been characterized by these methods.

To extend prediction approaches based on protease specificity and structure, further types of data can be used to predict protein–protein interactions in the protease web, including coexpression, colocalization, coannotation/functional similarity, and evolutionary similarity. Coexpression is based on the assumption that two interacting proteins are likely expressed in the same tissues or cells at the same time, such that they may interact. Coexpression is often measured by correlating gene expression patterns. This correlation was consistently observed in yeast [105–107], especially in protein complexes [108], but with weaker signal in humans [107, 109]. Similar to coexpression, colocalization of two proteins in the same cellular compartment is thought to suggest an interaction [110]. Annotations of cellular localization are stored in databases such as LocDB [111], the Human Protein Atlas [112], and Gene Ontology [113]. However, such annotations are incomplete and biased, making it difficult to predict interactions for the large number of sparsely annotated proteins. Beyond annotation of cellular localization, coannotation of two genes in similar biological processes is also used to predict interactions. Coannotation suggests functional similarity (i.e., activity in the same biological process) and is thought to increase the probability that two proteins interact [114]. Although coannotation may appear promising [115], it is confounded with protein interaction annotation [7], as both can be based on the same experiment, making the assessment of prediction accuracy overoptimistic. Finally, evolutionary (or phylogenetic) similarity is used to predict interactions: If two genes show similar evolution (i.e., their absence or presence in genomes of different organisms is coordinated), this may indicate that both are required to fulfill certain functions. Evolutionary similarity is calculated by correlating phylogenetic profiles (measures of absence/presence) of genes across multiple organisms [116], which is possible across all genes. Importantly, none of the types of data explained above are expected to perfectly predict interaction, but rather increase the probability that two proteins interact. To further boost prediction power, the types of data mentioned above are often combined using machine learning [117–119]. The resulting predictors often report high accuracy [119], but similar to predictions based on protease specificity, these estimates are likely overoptimistic [120] and generally not validated biochemically.

The above-mentioned measures are well studied in the prediction of general protein–protein interactions, and could also be applied to predict interactions of proteases and their substrates. However, to date research in this field is limited. Similarly, proteases and their inhibitors are expected to be coexpressed, colocalized, active in similar pathways, and also to coevolve. While comprehensive studies of the prediction power of these approaches in the protease web (as they exist for protein–protein interactions) are limited, a first attempt in the prediction of protease-inhibitor interactions from a large set of different data demonstrated limitations in these predictions [81]. Interestingly, this analysis demonstrated a lack of coexpression between proteases and their inhibitors—an often assumed feature based on biological rationale. Indeed, it was found that inhibitions occur between proteases in the blood and inhibitors localized in a given tissue, which limits the power of coexpression- and colocalization-based predictions. Predictions of novel protease web interactions thus remain challenging, both for sequence-based and for complementary approaches. Nevertheless, in silico predictions are a promising addition to biochemical experiments, which are much more labor-intensive.

Outlook

Proteases are major regulatory enzymes and consequently drug targets—in humans as well as viral diseases such as AIDS and COVID-19. For both basic science and drug development, proteolytic interactions in the protease web remain a large source of uncertainty. The lack of a comprehensive understanding of how proteases interact on a large scale makes it difficult to identify the specific functions of each protease and consequently hinders protease drug targeting. Network models have shed light onto the highly connected structure of the protease web and enabled the prediction of novel proteolytic pathways connecting protease classes and cascades, where protease inhibitors often serve as connecting hubs between different pathways. Databases of proteolytic interactions relevant for the study of the protease web exist, but data are likely incomplete. In the future, data on protease interactions will become more comprehensive due to developments in biochemical and computational methods to identify protease substrates, improving the model of the protease web and our understanding of its function. Beyond the protease web, an integration of protease interactions with other signaling interactions (such as phosphorylation and physical complexes) will further shed light onto the functions of proteases and their embedding in the general landscape of regulatory biology.

Acknowledgment

The authors thank Chris Overall for critical reading of the manuscript.

References

[1] Fortelny N, Cox JH, Kappelhoff R, et al. Network analyses reveal pervasive functional regulation between proteases in the human protease web. PLoS Biol 2014;12(5). https://doi.org/10.1371/journal.pbio.1001869, e1001869.
[2] Puente XS, Sánchez LM, Overall CM, López-Otín C. Human and mouse proteases: a comparative genomic approach. Nat Rev Genet 2003;4(7):544–58. https://doi.org/10.1038/nrg1111.
[3] Barrett AJ, McDonald JK. Nomenclature: protease, proteinase and peptidase. Biochem J 1986;237(3):935. https://doi.org/10.1042/bj2370935.
[4] Rawlings ND, Barrett AJ, Finn R. Twenty years of the MEROPS database of proteolytic enzymes, their substrates and inhibitors. Nucleic Acids Res 2015;44(D1):D343–50. https://doi.org/10.1093/nar/gkv1118.
[5] Fuchs JE, von Grafenstein S, Huber RG, Kramer C, Liedl KR. Substrate-driven mapping of the degradome by comparison of sequence logos. PLoS Comput Biol 2013;9(11). https://doi.org/10.1371/journal.pcbi.1003353, e1003353.
[6] Overall CM. Molecular determinants of metalloproteinase substrate specificity. Mol Biotechnol 2002;22(1):51–86. https://doi.org/10.1385/MB:22:1:051.
[7] Gillis J, Pavlidis P. Assessing identity, redundancy and confounds in gene ontology annotations over time. Bioinformatics 2013;29(4):476–82. https://doi.org/10.1093/bioinformatics/bts727.
[8] Gillis J, Pavlidis P. The impact of multifunctional genes on "guilt by association" analysis. PLoS One 2011;6(2). https://doi.org/10.1371/journal.pone.0017258, e17258.
[9] Doucet A, Butler GS, Rodríguez D, Prudova A, Overall CM. Metadegradomics toward in vivo quantitative degradomics of proteolytic post-translational modifications of the cancer proteome. Mol Cell Proteomics 2008;7(10):1925–51. https://doi.org/10.1074/mcp.R800012-MCP200.
[10] Butler GS, Overall CM. Updated biological roles for matrix metalloproteinases and new "intracellular" substrates revealed by degradomics. Biochemistry 2009;48(46):10830–45. https://doi.org/10.1021/bi901656f.
[11] Butler GS, Overall CM. Proteomic identification of multitasking proteins in unexpected locations complicates drug targeting. Nat Rev Drug Discov 2009;8(12):935–48. https://doi.org/10.1038/nrd2945.
[12] Goulet B, Baruch A, Moon N-S, et al. A cathepsin L isoform that is devoid of a signal peptide localizes to the nucleus in S phase and processes the CDP/Cux transcription factor. Mol Cell 2004;14(2):207–19. https://doi.org/10.1016/S1097-2765(04)00209-6.
[13] Kwan JA, Schulze CJ, Wang W, et al. Matrix metalloproteinase-2 (MMP-2) is present in the nucleus of cardiac myocytes and is capable of cleaving poly (ADP-ribose) polymerase (PARP) in vitro. FASEB J 2004;18(6):690–2. https://doi.org/10.1096/fj.02-1202fje.
[14] Golubkov VS, Boyd S, Savinov AY, et al. Membrane type-1 matrix metalloproteinase (MT1-MMP) exhibits an important intracellular cleavage function and causes chromosome instability. J Biol Chem 2005;280(26):25079–86. https://doi.org/10.1074/jbc.M502779200.
[15] Marchant DJ, Bellac CL, Moraes TJ, et al. A new transcriptional role for matrix metalloproteinase-12 in antiviral immunity. Nat Med 2014;20(5):493–502. https://doi.org/10.1038/nm.3508.
[16] Stennicke HR, Salvesen GS. Properties of the caspases. Biochim Biophys Acta BBA Protein Struct Mol Enzymol 1998;1387(1–2):17–31. https://doi.org/10.1016/S0167-4838(98)00133-2.
[17] Thornberry NA, Lazebnik Y. Caspases: enemies within. Science 1998;281(5381):1312–6. https://doi.org/10.1126/science.281.5381.1312.
[18] Slee EA, Adrain C, Martin SJ. Serial killers: ordering caspase activation events in apoptosis. Cell Death Differ 1999;6(11):1067–74. https://doi.org/10.1038/sj.cdd.4400601.
[19] Ashkenazi A, Salvesen G. Regulated cell death: signaling and mechanisms. Annu Rev Cell Dev Biol 2014;30(1):337–56. https://doi.org/10.1146/annurev-cellbio-100913-013226.

[20] Davie EW, Ratnoff OD. Waterfall sequence for intrinsic blood clotting. Science 1964;145(3638):1310–2. https://doi.org/10.1126/science.145.3638.1310.
[21] Macfarlane RG. An enzyme cascade in the blood clotting mechanism, and its function as a biochemical amplifier. Nature 1964;202(4931):498–9. https://doi.org/10.1038/202498a0.
[22] Adams RLC, Bird RJ. Review article: coagulation cascade and therapeutics update: relevance to nephrology. Part 1: overview of coagulation, thrombophilias and history of anticoagulants. Nephrol Ther 2009;14(5):462–70. https://doi.org/10.1111/j.1440-1797.2009.01128.x.
[23] Sim RB, Reboul A, Arlaud GJ, Villiers CL, Colomb MG. Interaction of 125I-labelled complement subcomponents C1r and C1s with protease inhibitors in plasma. FEBS Lett 1979;97(1):111–5. https://doi.org/10.1016/0014-5793(79)80063-0.
[24] Muller-Eberhard HJ. Molecular organization and function of the complement system. Annu Rev Biochem 1988;57(1):321–47. https://doi.org/10.1146/annurev.bi.57.070188.001541.
[25] Matsushita M, Thiel S, Jensenius JC, Terai I, Fujita T. Proteolytic activities of two types of mannose-binding lectin-associated serine protease. J Immunol 2000;165(5):2637–42.
[26] Sarma JV, Ward PA. The complement system. Cell Tissue Res 2011;343(1):227–35. https://doi.org/10.1007/s00441-010-1034-0.
[27] Ehrnthaller C, Ignatius A, Gebhard F, Huber-Lang M. New insights of an old defense system: structure, function, and clinical relevance of the complement system. Mol Med 2011;17(3–4):317–29. https://doi.org/10.2119/molmed.2010.00149.
[28] Goldberg AL. Protein degradation and protection against misfolded or damaged proteins. Nature 2003;426(6968):895–9. https://doi.org/10.1038/nature02263.
[29] Ciechanover A. Intracellular protein degradation: from a vague idea through the lysosome and the ubiquitin–proteasome system and onto human diseases and drug targeting. Cell Death Differ 2005;12(9):1178–90. https://doi.org/10.1038/sj.cdd.4401692.
[30] Rubinsztein DC. The roles of intracellular protein-degradation pathways in neurodegeneration. Nature 2006;443(7113):780–6. https://doi.org/10.1038/nature05291.
[31] Morel S, Lévy F, Burlet-Schiltz O, et al. Processing of some antigens by the standard proteasome but not by the immunoproteasome results in poor presentation by dendritic cells. Immunity 2000;12(1):107–17. https://doi.org/10.1016/S1074-7613(00)80163-6.
[32] Luzio JP, Pryor PR, Bright NA. Lysosomes: fusion and function. Nat Rev Mol Cell Biol 2007;8(8):622–32. https://doi.org/10.1038/nrm2217.
[33] Turk V, Turk B, Turk D. Lysosomal cysteine proteases: facts and opportunities. EMBO J 2001;20(17):4629–33. https://doi.org/10.1093/emboj/20.17.4629.
[34] Yasuda Y, Kaleta J, Brömme D. The role of cathepsins in osteoporosis and arthritis: rationale for the design of new therapeutics. Adv Drug Deliv Rev 2005;57(7):973–93. https://doi.org/10.1016/j.addr.2004.12.013.
[35] Mohamed MM, Sloane BF. Cysteine cathepsins: multifunctional enzymes in cancer. Nat Rev Cancer 2006;6(10):764–75. https://doi.org/10.1038/nrc1949.
[36] López-Otín C, Overall CM. Protease degradomics: a new challenge for proteomics. Nat Rev Mol Cell Biol 2002;3(7):509–19. https://doi.org/10.1038/nrm858.
[37] Kassell B, Kay J. Zymogens of proteolytic enzymes. Science 1973;180(4090):1022–7.
[38] Khan AR, James MNG. Molecular mechanisms for the conversion of zymogens to active proteolytic enzymes. Protein Sci 1998;7(4):815–36. https://doi.org/10.1002/pro.5560070401.
[39] Schatz G, Dobberstein B. Common principles of protein translocation across membranes. Science 1996;271(5255):1519–26. https://doi.org/10.1126/science.271.5255.1519.
[40] Paetzel M, Karla A, Strynadka NCJ, Dalbey RE. Signal peptidases. Chem Rev 2002;102(12):4549–80. https://doi.org/10.1021/cr010166y.
[41] Rapoport TA. Protein translocation across the eukaryotic endoplasmic reticulum and bacterial plasma membranes. Nature 2007;450(7170):663–9. https://doi.org/10.1038/nature06384.

[42] Welches WR, Bridget Brosnihan K, Ferrario CM. A comparison of the properties and enzymatic activities of three angiotensin processing enzymes: angiotensin converting enzyme, prolyl endopeptidase and neutral endopeptidase 24.11. Life Sci 1993;52(18):1461–80. https://doi.org/10.1016/0024-3205(93)90108-F.

[43] Turk B. Targeting proteases: successes, failures and future prospects. Nat Rev Drug Discov 2006;5(9):785–99. https://doi.org/10.1038/nrd2092.

[44] Overall CM, Kleifeld O. Validating matrix metalloproteinases as drug targets and anti-targets for cancer therapy. Nat Rev Cancer 2006;6(3):227–39. https://doi.org/10.1038/nrc1821.

[45] Lange PF, Overall CM. Protein TAILS: when termini tell tales of proteolysis and function. Curr Opin Chem Biol 2013;17(1):73–82. https://doi.org/10.1016/j.cbpa.2012.11.025.

[46] McQuibban GA, Gong J-H, Tam EM, McCulloch CAG, Clark-Lewis I, Overall CM. Inflammation dampened by gelatinase a cleavage of monocyte chemoattractant protein-3. Science 2000;289(5482):1202–6. https://doi.org/10.1126/science.289.5482.1202.

[47] Giannelli G, Falk-Marzillier J, Schiraldi O, Stetler-Stevenson WG, Quaranta V. Induction of cell migration by matrix metalloprotease-2 cleavage of laminin-5. Science 1997;277(5323):225–8. https://doi.org/10.1126/science.277.5323.225.

[48] Fukai F, Ohtaki M, Fujii N, et al. Release of biological activities from quiescent fibronectin by a conformational change and limited proteolysis by matrix metalloproteinases. Biochemistry 1995;34(36):11453–9. https://doi.org/10.1021/bi00036a018.

[49] López-Otín C, Bond JS. Proteases: multifunctional enzymes in life and disease. J Biol Chem 2008;283(45):30433–7. https://doi.org/10.1074/jbc.R800035200.

[50] López-Otín C, Hunter T. The regulatory crosstalk between kinases and proteases in cancer. Nat Rev Cancer 2010;10(4):278–92. https://doi.org/10.1038/nrc2823.

[51] Kötzler MP, Withers SG. Proteolytic cleavage driven by glycosylation. J Biol Chem 2016;291(1):429–34. https://doi.org/10.1074/jbc.C115.698696.

[52] Rawlings ND, Tolle DP, Barrett AJ. Evolutionary families of peptidase inhibitors. Biochem J 2004;378(3):705. https://doi.org/10.1042/BJ20031825.

[53] Jensen PEH, Stigbrand T. Differences in the proteinase inhibition mechanism of human α2-macroglobulin and pregnancy zone protein. Eur J Biochem 1992;210(3):1071–7. https://doi.org/10.1111/j.1432-1033.1992.tb17513.x.

[54] Luo L-Y, Jiang W. Inhibition profiles of human tissue kallikreins by serine protease inhibitors. Biol Chem 2006;387(6):813–6. https://doi.org/10.1515/BC.2006.103.

[55] Overall CM, Blobel CP. In search of partners: linking extracellular proteases to substrates. Nat Rev Mol Cell Biol 2007;8(3):245–57. https://doi.org/10.1038/nrm2120.

[56] Drag M, Salvesen GS. Emerging principles in protease-based drug discovery. Nat Rev Drug Discov 2010;9(9):690–701. https://doi.org/10.1038/nrd3053.

[57] Stopsack KH, Mucci LA, Antonarakis ES, Nelson PS, Kantoff PW. TMPRSS2 and COVID-19: serendipity or opportunity for intervention? Cancer Discov 2020;10(6):779–82. https://doi.org/10.1158/2159-8290.CD-20-0451.

[58] Dufour A, Overall CM. Missing the target: matrix metalloproteinase antitargets in inflammation and cancer. Trends Pharmacol Sci 2013;34(4):233–42. https://doi.org/10.1016/j.tips.2013.02.004.

[59] Kleifeld O, Doucet A. Auf dem Keller U, et al. isotopic labeling of terminal amines in complex samples identifies protein N-termini and protease cleavage products. Nat Biotechnol 2010;28(3):281–8. https://doi.org/10.1038/nbt.1611.

[60] Keller Uad, Prudova A, Eckhard U, Fingleton B, Overall CM. Systems-level analysis of proteolytic events in increased vascular permeability and complement activation in skin inflammation. Sci Signal 2013;6(258):rs2. https://doi.org/10.1126/scisignal.2003512.

[61] Lange PF, Huesgen PF, Nguyen K, Overall CM. Annotating N termini for the human proteome project: N termini and Nα-acetylation status differentiate stable cleaved protein species from degradation remnants in the human erythrocyte proteome. J Proteome Res 2014;13(4):2028–44. https://doi.org/10.1021/pr401191w.

[62] Prudova A, Serrano K, Eckhard U, Fortelny N, Devine DV, Overall CM. TAILS N-terminomics of human platelets reveals pervasive metalloproteinase dependent proteolytic processing in storage. Blood 2014;124(26):e49–60. https://doi.org/10.1182/blood-2014-04-569640.

[63] Eckhard U, Marino G, Abbey SR, Tharmarajah G, Matthew I, Overall CM. The human dental pulp proteome and N-terminome: levering the unexplored potential of semitryptic peptides enriched by TAILS to identify missing proteins in the human proteome project in underexplored tissues. J Proteome Res 2015;14(9):3568–82. https://doi.org/10.1021/acs.jproteome.5b00579.

[64] Wan J, Qian S-B. TISdb: a database for alternative translation initiation in mammalian cells. Nucleic Acids Res 2014;42(Database issue):D845–50. https://doi.org/10.1093/nar/gkt1085.

[65] Flicek P, Amode MR, Barrell D, et al. Ensembl 2014. Nucleic Acids Res 2014;42(D1):D749–55. https://doi.org/10.1093/nar/gkt1196.

[66] Fortelny N, Pavlidis P, Overall CM. The path of no return—truncated protein N-termini and current ignorance of their genesis. Proteomics 2015;15(14):2547–52. https://doi.org/10.1002/pmic.201500043.

[67] Desrochers PE, Jeffrey JJ, Weiss SJ. Interstitial collagenase (matrix metalloproteinase-1) expresses serpinase activity. J Clin Invest 1991;87(6):2258–65.

[68] Mast AE, Enghild JJ, Nagase H, Suzuki K, Pizzo SV, Salvesen G. Kinetics and physiologic relevance of the inactivation of alpha 1-proteinase inhibitor, alpha 1-antichymotrypsin, and antithrombin III by matrix metalloproteinases-1 (tissue collagenase), -2 (72-kDa gelatinase/type IV collagenase), and -3 (stromelysin). J Biol Chem 1991;266(24):15810–6.

[69] Overall C, Dean R. Degradomics: systems biology of the protease web. Pleiotropic roles of MMPs in cancer. Cancer Metastasis Rev 2006;25(1):69–75. https://doi.org/10.1007/s10555-006-7890-0.

[70] Beaufort N, Plaza K, Utzschneider D, et al. Interdependence of kallikrein-related peptidases in proteolytic networks. Biol Chem 2010;391(5):581–7. https://doi.org/10.1515/bc.2010.055.

[71] Mason SD, Joyce JA. Proteolytic networks in cancer. Trends Cell Biol 2011;21(4):228–37. https://doi.org/10.1016/j.tcb.2010.12.002.

[72] Stelzl U, Worm U, Lalowski M, et al. A human protein-protein interaction network: a resource for annotating the proteome. Cell 2005;122(6):957–68. https://doi.org/10.1016/j.cell.2005.08.029.

[73] Jeong H, Tombor B, Albert R, Oltvai ZN, Barabási A-L. The large-scale organization of metabolic networks. Nature 2000;407(6804):651–4. https://doi.org/10.1038/35036627.

[74] Barabási A-L, Albert R. Emergence of scaling in random networks. Science 1999;286(5439):509–12.

[75] Wagner A, Fell DA. The small world inside large metabolic networks. Proc R Soc Lond B Biol Sci 2001;268(1478):1803–10. https://doi.org/10.1098/rspb.2001.1711.

[76] Milo R, Shen-Orr S, Itzkovitz S, Kashtan N, Chklovskii D, Alon U. Network motifs: simple building blocks of complex networks. Science 2002;298(5594):824–7. https://doi.org/10.1126/science.298.5594.824.

[77] Shen-Orr SS, Milo R, Mangan S, Alon U. Network motifs in the transcriptional regulation network of Escherichia coli. Nat Genet 2002;31(1):64–8. https://doi.org/10.1038/ng881.

[78] Arita M. The metabolic world of Escherichia coli is not small. Proc Natl Acad Sci U S A 2004;101(6):1543–7. https://doi.org/10.1073/pnas.0306458101.

[79] Lima-Mendez G, van Helden J. The powerful law of the power law and other myths in network biology. Mol Biosyst 2009;5(12):1482–93. https://doi.org/10.1039/B908681A.

[80] Gillis J, Ballouz S, Pavlidis P. Bias tradeoffs in the creation and analysis of protein–protein interaction networks. J Proteomics 2014;100:44–54. https://doi.org/10.1016/j.jprot.2014.01.020.

[81] Fortelny N, Butler GS, Overall CM, Pavlidis P. Protease-inhibitor interaction predictions: Lessons on the complexity of protein-protein interactions. Mol Cell Proteomics 2017. https://doi.org/10.1074/mcp.M116.065706, mcp.M116.065706.

[82] Uetz P, Giot L, Cagney G, et al. A comprehensive analysis of protein–protein interactions in Saccharomyces cerevisiae. Nature 2000;403(6770):623–7. https://doi.org/10.1038/35001009.

[83] Mostafavi S, Ray D, Warde-Farley D, Grouios C, Morris Q. GeneMANIA: a real-time multiple association network integration algorithm for predicting gene function. Genome Biol 2008;9(Suppl 1):S4. https://doi.org/10.1186/gb-2008-9-s1-s4.

[84] Vinayagam A, Stelzl U, Foulle R, et al. A directed protein interaction network for investigating intracellular signal transduction. Sci Signal 2011;4(189):rs8. https://doi.org/10.1126/scisignal.2001699.

[85] Kim BJ, Yoon CN, Han SK, Jeong H. Path finding strategies in scale-free networks. Phys Rev E 2002;65(2). https://doi.org/10.1103/PhysRevE.65.027103, 027103.

[86] Tester AM, Cox JH, Connor AR, et al. LPS responsiveness and neutrophil chemotaxis in vivo require PMN MMP-8 activity. PLoS One 2007;2(3). https://doi.org/10.1371/journal.pone.0000312, e312.

[87] Fortelny N, Yang S, Pavlidis P, Lange PF, Overall CM. Proteome TopFIND 3.0 with TopFINDer and PathFINDer: database and analysis tools for the association of protein termini to pre- and post-translational events. Nucleic Acids Res 2015;43(D1):D290–7. https://doi.org/10.1093/nar/gku1012.

[88] Igarashi Y, Eroshkin A, Gramatikova S, et al. CutDB: a proteolytic event database. Nucleic Acids Res 2007;35(Database issue):D546. https://doi.org/10.1093/nar/gkl813.

[89] Crawford ED, Seaman JE, Agard N, et al. The DegraBase: a database of proteolysis in healthy and apoptotic human cells. Mol Cell Proteomics 2013;12(3):813–24. https://doi.org/10.1074/mcp.O112.024372.

[90] Lüthi AU, Martin SJ. The CASBAH: a searchable database of caspase substrates. Cell Death Differ 2007;14(4):641–50. https://doi.org/10.1038/sj.cdd.4402103.

[91] Colaert N, Maddelein D, Impens F, et al. The online protein processing resource (TOPPR): a database and analysis platform for protein processing events. Nucleic Acids Res 2013;41(D1):D333–7. https://doi.org/10.1093/nar/gks998.

[92] Lange PF, Overall CM. TopFIND, a knowledgebase linking protein termini with function. Nat Methods 2011;8(9):703–4. https://doi.org/10.1038/nmeth.1669.

[93] Edman P, Begg G. A protein sequenator. Eur J Biochem 1967;1(1):80–91. https://doi.org/10.1111/j.1432-1033.1967.tb00047.x.

[94] von Mering C, Krause R, Snel B, et al. Comparative assessment of large-scale data sets of protein–protein interactions. Nature 2002;417(6887):399–403. https://doi.org/10.1038/nature750.

[95] Braun P, Tasan M, Dreze M, et al. An experimentally derived confidence score for binary protein-protein interactions. Nat Methods 2009;6(1):91–7. https://doi.org/10.1038/nmeth.1281.

[96] Song J, Tan H, Boyd SE, et al. Bioinformatic approaches for predicting substrates of proteases. J Bioinform Comput Biol 2011;09(01):149–78. https://doi.org/10.1142/S0219720011005288.

[97] Gasteiger E, Hoogland C, Gattiker A, et al. Protein identification and analysis tools on the ExPASy server. In: Walker John M, editor. The proteomics protocols handbook. Humana Press; 2005. p. 571 607. https://doi.org/10.1385/1-59259-890-0:571.

[98] duVerle DA, Mamitsuka H. A review of statistical methods for prediction of proteolytic cleavage. Brief Bioinform 2012;13(3):337–49. https://doi.org/10.1093/bib/bbr059.

[99] Backes C, Kuentzer J, Lenhof H-P, Comtesse N, Meese E. GraBCas: a bioinformatics tool for score-based prediction of caspase- and granzyme B-cleavage sites in protein sequences. Nucleic Acids Res 2005;33(suppl 2):W208–13. https://doi.org/10.1093/nar/gki433.

[100] Barkan DT, Hostetter DR, Mahrus S, et al. Prediction of protease substrates using sequence and structure features. Bioinformatics 2010;26(14):1714–22. https://doi.org/10.1093/bioinformatics/btq267.

[101] Song J, Tan H, Shen H, et al. Cascleave: towards more accurate prediction of caspase substrate cleavage sites. Bioinforma Oxf Engl 2010;26(6):752–60. https://doi.org/10.1093/bioinformatics/btq043.

[102] du Verle DA, Ono Y, Sorimachi H, Mamitsuka H. Calpain cleavage prediction using multiple kernel learning. PLoS One 2011;6(5). https://doi.org/10.1371/journal.pone.0019035.

[103] Schilling O, Barré O, Huesgen PF, Overall CM. Proteome-wide analysis of protein carboxy termini: C terminomics. Nat Methods 2010;7(7):508–11. https://doi.org/10.1038/nmeth.1467.

[104] O'Donoghue AJ, Eroy-Reveles AA, Knudsen GM, et al. Global identification of peptidase specificity by multiplex substrate profiling. Nat Methods 2012;9(11):1095–100. https://doi.org/10.1038/nmeth.2182.

[105] Ge H, Liu Z, Church GM, Vidal M. Correlation between transcriptome and interactome mapping data from Saccharomyces cerevisiae. Nat Genet 2001;29(4):482–6. https://doi.org/10.1038/ng776.

[106] Grigoriev A. A relationship between gene expression and protein interactions on the proteome scale: analysis of the bacteriophage T7 and the yeast Saccharomyces cerevisiae. Nucleic Acids Res 2001;29(17):3513–9. https://doi.org/10.1093/nar/29.17.3513.

[107] Bhardwaj N, Lu H. Correlation between gene expression profiles and protein–protein interactions within and across genomes. Bioinformatics 2005;21(11):2730–8. https://doi.org/10.1093/bioinformatics/bti398.

[108] Dezső Z, Oltvai ZN, Barabási A-L. Bioinformatics analysis of experimentally determined protein complexes in the yeast Saccharomyces cerevisiae. Genome Res 2003;13(11):2450–4. https://doi.org/10.1101/gr.1073603.

[109] Rual J-F, Venkatesan K, Hao T, et al. Towards a proteome-scale map of the human protein–protein interaction network. Nature 2005;437(7062):1173–8. https://doi.org/10.1038/nature04209.

[110] Shin CJ, Wong S, Davis MJ, Ragan MA. Protein-protein interaction as a predictor of subcellular location. BMC Syst Biol 2009;3(1):28. https://doi.org/10.1186/1752-0509-3-28.

[111] Rastogi S, Rost B. LocDB: experimental annotations of localization for Homo sapiens and Arabidopsis thaliana. Nucleic Acids Res 2011;39(suppl 1):D230–4. https://doi.org/10.1093/nar/gkq927.

[112] Uhlen M, Oksvold P, Fagerberg L, et al. Towards a knowledge-based human protein atlas. Nat Biotechnol 2010;28(12):1248–50. https://doi.org/10.1038/nbt1210-1248.

[113] Ashburner M, Ball CA, Blake JA, et al. Gene ontology: tool for the unification of biology. Nat Genet 2000;25(1):25–9. https://doi.org/10.1038/75556.

[114] King OD, Foulger RE, Dwight SS, White JV, Roth FP. Predicting gene function from patterns of annotation. Genome Res 2003;13(5):896–904. https://doi.org/10.1101/gr.440803.

[115] Maetschke SR, Simonsen M, Davis MJ, Ragan MA. Gene ontology-driven inference of protein–protein interactions using inducers. Bioinformatics 2012;28(1):69–75. https://doi.org/10.1093/bioinformatics/btr610.

[116] Pellegrini M, Marcotte EM, Thompson MJ, Eisenberg D, Yeates TO. Assigning protein functions by comparative genome analysis: protein phylogenetic profiles. Proc Natl Acad Sci U S A 1999;96(8):4285–8.

[117] Jansen R, Yu H, Greenbaum D, et al. A Bayesian networks approach for predicting protein-protein interactions from genomic data. Science 2003;302(5644):449–53. https://doi.org/10.1126/science.1087361.

[118] Rhodes DR, Tomlins SA, Varambally S, et al. Probabilistic model of the human protein-protein interaction network. Nat Biotechnol 2005;23(8):951–9. https://doi.org/10.1038/nbt1103.

[119] Zhang QC, Petrey D, Deng L, et al. Structure-based prediction of protein-protein interactions on a genome-wide scale. Nature 2012;490(7421):556–60. https://doi.org/10.1038/nature11503.

[120] Moon S, Han D, Kim Y, Jin J, Ho W-K, Kim Y. Interactome analysis of AMP-activated protein kinase (AMPK)-$\alpha 1$ and -$\beta 1$ in INS-1 pancreatic beta-cells by affinity purification-mass spectrometry. Sci Rep 2014;4:4376.

CHAPTER 12

The puzzle of proteolytic effects in hemorrhage induced by Viperidae *snake venom metalloproteinases*

Dilza Trevisan-Silva[a], Jessica de Alcantara Ferreira[b], Milene Cristina Menezes[b], and Daniela Cajado-Carvalho[b]

[a]Centre of Excellence in New Target Discovery (CENTD), Butantan Institute, São Paulo, Brazil, [b]Special Laboratory for Applied Toxinology, Butantan Institute/Center of Toxins, Immune-Response and Cell Signaling (CeTICS), São Paulo, Brazil

Introduction

Overview of snakebite envenomation

Snakebite envenomation epidemiology

Accidents with snakes represent a public health problem worldwide, which can lead victims to temporary or permanent injury, or even to death [1]. According to the World Health Organization (WHO), 4.5–5.4 million people get bitten by snakes every year, from which 138.000 progressed to death [2]. However, the number of accidents and deaths is probably much higher, since many countries do not have a reliable notification system for snake envenomation [3], or even do not consider it as a reportable disease [4], which also imply in underreporting of long-term or permanent injuries [5]. The specific treatment for snakebite is the use of antivenoms, which consist of isolated hyperimmune globulin from animals that were immunized with target snake venom [6].

Cases of snake envenomation are more common in developing and under-developing countries located in tropical and subtropical regions [7], especially in rural areas. Snakebite mortality has been linked to poverty since there is a strong correlation between snake accidents and fatality with Human Development Index (HDI) and Gross domestic product (GDP)/capita [8]. The majority of snakebites occur in sub-Saharan Africa, South to Southeast Asia, Papua New Guinea, and Latin America. According to Kasturiratne and collaborators (2008) [5], the most affected countries are from South Asia, such as India (81,000 deaths), Sri Lanka (33,000 deaths), Vietnam (30,000 death), and Nepal (20,000 death), followed

Proteolytic Signaling in Health and Disease. https://doi.org/10.1016/B978-0-323-85696-6.00012-3

by American countries like Brazil (30,000 death) and Mexico (28,000). Although Sub-Saharan Africa was registered as one of the main affected areas with 43,000 deaths, the lack of epidemiological data indicates that the number of cases and death could be even more significant [9].

Despite the risk associated, the global public health community and government agencies underrated snakebite as a relevant health issue and, due to this fact, in 2009, WHO included accidents with snakes as a neglected tropical disease (NTD). However, it was soon removed from NTD list in 2013 and reclassified as a neglected condition. After a massive campaign by concerned stakeholders, including Non-Government Organizations (NGOs), researchers, health-care providers, and antivenom producers, along with a detailed submission by more than 20 countries, in 2017, WHO re-included the snakebite as a top priority neglected tropical disease in category A. The inclusion was crucial to gain relevant attention to this disease, which helped to elaborate strategies to prevent and control snakebite envenomation [10]. The biggest challenge is that snakebite mostly affected emerging countries, and antivenom is a high-cost product. In most cases, it does not promote profit to manufacturers, which causes a low production scale and, therefore, a shortage in antivenom storage [11]. Thankful to the inclusion in the NTD list, WHO established objectives globally to reduce death and disability by 50% until 2030, which includes (1) ensure that a safe and effective treatment is accessible and affordable to all, (2) help to strengthen health systems, (3) empower communities to take proactive actions, and (4) build a robust global coalition to mobilize resources and partnerships as a coordinate action [12].

Brief overview of medically relevant venomous snakes

Snakes are a specialized group of limbless lizards that belong to the second most speciose group of reptiles [13]. Although these animals are commonly associated with danger due to their venom, only 10% of the 3700 snake species are venomous [14]. Taxonomically, venomous snakes are distributed in four families, but most species are distributed in only two of them: the *Elapidae*, which contains coral snakes, cobras, and sea snakes, or the *Viperidae*, whose members are vipers and pit vipers. With fewer venomous taxa, the family *Colubridae*, known as back-fanged snakes, also has medically relevant species [14]. Moreover, the subfamily *Atractaspididae*, known as *mole snakes* or *burrowing asps*, has some species that were registered to cause accidents with humans [15].

The *Viperidae* is considered a successful group of venomous snakes due to the presence of a sophisticated venom apparatus, which consists of long fangs positioned into a mobile maxillary bone that is able to rotate and fold the fangs on the roof of the mouth when it is closed [16]. The long fangs, whose length may exceed 2 cm, allows an effective delivering of the venom into the prey [17]. The *Viperidae* family is divided into three subfamilies—*Azemiopinae*, *Viperinae*, and *Crotalinae* [18]. *Viperinae*, also called "true vipers," has 101 species cataloged and is found in the "Old World" (Africa, Europe, and Asia). The "pit vipers," *Crotalinae* subfamily, comprises the higher diversity, with 261 species, and has a wider localization in both Old and New World, which includes Americas. *Azemiopinae* is a small group with only two species exclusively found in the "Old World," and phylogenetic data indicate that it may be a sister taxon of *Crotalinae* [18]. In general, the viper venoms present hemotoxic and necrotic activities [19].

Clinical manifestations of snakebites: Envenomation symptoms

Envenomation caused by snakebites consists of physiopathological manifestations in the victim caused by multiple factors that can affect several organs. Noteworthy, it is estimated that 10%–50% of all snakebites are "dry bites," that are characterized by the absence of local and systemic manifestations as a consequence of a lack of venom injection, even though the presence of fang marks is observed in the bite site [20–22]. A number of factors contribute to the severity of envenomation, such as victim's comorbidity, body structure (amount of toxin per kg of body weight), amount of venom injected, and time elapsed of first assistance after the snakebite [23]. Even after recovery, physical injuries can prevail permanently, such as amputations or loss of function of the bitten body member, kidney damage, stroke, blindness, and chronic pain [21].

Although there is a specific symptomatology that comprises multiple clinical effects for envenomations induced by each snake species that is closely related with venom composition, four clinical patterns are classically observed: cytotoxic, neurotoxic, myotoxic, and hemotoxic/hemorrhagic envenomation [22, 24, 25]. The present review will focus on hemorrhagic effects and the other clinical patterns will be briefly described.

Cytotoxic envenomation is commonly associated with viper snake venoms and is characterized by the presence of edema, which grows progressively and painfully, and of local tissue destruction such as necrosis, blistering, and bruising [21, 24]. Notably, after local effects are established, antivenom treatment cannot revert tissue damage but prevents further damage by neutralizing active toxins [21].

Some snake toxins can interact with pre- and/or post-synaptic junctions, causing neurotoxic symptoms [24, 26]. The bitten patient can develop a descending paralysis, starting with manifest in eyes by falling of the upper eyelid (ptosis) and dilation of the pupil (mydriasis) and, without proper treatment, can progress to life-threatening complications such as bulbar and respiratory paralysis [26]. In general, neurotoxicity is related to viper and elapid snake species [21].

Myotoxicity induced by snakebite affects mainly muscles on the bite site, but rhabdomyolysis, generalized myalgia and increase of enzymes creatine kinase (CK) and myoglobinuria associated with muscle toxicity are reported [26]. Sea snakes, elapids, and some viper venoms are considered myotoxic.

Snake envenomation can also promote disturbance of the hemostatic system locally or systemically [27]. *Viperidae* snakes had the most species associated with these effects, but some *Colubridae* and *Elapidae* venoms can also lead to this clinical pattern [26]. Venom toxins can interact with distinct proteins of the blood or tissue components of the victim and can induce coagulopathy and systemic bleeding in multiple body areas besides the bitten site, including gum and gastrointestinal urinary tract [21, 27]. Victims can also develop secondary effects like vomiting (hematemesis) or coughing blood (hemoptysis) [27].

The hemorrhage is mainly associated with the proteolytic activity of metalloproteases, named in the Toxinology field as Snake Venom Metalloproteinases (SVMP) [28]. Noteworthy, some snake venom proteases possess a procoagulant activity by consuming coagulation factors, such as FII, Factor V, Factor XIII, and fibrinogen [29, 30]. Although some SVMP can also hydrolyze these coagulation factors, mainly Thrombin-like Snake Venom Serine Proteinases (TL-SVSP) are related to this activity. The shortage of the proteins associated with blood coagulation may lead to one significant clinical manifestation called Venom Induced Consumption Coagulopathy (VICC) [31]. Snake venoms with a high abundance of hemorrhagic SVMP are linked to worsening clinical conditions of VICC, causing systemic bleeding such as intracranial and pulmonary hemorrhages [30, 32].

Nonetheless, although SVMPs are the main effector agents, other toxins presented in the crude venom can act as supporting actors on the hemorrhagic event and contribute to this symptom [33].

Snake venom metalloproteinases related to hemorrhagic effects

Snake venoms are complex mixtures of components, consisting mainly of proteins and peptides, but non-protein molecules like nucleotides, carbohydrates, lipids, and metal ions are also present [34]. The composition of venoms has a high percentage of variability, and toxins are distinct among species or even possess different protein abundance among specimens depending on their sex [35] or location [36, 37]. It is hypothesized that, since the primary function of the venoms is ecological and used by snakes for both defense and feeding, the variability could be associated with biological pressure [38]. Furthermore, the complexity of venom composition on the ontogenetic level was extensively reported by transcriptomic and proteomic reports, where toxins abundance can change according to prey or even snake ages [39–41]. A database of venom proteases revealed that four classes of proteins are most commonly found in all four main families of venomous snakes—phospholipase A2s, three-finger toxins, serine proteases, and metalloproteases [42].

Snake venom metalloproteinases are zinc-dependent peptidases that belong to the Metzincins superfamily of proteases, and subfamily of reprolysins (Clan MA, subfamily M12). The term "reprolysin" represents both metallopeptidases isolated from snake venoms (reptilia) and some reproductive proteins from mammals, all included in this subfamily. These peptidases have their biological activity similar to human metallopeptidases that degrade matrix components (MMPs), and also structurally resemble to adamlysins (ADAM, A Disintegrin and Metalloproteinase) [43].

Structurally, the proteolytic domain of SVMPs usually has 200 to 210 residues in length and possesses a consensus sequence on a catalytic site—HExxHxxG (where H = histidine, E = glutamic acid, G = glycine and x = any amino acid)—which is responsible for the zinc coordination [44]. Besides its catalytic domain, SVMPs could also have extra domains: disintegrin (Dis) or disintegrin-like (Dis-like), and cysteine-rich (Cys-rich). Based on the presence or absence of these domains, SVMPs can be classified as follows: PI (metallopeptidase domain only); PIIa-PIIe (metallopeptidase domain and disintegrin domain, present in the form of dimers); PIIIa-PIIIc (metallopeptidase domain, disintegrin-like domain and cysteine-rich domain, can also form dimers), and PIIId, or PIV (have a PIII-type structure and two type-C lectins connected by disulfide bridges) [43, 45].

Several SVMPs have already been isolated and characterized for having some level of hemorrhagic activity. Moreover, hemorrhagic SVMP can possess other targets, such as platelet aggregation inhibition, activation of coagulation factors such as Factor II and Factor X, and the ability to degrade both fibrinogen and/or fibrin. Apart from the hydrolysis of substrates, the hemorrhagic potency of venom components can be assessed by determining the Minimum Hemorrhagic Dose (MHD). This measurement consists in the amount of venom/toxin that is able to generate a hemorrhagic spot of 10 mm diameter after 24 h of injection into the depilated back skin of rabbits [46]. Therefore, a lower value of MHD indicates a higher hemorrhagic potency. Table 1 summarizes SVMPs described so far in *Viperidae* snakes related to hemorrhagic effects.

In the following topics, the in vivo and in vitro findings that contributed to the hypothesis of mechanisms by which SVMPs induce hemorrhage are discussed and cover crucial studies for this understanding as well as the most investigated hemorrhagic SVMPs.

Tissue effects and mechanisms evolving SVMP in hemorrhage

The pathophysiology of hemorrhage induced by SVMP is experimentally studied using a number of purified proteases and ex vivo tissues or animal models (for reviews, see Refs. [47, 48]). SVMPs promote high or weak local hemorrhage, and some are also able to induce bleeding in distant tissues from the local site of the injection or the snakebite. These differences have been related to the enzyme structures since PIII SVMPs, which possess domains involved in their interactions with substrates in addition to the metalloprotease catalytic domain, are in general more hemorrhagic than PI SVMP, which lack non-catalytic domains and even a number of non-hemorrhagic PI SVMPs is reported (Table 1). These observations have support on accumulated evidence in the literature that investigates the relationship between function and structure of distinct classes of isolated SVMPs from *Bothrops* and *Crotalus* venoms [49–55].

Table 1: Viperidae snake toxins related to hemorrhagic effects.

SVMP class	Toxin name (alternative names)	Hemorrhagic potential	Minimal hemorrhagic dose (MHD) and/or minimal pulmonary hemorrhagic dose (MPHD)	Main protein substrates	Main in vivo effects	Snake subfamily	Snake Species	References
PIII	Atrolysin A (HT-A)	Ultra-High	0.04 μg, mice skin	Plasma fibronectin, type-I collagenase, type-IV collagenase, fibrinogenase, gelatinase, sheddase of Type VI collagen, and Anexin V Azocoll	Local hemorrhage	Crotalinae	*Crotalus atrox*	[51, 140]
PIII	BaH1	High	0.18 μg, mice dorsal skin		Endothelial cells detachment from basal lamina	Crotalinae	*Bothrops asper*	[82]
PIII	BaH4	Moderate	2 μg, mice skin	Fibronectin, partial hydrolysis of laminin and Type IV collagen	Hemorrhage in local injection (skin and grastrocnemius muscle) and lung hemorrhagic spots	Crotalinae	*Bothrops asper*	[59]
PIII	Baltergin	Moderate	MPDH = 10 μg in mice lung	Type IV collagen	Lung bleeding, kidney and liver injuries (evaluated in mice)	Crotalinae	*Bothrops alternatus*	[141]
PIII	Batroxrhagin	Moderate	1.24 μg, mice dorsal skin	Fibrinogen, fibrin	Inhibited collagen-dependent platelet-aggregation	Crotalinae	*Bothrops atrox*	[142]
PIII	Bothropasin	Moderate	1 μg, rabbit skin; 20 μg, mice dorsal skin	Fibrinogen	Hemorrhage followed by myonecrosis and arterial necrosis	Crotalinae	*Bothrops jararaca*	[143–145]
PIII	Cc HSM-III	Moderate	5 μg, mice dorsal skin	Fibrinogen, Type IV collagen, laminin, gelatin	Inflammatory reaction (presence of erythrocytes and polynuclears leukocyte infiltrate in hypodermis and in the skeletal muscle)	Viperinae	*Cerastes cerastes*	[146]

PIII	CsH1	Moderate	2.2 μg, mice dorsal skin	Type IV collagen, laminin, nidogen	Local hemorrhage	Crotalinae	*Crotalus simus*	[76]
PIII	Daborhagin-K	High	0.82 μg, mice dorsal skin	Fibrinogen, fibronectin, Type IV collagen, Insulin B Chain, Casein	Local hemorrhage	Viperinae	*Daboia siamensis*	[147]
PIII	Daborhagin-M	High	0.86 μg, mice dorsal skin	Fibrinogen, fibronectin, Type IV Collagen, Insulin B Chain, Casein	Local hemorrhage	Viperinae	*Daboia siamensis*	[148]
PIII	HF1	High	0.10 μg, rabbit skin	Insulin B Chain	Local hemorrhage	Crotalinae	*Bothrops jararaca*	[149]
PIII	HF2	Ultra-High	0.02 μg, rabbit skin	Casein, Insulin B Chain	Local hemorrhage	Crotalinae	*Bothrops jararaca*	[149, 150]
PIII	HF3	Ultra-High	0.015 μg, rabbit skin	Casein. Extracellular proteins: fibronectin, vitronectin, fibrinogen, von Willebrand factor, and Type IV and VI collagens. Plasma proteins inhibitors. Endothelial glycocalyx proteoglycans. PDGFR.	Local hemorrhage	Crotalinae	*Bothrops jararaca*	[54, 64, 65, 143, 149, 151]
PIII	HR1A	Ultra-High	0.011 μg, rabbit skin	Casein	Local hemorrhage	Crotalinae	*Probothrops flavoviridis*	[147]
PIII	HR1B	Ultra-High	0.0067 μg, rabbit skin	Casein (very low activity)	Local hemorrhage	Crotalinae	*Probothrops flavoviridis*	[152]
PIII	HR-Ele1	Ultra-High	0.05 mg in the guinea pig	Hemorrhagic, αβ-fibrinogenase, type I collagenase	Local hemorrhage	Crotalinae	*Protobothrops elegans*	[153]
PIII	Jararhagin	Moderate	20 μg (22); 1.5 μg after optimizing purification (23); MPHD = 10 μg, mice (24)	Fibrinogen, von Willebrand factor, alpha-2/beta-1 integrin (subunity beta-1)	Systemic hemorrhage in lungs and kidneys (lesser extent). Inhibits both risocetin- and collagen-induced platelet-aggregation	Crotalinae	*Bothrops jararaca*	[72, 153–155]
PIII	Leucurolysin-B	Ultra-High	0.03 μg, rabbit skin	Fibrinogen, Fibrin, Insulin B Chain,	Local hemorrhage	Crotalinae	*Bothrops. leucurus*	[152]

Continued

Table 1: Viperidae snake toxins related to hemorrhagic effects—cont'd

SVMP class	Toxin name (alternative names)	Hemorrhagic potential	Minimal hemorrhagic dose (MHD) and/or minimal pulmonary hemorrhagic dose (MPHD)	Main protein substrates	Main in vivo effects	Snake subfamily	Snake Species	References
PIII	VaH4	Moderate	18 μg hemorrhagic spot of 12.88 + 0.88 mm, rat skin	Hydrolysis of fibronectin, fibrinogen, nidogen, and Factor X, Limited proteolysis of Collagen IV, Laminin all basal membrane proteins. Cytotoxic activities against HeLa cells	Local hemorrhage	Viperinae	*Vipera ammodytes ammodytes*	[156]
PIII	VAP1	Weak	8 μg, mice dorsal skin	Fibrinogen	Local hemorrhage	Crotalinae	*Crotalus atrox*	[157]
PIII	VAP2 (CAMP)	Weak	15 μg, mice dorsal skin	Fibrinogen	Systemic hemorrhage in kidney and intestine, Severe hemorrhage in lungs, bleeding in the heart and liver	Crotalinae	*Crotalus atrox*	[157]
PIII	VLH2	Moderate	3 μg, mice dorsal skin	Fibrinogen (Aα-and Bβ-chains)	Local bleeding, interfibrillar hemorrhage with polymorphonuclear inflammatory infiltrate, late increase in serum CK, necrose on muscle cells	Viperinae	*Vipera lebetina*	[158, 159]
PIII	EoVMP2	Weak	10 μg, mice dorsal skin	Fibrinogen, Prothrombin	Local hemorrhage	Viperinae	*Echis ocellatus*	[160]
PII	Ammodytagin	Weak (rat)/ High (mice)	10.86 μg ,rat; 0.65 μg, mice dorsal skin	Fibrinogen	Local hemorrhage	Viperinae	*Vipera ammodytes ammodytes*	[161]

PII	BlaH1	Moderate	5 μg, rats dorsal skin	Casein, Gelatin fibrinogen	Local and systemic hemorrhagic, platelet aggregation inhibition	Crotalinae	*Bothrops lanceolatus*	[162]
PII	BlatH1	High	0.23 μg; MHPD = 1.32 μg/g	Azocasein, Gelatin, and fibrinogen	Local and pulmonary hemorrhagic activity	Crotalinae	*Bothriechis lateralis*	[159]
PII	Bilitoxin-1	Ultra-High	0.008 μg/g, mice dorsal skin	Fibrinogen, Casein	Local hemorrhage	Crotalinae	*Agkistrodon bilineatus*	[163]
PII	CcMP-II	Moderate	5.5 μg/20 g	Casein, Fibrinogen, Fibrin, Laminin, Type IV collagen	Local hemorrhage	Viperinae	*Cerastes cerates*	[164, 165]
PI	Acutolysin A (Aa-hemorrhagin I/AaH I)	High	0.4 μg, rabbit skin	Casein, Fibrinogen	Local hemorrhage	Crotalinae	*Agkistrodon acutus*	[166]
PI	Acutolysin C	Weak	12.0 μg, rabbit skin	Casein	Local hemorrhage	Crotalinae	*Agkistrodon acutus*	[167]
PI	Atrolysin C (Ht-d)	Weak	11 μg, mice dorsal skin	Type IV collagen, gelatin	Local hemorrhage	Crotalinae	*Crotalus atrox*	[168]
PI	Atroxlysin Ia	Moderate	1.8 μg, mice dorsal skin	Fibronectin, laminin, Type IV collagen, nidogen	Hemorrhage and dermonecrosis	Crotalinae	*Bothrops atrox*	[169]
PI	Batroxase	Weak	10 μg, mice dorsal skin	Fibrinogen, laminin, plasminogen	Local hemorrhage	Crotalinae	*Bothrops atrox*	[170]
PI	BaP1	Extremely weak	Bleeding observed in mice cremaster muscle during intravital experiments	Perlecan, Type IV Collagen, Laminin	Mild local tissue damage (Necrosis). Do not cause systemic bleeding	Crotalinae	*Bothrops asper*	[47, 171]
PI	Batx-1	Weak	17 μg/20 g, mice dorsal skin	Fibrinogen	Mild myotoxicity due to a slightly increased CK activity	Crotalinae	*Bothrops atrox*	[172]
PI	BmHF-1	Extremely weak	41.14 μg, mice dorsal skin	Fibrinogen, casein	Hemorrhage and edema in mouse footpad (Edema-forming dose = 10.28 μg)	Crotalinae	*Bothrops marajoensis*	[93]
PI	BnP1	Extremely weak	50 μg, mice dorsal skin	Fibrinogen	Local hemorrhage	Crotalinae	*Bothrops neuwiedi*	[173, 174]

Continued

Table 1: Viperidae snake toxins related to hemorrhagic effects—cont'd

SVMP class	Toxin name (alternative names)	Hemorrhagic potential	Minimal hemorrhagic dose (MHD) and/or minimal pulmonary hemorrhagic dose (MPHD)	Main protein substrates	Main in vivo effects	Snake subfamily	Snake Species	References
PI	BpirMP	Extremely weak	50 μg, mice dorsal skin	Azocasein, fibrinogen, Fibrin, laminin, nidrogen	Dose- and time-dependent edematogenic response	Crotalinae	*Bothrops pirajai*	[175]
PI	BtaMP-1	Weak	20.32 μg, mice dorsal skin	Casein, Fibrinogen	Pulmonary hemorrhage, alveolar septum thickening and neutrophil infiltration in mouse pulmonary tissue; edema formation and dermonecrosis	Crotalinae	*Bothriopsis taeniata*	[176]
PI	BthMP	Extremely weak	N.D. (Weak hemorrhage using 30 μg of purified enzyme)	Fibrinogen (Aα-and Bβ-chains), Fibrin and casein	Anticoagulant activity (prolong prothrombin time and partial thromboplastin time)	Crotalinae	*Bothrops moojeni*	[163, 177]
PI	CcH1	Moderate	3 μg/20 g, mice dorsal skin	Casein, Fibrinogen, laminin, and type IV collagen	Local hemorrhage	Viperinae	*Cerastes cerastes*	[178]
PI	HR2A	High	0.068 μg, rabbit skin	Fibrinogen, FIbronectin	Local hemorrhage	Crotalinae	*Trimeresurus flavoviridis*	[179, 180]
PI	HR2B	High	0.066 μg, rabbit skin	Fibrinogen, FIbronectin	Local hemorrhage	Crotalinae	*Trimeresurus flavoviridis*	[179, 180]
PI	HT-1	High	0.17 μg, mice dorsal skin	Fibrinogen, Insulin B Chain	Severe hemorrhage in lung and heart	Crotalinae	*Crotalus ruber*	[180]
PI	HT-2	High	0.27 μg, mice dorsal skin	Fibrinogen	Low hemorrhage activity in lung and heart	Crotalinae	*Crotalus ruber*	[180]

PI	HT-3	High	0.43 μg, mice dorsal skin	Fibrinogen	Low hemorrhage activity in lung and heart	Crotalinae	*Crotalus ruber*	[181]
PI	Jararafibrase I	Moderate	8 μg, mice dorsal skin	Fibrinogen, fibrin, fibronectin, gelatin, type IV collagen, laminin, and entactin	Local hemorrhage	Crotalinae	*Bothrops jararaca*	[181]
PI	Jararafibrase II	Weak	27 μg, mice dorsal skin	Fibrinogen, fibrin, fibronectin, gelatin, type IV collagen, laminin, and entactin	Local hemorrhage	Crotalinae	*Bothrops jararaca*	[181]
PI	Jararafibrase III	Weak	31 μg, mice dorsal skin	Fibrinogen, fibrin, fibronectin, gelatin, type IV collagen, laminin, and entactin	Local hemorrhage	Crotalinae	*Bothrops jararaca*	[181]
PI	Jararafibrase IV	Weak	34 μg, mice dorsal skin	Fibrinogen, fibrin, fibronectin, gelatin, type IV collagen, laminin, and entactin	Local hemorrhage	Crotalinae	*Bothropsjararaca*	[182]
PI	LHF-II	Weak	16.2 μg, rabbit skin	fibrinogen, fibrin, casein, insulin B chain	Hemorrhagic and myotoxic, local tissue damage, edema formation	Crotalinae	*Lachesis muta*	[183, 184]
PI	Neuwiedase	Extremely weak	N.D. (40 μg induce muscle hemorrhage, lung hemorrhage in high doses)	Fibrinogen, Insulin B Chain, laminin, fibronectin and Type I collagen	Hemorrhage in gastrocnemius and lung (i.v. injection)	Crotalinae	*Bothrops neuwiedi*	[185]
PI	Porthidin-1	Weak	11.12 μg, mice dorsal skin	Fibrinogen, Fibrin, Casein, and Gelatin	Edema in mouse footpad (Edema-forming dose = 17.36 μg)	Viperinae	*Porthidium lansbergii hutmanni*	[185]

N.D., not determined; M.H.D., minimal hemorrhagic dose.

Since the late 1970–90s, the enzymatic degradation of Basement Membrane (BM) components of capillary vessels is suggested to be a key event on hemorrhage induced by SVMPs [56, 57] and during the early 2000s, it was added that SVMPs could cleave specific peptide bonds of BM components, disturbing the interaction between endothelial cells and BM, resulting in bleeding [58–60]. The proposed mechanism, *per rhexis* hemorrhage, included rapid damage of endothelial cells of capillary blood vessels, generating lesions or gaps of the endothelium, allowing erythrocyte and other blood components to escape to the interstitial space [33, 60]. The hemorrhage induced by SVMPs is an extremely rapid noticeable effect in vivo, indeed an explosive hemorrhage is observed after a few minutes of toxin injection. However, in vitro analysis of purified BM components degradation and changes in cell morphology are observed only hours after incubation with distinct SVMPs from *Crotalinae* and *Viperinae* venoms. The presence of biophysical factors in the microvasculature is proposed to explain these apparent controversial observations, and a two-step mechanism was proposed, briefly: (1) Cleavage of specific regions of BM proteins and of membrane proteins of endothelial cell-related to cell-matrix interaction, that probably reduces capillary wall resistance to distention and separation of endothelial cells from BM and (2) increase of wall tension by distention of capillary wall, leading to decrease the endothelium thickness, and blood flow-dependent shear stress may cause further mechanical damage to the weakened endothelial cells. This process ends in microvessel rupture and extravasation of blood components (for a complete description of this hypothesis, check Ref. [61]). An experimental support of this proposed mechanism is derived from experiments in which the blood flow in muscle tissue was interrupted, excluding the biophysical forces associated with blood flow, and the injection of BaP1 did not induce endothelial cell damage [62].

A comparison of the proteolytic effect upon BM components in vitro and in vivo was performed using BaP1 (PI SVMP) and Jararhagin (PIII SVMP) [63]. The in vitro results show a distinct pattern of proteolysis of BM components. In vivo analysis showed that when BaP1 was injected in conditions where hemorrhage did not occur (5 μg), there was no evidence of loss of endothelial cell markers or capillary basement membrane antigens. In contrast, in conditions where the hemorrhagic effects were observed (30 μg of BaP1 or 5 μg of Jararhagin), the endothelial cell markers and capillary BM antigens were diminished. The BaP1 and Jararhagin present a distinct cleavage specificity site on nidogen, generating different N-terminal peptide sequences from nidogen after incubation with these enzymes and this could also play a role in the hemorrhagic process [63]. A variety of studies have compared SVMPs from the PI, PII, and PIII classes in terms of their ability to induce hemorrhage and to hydrolyze specific proteins in vitro and in vivo. The intradermal injection of *B. jararaca* SVMPs HF3 (PIII SVMP—Table 1), Bothropasin (PIII—Table 1), BJ-PI (non-hemorrhagic PI class), or DC protein (non-catalytic domain generated by the hydrolysis of Bothropasin and Jararhagin in the venom of *B. jararaca*) in mice skin generated distinct hemorrhagic effects. HF3 induced hemorrhage in low dose and a short period of time

(1 mg–5 min), Bothropasin is much less hemorrhagic (5 mg–2 h) while BJ-PI and DC protein did not induce hemorrhage even in high dose (20 mg–2 h) [54]. It is important to notice that the evaluated enzymes present different domain organizations and distinct glycosylation levels. N-deglycosylation of Bothropasin and BJ-PI caused the loss of structural stability, but HF3 remained intact. N-deglycosylated HF3 presented reduced fibrinogenolytic and hemorrhagic activities, higher affinity to bind to type I and VI collagens but no difference in proteolytic activities on these substrates [54]. This study suggests that carbohydrate moieties of hemorrhagic SVMPs could be related to enzyme stability, to the interaction with its substrates, and to the hemorrhagic potency. This study also exemplifies a conclusion from a variety of other studies that in vitro proteolytic comparison of SVMPs proteolytic activities does not show a direct correlation to the hemorrhagic potency observed in vivo. However, in vitro biochemical analyses have largely contributed to understanding the PI-SVMP and PIII-SVMP differences.

Distinct analyses of hemorrhagic mice skin after injection of HF3 (Table 1) have demonstrated that this isolated PIII-SVMP is able to in vivo: (1) hydrolyze extracellular matrix proteins and glycocalyx proteoglycans: type I, III, and IV collagens, decorin, lumican, mimecan, biglycan, and syndecan [64, 65]; (2) activate tissue collagenases; (3) hydrolyze plasma proteins: alpha 1-antitrypsin, fibrinogen, apolipoprotein AII, fibronectin, hemoglobin [64]; and (4) hydrolyze platelet-derived growth factor receptor (PDGFR) [65]. The identification of the new substrates generated in vivo by HF3 suggests the involvement of these substrates in hemorrhage [65]. Moreover, analysis of muscle samples from mice injected with BaP1 (PIII-SVMP) or Leucurolysin-a (non-hemorrhagic PI-SVMP) (Table 1) showed that only BaP1 was able to degrade perlecan and type VI collagen, indicating that PIII-SVMP could reach and hydrolyze these BM components in the tissue [47]. Interestingly, it was shown that despite BM peptides generated by the hydrolysis of Atroxlysin-Ia (PI-SVMP) or Batroxrhagin (PIII-SVMP) presented different sizes, they were able to induce a significant edema reaction after injection in mice paw [66]. In agreement, an in vitro study demonstrated that the Bothropasin could generate bioactive peptides from proteolysis of two frequent SVMP substrates, fibrinogen and fibronectin [67] and these peptides can interact with diverse functions of endothelial cells [68]. Nonetheless, the interaction with some molecules is not exclusively performed by catalytic mechanism, but PIII-SVMP can also bind to some substrates like von Willebrand factor, matrilins and type XIV and XII collagens via other domains, which may contribute to the destabilization of basement membranes [50].

By using a proteome derived peptide library of albumin-depleted plasma as a substrate and mass spectrometry analysis, the cleavage site specificity of SVMPs was compared and the results demonstrate that both PI and PIII SVMPs possess a clear preference of leucine residues at P1′ site, but PI-SVMPs showed preferences across the full P4 to P4′ range whereas the PIII-SVMP showed a more specific range of amino acid preferences across the sites [69]. The analysis of the peptidome and the N-terminome of plasma and tissue samples from mice after injection with HF3 have contributed to the discovery of the great variety of substrates of this enzyme and showed that the generation of a great set of peptides in vivo does not resemble to the HF3 cleavage specificity site (Leu at P1′) and indicates the activation of tissue proteases on the analyzed time-points after HF3-treatment [64, 65, 70, 71]. Therefore, the detailed molecular analysis of generated peptide sequences suggests two additional pieces in the hemorrhagic effects puzzle beyond the key events to the onset of hemorrhage: (1) SVMPs from distinct class induce different patterns of proteolysis of BM components, generating a distinct set of peptides and the functional effects of these proteolytic products could be a key event to distinguish the molecular signaling induced during the hemorrhage process and (2) SVMPs in vivo are suggested to induce the activation of tissue/endogenous proteases that contribute to the extent of the hemorrhagic process and could also help to explain the partial effectiveness of antivenom therapy on the local effects of envenomation. This hypothesis requires further investigation.

It has also been proposed that the ability of PIII enzymes to assess and accumulate on specific sites of tissue microvasculature may play a role in the observed hemorrhagic differences and also the ability to escape the plasma proteinase inhibitor, a2-macroglobulin [47, 72, 73]. The distribution of Jararhagin (moderate hemorrhagic SVMP) and BnP1 (weakly hemorrhagic SVMP) was evaluated in mice skin after intradermal injection and the induced effects on the skin tissue were also described [74]. The histological and immunofluorescence analyses of the injected mice skin revealed that Jararhagin induced: (1) a huge hemorrhage in the hypodermis and in the adjacent skeletal muscle, (2) a huge disorganization of collagen fibers in the dermis, massive degradation of collagen in the hypodermis, and promoted degradation of type IV collagen in BM of blood vessels and skeletal muscle, and (3) a slight alteration in laminin from BM. While BnP1 induced: (1) an enhancement of the skin thickness (edema) and even in a higher dose, a persistent edema and only sparse hemorrhage spots in the hypodermis were observed, (2) a moderated disorganization of collagen fibers throughout the dermis and of type IV collagen on BM, and (3) a slight alteration in laminin from BM. Besides these enzymes induce distinct tissue effects, Jararhagin binds selectively to microvessels and co-localizes to type IV collagen. It is proposed that the observed selective binding sites depend on non-catalytic domains, Dis-like and Cys-rich domains, of PIII SVMPs [74]. A variety of in vitro studies support the importance of non-catalytic domains for the interaction between PIII-SVMPs and their specific substrates [49–51, 54, 55, 75].

Patterns of tissue localization of BaP1 (PI class–Table 1), BlatH1 (PII class–Table 1), and CsH1 (PIII class–Table 1) SVMPs have been compared using an ex vivo model of cremaster muscle of mice. The results showed that the PIII CsH1 and PII BlatH1 are preferentially localized in the BM of the microvasculature and PIII is co-localized to type IV collagen while PI BaP1 presented a wide distribution in the tissue [76]. Moreover, a comparison of tissue localization between Basparin A, a pro-coagulant PIII-SVMP, and CsH1, a hemorrhagic PIII-SVMP, demonstrated that Basparin A did not bind to the microvasculature, the deglycosylation of this enzyme did not promote a change in its binding pattern, and it suggests the possibility of Basparin A to present amino acid substitutions in non-catalytic domains that are necessary to microvessel binding [77]. Therefore, these results reinforce that SVMPs with distinct domain structure possess different patterns of tissue localization that directly correlates to the hemorrhagic potency of PI, PII, and PIII SVMPs and it is likely to be the main reason of the higher hemorrhagic activity characteristic of PIII-SVMPs. Moreover, the hydrolytic pattern of BM components of mice skin injected with BaP1, BlatH1 and CsH1, and BM components in the exudates collected from the gastrocnemius were evaluated and revealed similarities and differences among the three SMVPs. The results corroborate that hydrolysis of type IV collagen is a key event on the destabilization and damage of microvessels leading to hemorrhage [77]. Although analysis of tissue effects induced by SVMPs largely contributed to the scientific advances on the understanding of the mechanism of action of hemorrhagic SVMPs, the precise cellular and molecular bases of microvessels injuries still remain to be investigated.

The hemorrhagic effects of SVMPs in organs/tissues distant from the snakebite site or the experimentally injected toxin site are less investigated. It has been reported that PIII SVMPs induce hemorrhage in various organs [78–81] and PI-SVMPs do not induce systemic hemorrhage despite the ability to induce local bleeding [82]. Pulmonary hemorrhage is probably the best evaluated/known systemic hemorrhage induced by SVMPs, since one of the methods to determine the minimum hemorrhagic dose relies on an analysis of hemorrhagic spots in lungs (Minimum Pulmonary Hemorrhagic Dose). Jararhagin induces hemorrhage in the lungs, while hemorrhage in kidneys, liver, and brain was not observed. In contrast, the distribution of toxin was more prominent in kidneys and liver other than in lungs. The histological and ultrastructure analyses of lungs after toxin injection showed erythrocyte extravasation in alveolar space, cellular debris, edema on some alveolar septa, distinct morphological damage on endothelial cells, and type I pneumocytes. Therefore, the pulmonary hemorrhage induced by Jararhagin is suggested to depend on the proteolytic activity on endothelial cells and BM proteins and its partial resistance to plasma proteinase inhibitors. There was no direct correlation to the accumulation of toxin in the pulmonary microvasculature and it is proposed that the rapid exposure of lung microvasculature to the toxin could also explain that lung bleeding or the lung microvasculature could be more susceptible to SVMPs than microvessels from other organs/tissues [72]. Moreover, it was also reported that pulmonary hemorrhagic activity of metalloproteinases BaP1 (PI-SVMP) and Jararhagin (PIII-SVMP) could be potentiated by a pre-treatment of mice with Aspercetin, a C-type lectin from *B asper* venom which induces von Willebrand factor-mediated platelet aggregation in vitro and thrombocytopenia in vivo [83]. This result shows that hemostasis disturbance, such as moderate to severe thrombocytopenia, could contribute to the hemorrhagic mechanism of SVMPs and even a PI-SVMP, which alone does not promote pulmonary bleeding, is able to induce systemic hemorrhage if encounters a system with impaired hemostasis. This clearly exemplifies that hemorrhagic syndrome observed in envenoming is a much more complex event since venoms are a mixture of bioactive molecules and a diverse of hemostatic disturbance is observed and all these factors could potentiate the hemorrhagic effects induced by SVMPs.

Local myonecrosis is a common manifestation in viperid envenomings and is mainly caused by the direct action of PLA2, known as myotoxins [84]. A variety of SVMPs are described to induce muscle cell degeneration [85] and it was proposed that SVMPs contribute generate ischemia in the muscle tissue, as a consequence of extravasation that occurs after the microvasculature impairment induced by hemorrhagic SVMPs [86]. Deficient skeletal muscle regeneration after myonecrosis in viperid envenomings is an intriguing phenomenon. It has been proposed that inflammatory responses to venom toxins are key events on regeneration (for review, see [86]). Recently, a study has proposed a mechanism by which PIII-SVMPs induce permanent muscle damage. Muscle effects of an isolated *Crotalus atrox* Metalloproteinase, CAMP, also known as VAP2 (PIII-SVMP—Table 1), were compared with the effects induced by cardiotoxin (a three-finger toxin, non-hemorrhagic) from an elapid venom after 5 and 10 days of toxin injection in the tibia anterior muscle of mice [87]. Microscopy analysis showed that CAMP induces a decrease of muscle fibers containing centrally located nuclei and sparse space between fibers was observed. After 10 days of injection, the analysis showed robust muscle regeneration in muscle treated with CTX but not in CAMP-treated samples. Moreover, immunohistochemistry analyses demonstrated that CAMP degrades collagen scaffold of the fibers, the expression of embryonic myosin heavy chain (a protein important for muscle regeneration) protein was observed in smaller number and non-uniform fibers in CAMP-treated mice, the number of capillaries serving each regenerating fiber was greater in CTX-treated mice (CD31 positive labeling), and the density of macrophage infiltration that is a key for effective muscle regeneration was lower in CAMP treated mice. Moreover, CAMP reduced the number of satellite cells observed in isolated single muscle myofiber and reduced the migration and proliferation of these cells. A simple mechanism was proposed in which SVMPs contribute to permanent muscle damage: (1) SVMPs prevent muscle regeneration by degrading ECM components that serve as a scaffold for the formation of new muscle fibers and surrounding blood vessels and (2) SVMPs affect satellite cells that are essential for muscle regeneration [87]. In order to establish a more comprehensive scenario on muscle regeneration, more experimental evidence supporting the proposed hypothesis is needed.

Therefore, analysis of tissue effects induced by SVMPs largely contributed to the scientific advances on the understanding of the mechanism of action of hemorrhagic SVMPs, but the detailed cellular and molecular bases of microvessels injuries remain to be investigated. For a review on unresolved issues of local tissue damage induced by snake venoms, see Ref. [88].

Cell effects and signaling events that might be evolved in the hemorrhage induced by SVMPs

For many years, the in vitro assays have contributed to the development of hypotheses and unraveling possible mechanisms of actions of toxin venoms. There are many advantages of in vitro analysis compared to animal models as most of these methods are low cost, fast, allow a greater control of variables, and are relatively high-throughput strategies. The recognized drawback of in vitro activities associated with SVMPs is that the findings partially recapitulate the in vivo function of hemorrhagic toxins as they lack the complexity of tissue organization, biophysical forces, and the crosstalk among all these components. Therefore, extrapolations of in vitro findings in the in vivo context where SVMPs exert their functions should be made with caution and in vitro hypotheses need to be validated in a complex scenario whenever possible. With this in mind, herein the possible mechanisms involved in hemorrhage-induced SVMPs that have been assessed using cell cultures are discussed (for review, check Ref. [89]). Most of the cell biology experiments related to the effects of snake venoms proteolytic enzymes are appraised by cell viability (MTT assay or similars), cytotoxicity to verify membrane integrity and possible DNA fragmentation (Cell cytometry technique), cell adhesion, angiogenesis, and cell migration [90]. Only a few studies have assessed the cell signaling events of hemorrhagic SVMPs, most of the data focus on the investigation of the hydrolysis of BM or ECM components (Table 1) and in-depth insights of molecular signaling cascade induced by SVMPs are still to be elucidated.

The adhesion of cells to ECM can be mediated by the integrins signal transduction [91], and it has been demonstrated that Focal Adhesion Kinase (FAK) is involved in the enhancement of adhesive forces in this pathway [92]. Both in vivo [83] and in vitro evidence [93–96] shows that SVMPs can disestablish cell–cell contact, promoting cell detachment. The interaction of SVMPs with integrins by the disintegrin-like non-catalytic domain was also evaluated, demonstrating that the binding on this receptor inhibits its interaction to collagens [97]. For example, the modification of endothelial cells (tEND) shape is concomitant to the rearrangement of the actin cytoskeleton and a decrease in FAK association caspase activation, and this event was not inhibited by EDTA, indicating the participation of non-catalytic domains [95].

The catalytic effects of Jararhagin (PIII-SVMP) and BnP1 (PI-SVMP) induced alterations in cell shape to roundness, followed by cell detachment in a concentration- and time-dependent manner [93, 95]. In Human Umbilical Vascular Endothelial Cells (HUVEC) cultures, the Jararhagin mechanism of cell detachment also involves FAK signaling pathway [94]. Curiously, this phenomenon seems to be associated specifically with endothelial cells, as fibroblasts (A31, HS-68), myoblasts (C2C12), and murine peritoneal adherent cells (MPACs) showed few or no effects in cell adhesion after incubation with Jararhagin [93, 95, 98]. On the other hand, a high concentration of HF3 (400 nM) was able to induce detachment of C2C12 cultures [99].

Another possible signaling pathway that could explain endothelial cell detachment by hemorrhagic SVMPs was recently explored. The activation of LPR5/6 modulates the expression of proteins of the tight cell junctions, such as claudin-3 and PLVAP, which is related to endothelial cell permeability. Moreover, it has been hypothesized that the Wnt pathway seems to have a relevant contribution to vascular integrity, since reports indicate that this pathway is deactivated near bleeding sites when hemorrhagic strokes occur [100]. Recently, Seo and Collaborators [101] showed that the hemorrhagic SVMP from *C. atrox*, VAP1 (PIII SVMP, Table 1), mediates the disruption of cell–cell interaction through Wnt signaling pathway due to limited proteolysis of LRP6 and LPR5 receptors of HUVEC cells. VAP1 was also able to cleave the recombinant LRP6, surprisingly more efficiently than the classical SVMP substrate, fibrinogen [101]. Furthermore, the authors demonstrate that anti-LPR6 antibodies designed to bind specifically to regions that are hydrolyzed by VAP1 were able to both suppress cell–cell detachment and also significantly inhibit intracutaneous hemorrhage induced by VAP1. Therefore, these results indicate that the hemorrhage induced by VAP1 is, at least in parts, promoted by the hydrolysis of LRP6, and probably LRP5, to trigger Wnt intracellular signaling events [101].

The loss of ECM interaction can lead cells to a specific apoptosis pathway called anoikis (a Greek word that means "loss of home") [102]. Tanjoni and collaborators [95] hypothesized that the apoptosis of tEND cells induced by Jararhagin after detachment might be through the anoikis pathway due to modulation of apoptosis-related genes Bax and Bcl-xL. In agreement, BaP1 (PI SVMP) also induces apoptosis in endothelial cells by modulating the expression of Bcl-xL and Bax and activation of caspase-8, which authors also suggest being an anoikis mechanism [103]. Although no significant morphological or cell adhesiveness alterations were observed, the analysis of gene expression of HS-68 fibroblast cultures after treatment with Jararhagin indicated that integrins pathway is being affected, and this might induce cells to anoikis pathway [98].

When HUVEC cell cultures were treated with Jararhagin, DNA fragmentation was related to generic apoptotic events [94]. Interestingly, the cell coating substrate has a significant role in apoptosis induced by Jararhagin due to a more expressive and faster DNA fragmentation on type I collagen and matrigel coating cultures than on cells seeding in fibronectin [94]. Curiously, although VAP1 causes severe endothelial cell fragmentation during apoptosis, there is no interference in ECM-cell interaction. Therefore, the involvement of integrins in apoptosis process induced by VAP1 suggests that the mechanism of cell death is not mediated by the anoikis pathway [104].

Non-catalytic domains involved in SVMPs effects: DC domain

Considering the domain composition of the PIII-SVMPs, it is suggested that these toxins may have more than one site of interaction with their targets and several studies indicate that the disintegrin-like and cysteine-rich domains of SVMPs, and also from ADAMs and ADAMTSs, are involved in its interaction with specific ligands [45, 105, 106]. Native and recombinant proteins, composed of the disintegrin-like and cysteine-rich domains (DC), are potent inhibitors of platelet aggregation by blocking the binding of collagen to platelet integrin $\alpha_2\beta_1$ [53, 55, 107, 108]. The interaction of DC domain to bind collagen by integrins can be explained by the presence of RGD (Arg-Gly-Asp) motif into the DC primary sequence, which resembles the interaction motif of collagens with integrins [97, 109]. The DC domains released from the processing of the PIIIb class metalloproteinases and those produced in the recombinant form (such as those present in the PIIIa class SVMPs, for example) have biological activities, such as the inhibition of collagen-induced platelet aggregation, modulation of adhesion, migration, and cell proliferation, which implies that these domains are also crucial in venom toxicity [45, 58].

Regarding ADAMs and ADAMTs, studies have demonstrated the importance of the cysteine-rich domain in the interaction with its targets. It is known, for example, that the cysteine-rich domain of ADAM12 promotes the adhesion of fibroblasts and myoblasts [110]. The same domain in ADAM13 cooperates with the metalloproteinase domain to regulate the enzyme function in vivo [111]. It was also shown that the disintegrin-like and cysteine-rich domains of ADAM-13 bind to fibronectin and integrins that contain subunit 1, and this binding can be inhibited by antibodies against the cysteine-rich domain, which is still the same domains that are involved in cell migration [112, 113]. Janes and collaborators [114], in order to analyze the acidic surface region located in the cysteine-rich domain of ADAM10, mutated three glutamic acid residues (E573A, E578A, and E579A) and revealed that this negatively charged region is essential for the interaction of ADAM10 with the EphA3/ephrinA5 complex.

Recombinant proteins containing mutations of charged residues in the V-loop (Variable loop—located before the hypervariable region, HVR) and HVR regions of ADAMTS13 dramatically decreased their proteolytic activity, suggesting that these residues play a crucial role in recognition of their main target, the von Willebrand factor (vWF) [105]. Recently, Groot and collaborators [115] corroborated the importance of ADAMTS13's cysteine-rich domain for vWF interaction and cleavage and identified important hydrophobic residues for these activities.

Therapeutic strategies targeting hemostatic disturbance of snake venoms: Antivenoms and new therapies

Antivenoms are the only therapy considered specific to the treatment of snakebite envenoming, and the search for alternative treatments is essential [116, 117]. Antivenoms consist of polyclonal immunoglobulins and, depending on the manufacturer, they may comprise intact IgGs, F (ab') 2, or Fab fragments. The production of antivenoms involves consecutive immunization (for months to years) of large mammals (in general, horses and sheep), followed by purification of the immunoglobulin G (IgG) antibodies from the hyper-immunized plasma [118]. The appropriate and fast use of the antivenom provides a greater chance of therapeutic success, minimizing or even avoiding the occurrence of pathologies such as hemorrhage, coagulopathy, neurotoxic effects, and even hypotensive shock [119]. Antivenoms can be classified as monovalent or polyvalent, depending on the immunogen used during production. Monovalent antivenoms are a result of immunizing animals with venom from a single snake species. In contrast, polyvalent antivenoms contain antibodies produced from a cocktail of several medically relevant snake venoms from a diverse geographical region [120, 121].

Besides, the use of antivenoms has led to a significant reduction in morbidity and mortality in the number of registered accidents involving complex animal venoms [122]. Therefore, antivenom efficacy is typically limited to those species whose venoms can be used as immunogens and, in several cases, closely related to snake species that share sufficient toxin overlap for the generated antibodies that recognize and neutralize the key toxic components [38, 121]. Moreover, the complexity and variability of the venom components contribute to the challenge of developing more assertive therapeutic alternatives to deal with the toxic effects of envenoming. Even though antivenom is the primary treatment for systemic effects of envenomation, it is ineffective in treating local injuries [123]. The pharmacokinetics of these immunobiological help to explain the ineffectiveness in some cases; due to the variation in molecular weight presented by different types of antibodies, there are many pharmacokinetic profiles, and these differences end up influencing the pharmacodynamics of antibodies so that they can be ineffective for specific actions [124]. Regarding hemorrhagic effects, studies have shown that antivenoms act only partially in the inhibition of local effects. da Silva and collaborators [125], showed a weak action of antibothropic Brazilian antivenom against the hemorrhagic activity of *B. jararaca* venom in a pre-treatment protocol. Thus, this weak neutralization effect against the hemorrhage can be explained by the difficulty of the antivenom in penetrating through the tissue barriers and reaching the injury site, which can be related to the above-mentioned pharmacokinetic properties [3, 124].

In the past few years, studies have shown pharmacological evidence regarding the benefits of various extracts and compounds isolated from different plant species against the local and systemic effects induced by a wide range of snake venoms, including lethality. In addition to natural sources, there are several small molecules with toxin-neutralizing mechanisms, and some of these therapeutic alternatives are shown below, with an emphasis on strategies targeting hemostatic disturbance therapies [85, 126, 127].

Molecules from natural products as alternative therapy methods for snake envenoming local effects

Over the years, the ethnobotanical studies have enabled, the identification of medicinal plants that provide compounds that inhibit the action of snake venom. It has been shown that secondary metabolites isolated mainly from plants can affect human homeostasis [128].

There are few mechanisms related to the therapeutic action of plant extracts and isolated compounds from plants: precipitation of proteins, ions chelation, inhibition of specific enzymes, and others. Concerning the secondary metabolites, the flavonoids are of great interest when it comes to action against snake venom-induced hemorrhage. This class of compounds embraces a range of possibilities of hydrogen bonding that is arranged around a small carbon skeleton, and these structures are capable of interacting with molecular targets [127]. Dietary flavonoids, such as catechin, quercetin, luteolin, kaempferol, myricetin, and apigenin, are examples of compounds that have antivenom capacity, which occurs through the inhibition of venom hyaluronidases [129]. Another well-known dietary flavonoid, rutin, proved to be a potent antioxidant and a hemostasis modulatory compound. It was recently demonstrated that rutin attenuates local hemorrhage and the increase in reactive species and prevents the fall in red blood cell counts and fibrinogen levels provoked by envenomation [130].

Another known mechanism of flavonoids is the ability to cause protein precipitation. An example is the action of an isolated flavanone named pinostrobin, obtained from the leaves of *Renealmia alpinia*, which is known for its anti-hemorrhagic and anti-myotoxic activities, herewith its ability to neutralize the *in vitro* activities against the venom of *B. asper* [131].

The chelating activity of compounds present in plants is also a mechanism to minimize the injuries caused by snake venom. As metalloproteinases play such a prominent role in local effects, their inhibition can drastically reduce the extension of local tissue damage. As proper coordination of metal ions is necessary for the appropriate activity of metalloproteases, metabolites can interfere in the metal ion coordination of catalytic sites, therefore, minimizing the damage induced by these enzymes [132].

Some compounds have no specific binding site to the enzymes that cause hemorrhagic effects, like tannins and other phenolic compounds that interact with the snake venom enzymes by non-specific binding proteins. Over the past few years, studies have concluded that the effect of these compounds is a result of the interaction between venom enzymes and the hydroxyl group of the above-mentioned compounds, resulting in the formation of hydrogen bonds that stabilize the complexes formation [133].

Small molecule inhibitors as alternative therapy methods

Small molecule inhibitors are small size compounds compared with conventional antibodies. They are regarded as promising alternatives for affordable snakebite treatment because their small size confers desirable drug-favorable properties, once they are inhibitors of high diffusibility and, depending on the pharmacokinetics and physicochemical characteristics that can be administered in the field and in isolated areas where medical access is limited [126, 134].

Many of the small molecules currently known to inhibit SVMP venom toxins are repurposed molecules. One example is the metalloprotease inhibitor batimastat and its water-soluble orally available prodrug marimastat. This molecule was initially developed to treat solid tumors, including pancreatic, lung, breast, colorectal, brain and prostate cancer [135], and was later repurposed to minimize the effects of envenoming [136]. Marimastat is a matrix metalloprotease inhibitor, and its action gives through the connection to the matrix metalloprotease active site, coordinating the metallic ion [137]. Arias et al. [138] showed that marimastat effectively inhibits hemorrhage and the coagulant and defibrinogenating effects caused by the venom of *Echis ocellatus*. Another excellent example of the use of small-molecule inhibitors for the neutralization of enzymatic toxins is varespladib and also its prodrug, methyl-varespladib. These molecules were also developed for other health problems and then were repurposed to be used to inhibit phospholipase A2 (PLA2) [139].

Xie and collaborators [126] investigated the effect of the small molecule inhibitors marimastat, dimercaprol, and DMPS that target snake venom metalloproteinases on inhibiting the activities of coagulopathic toxins found in *Viperidae* snake venoms, and they have concluded that these molecules are promising candidates for affordable treatment of snakebite envenomation.

Another interesting group of small molecule enzymatic inhibitors consists of metal chelators, and EDTA is one of the main ones. Rucavado and collaborators [85] showed that the injection of the $CaNa_2EDTA$ chelator or the batimastat inhibitor in situ right after the venom injection is able to abrogate the hemorrhagic and dermonecrotic activity of the venom of *B. asper*. Dimercaprol and its derivative 2,3-dimercapto-1-propane sulfonic acid (DMPS) are also ion chelators and were found to antagonize the activity of Zn^{2+} dependent snake venom metalloproteinases in vitro. Xie and collaborators [126] recently investigated the value of metal ion chelators as prehospital therapeutics for snakebite. They concluded that the potential clinical scenario of early oral DMPS therapy combined with a delayed, intravenous dose of conventional antivenom could provide more extended protection against effects of envenoming.

Therefore, it is clear that there is a wide variety of molecules that can minimize the effects of envenomation; however, although the results of the studies are promising, none of the small molecule inhibitors have yet progressed to clinical trials.

Conclusions remarks

The present review includes studies that revealed experimental evidence of distinct targets of hemorrhagic metalloproteinases from *Viperidae* venoms. The integration of accumulated data in the literature from detailed molecular analysis of cleavage site specificity of SVMPs to culture cell assays that somehow preclude the investigated in vivo effects of hemorrhage has enabled the construction of a possible mechanism of action of these hemorrhagic toxins.

However, there is still a long road to complete the puzzle of molecular targets and signaling events that fully explain this pathological effect. There is a brilliant list of unsolved issues regarding the pathogenesis of hemorrhage that was already published and is still up-to-date (for review, please check Ref. [88]). For few illustrative examples, the literature has extensively shown the differences in hemorrhage levels induced by PI and PIII classes, as well as the distinct proteolytic cleavage patterns of BM and ECM components. However, little is known about the hydrolysis products, which could generate bioactive molecules that can play important roles in the bleeding process. Moreover, the secretion of endogenous proteases stimulated by SVMPs on hemorrhagic sites may also explain the poor neutralization of local effects by antivenoms, but the participation of MMPs and other endogenous enzymes is still unknown. Another intriguing question is related to the systemic hemorrhage in tissues distant from the bitten site that can be reproduced by injection of purified SVMPs—How is the distribution of SVMP into the body? Is there a preference for specific organs, such as lungs and kidneys? Which are the mechanisms involved in systemic hemorrhage?

There is a consensus in the Toxinology field that an in-depth description of molecular pathways involved in hemorrhage events induced by SVMPs is pivotal to drive the development of more specific treatments. This kind of search for new treatment strategies was established by WHO as one of the objectives globally required to reduce death and disability by 50% until 2030. Therefore, the open questions are a stimulus for new integrative studies that can contribute to more effective and promised therapies.

References

[1] Gutiérrez JM, Williams D, Fan HW, Warrell DA. Snakebite envenoming from a global perspective: towards an integrated approach. Toxicon 2010;56(7):1223–35.

[2] World Health Organization (WHO). Snakebite [Internet]. [cited 2021 Jan 8]. Available from: https://www.who.int/health-topics/snakebite#tab=tab_1; 2021.

[3] Chippaux JP, Goyffon M. Venoms, antivenoms and immunotherapy. Toxicon 1998;36(6):823–46.

[4] Bagcchi S. Experts call for snakebite to be re-established as a neglected tropical disease. BMJ 2015;351(October):h5313.

[5] Kasturiratne A, Wickremasinghe AR, De Silva N, Gunawardena NK, Pathmeswaran A, Premaratna R, et al. The global burden of snakebite: a literature analysis and modelling based on regional estimates of envenoming and deaths. PLoS Med 2008;5(11):1591–604.

[6] Warrell DA. Snake bite. Lancet [Internet] 2010;375(9708):77–88. Available from https://doi.org/10.1016/S0140-6736(09)61754-2.

[7] Gutiérrez JM, Calvete JJ, Habib AG, Harrison RA, Williams DJ, Warrell DA. Snakebite envenoming. In: Vol. 3, Nature reviews. Disease primers; 2017. p. 17063.

[8] Harrison RA, Hargreaves A, Wagstaff SC, Faragher B, Lalloo DG. Snake envenoming: a disease of poverty. PLoS Negl Trop Dis 2009;3(12).

[9] Chippaux JP, Saz-Parkinson Z, Amate Blanco JM. Epidemiology of snakebite in Europe: comparison of data from the literature and case reporting. Toxicon 2013 Dec 15;76:206–13.

[10] Chippaux JP, Massougbodji A, Habib AG. The WHO strategy for prevention and control of snakebite envenoming: a sub-Saharan Africa plan. J Venom Anim Toxins Incl Trop Dis 2019;25(November 2019):4–9.

[11] Chippaux JP. Snakebite envenomation turns again into a neglected tropical disease! J Venom Anim Toxins Incl Trop Dis 2017;23(1):1–2.
[12] Williams DJ, Faiz MA, Abela-Ridder B, Ainsworth S, Bulfone TC, Nickerson AD, et al. Strategy for a globally coordinated response to a priority neglected tropical disease: snakebite envenoming. PLoS Negl Trop Dis 2019;13(2):12–4.
[13] Burbrink F, Crother B. Evolution and taxonomy of snakes. In: Reproductive biology and phylogeny of snakes, 2011; April 2011. p. 19–53.
[14] Perry G, Lacy M, Das I. May. Problematic wildlife II; 2020.
[15] Pyron RA, Burbrink FT, Wiens JJ. A phylogeny and revised classification of Squamata, including 4161 species of lizards and snakes. BMC Evol Biol 2013;13(1).
[16] Wüster W, Peppin L, Pook CE, Walker DE. A nesting of vipers: phylogeny and historical biogeography of the Viperidae (Squamata: Serpentes). Mol Phylogenet Evol [Internet] 2008;49(2):445–59. Available from: https://doi.org/10.1016/j.ympev.2008.08.019.
[17] White J, Persson H. Snakes. In: Human toxicology [Internet]. Elsevier; 1996. p. 757–802. [cited 2021 Feb 16]. Available from: https://linkinghub.elsevier.com/retrieve/pii/B9780444815576500344.
[18] Alencar LRV, Quental TB, Grazziotin FG, Alfaro ML, Martins M, Venzon M, et al. Diversification in vipers: phylogenetic relationships, time of divergence and shifts in speciation rates. Mol Phylogenet Evol [Internet] 2016;105:50–62. Available from: https://doi.org/10.1016/j.ympev.2016.07.029.
[19] Adukauskienė D, Varanauskienė E, Adukauskaitė A. Continuing medical education. In: Venomous snakebites, vol 47. Kaunas: Medicina; 2011.
[20] Naik BS. "Dry bite" in venomous snakes: a review. In: Toxicon, vol 133. Elsevier Ltd; 2017. p. 63–7.
[21] Benjamin JM, Abo BN, Brandehoff N. Review article: snake envenomation in Africa. Curr Trop Med Reports 2020;7(1).
[22] World Health Organization (WHO). Guidelines for the prevention and clinical management of snakebite in Africa. World Health Organization. Regional Office for Africa; 2010.
[23] Ahmed SM, Ahmed M, Nadeem A, Mahajan J, Choudhary A, Pal J. Emergency treatment of a snake bite: pearls from literature. J Emerg Trauma Shock [Internet] 2008 Jul;1(2):97–105. Available from: https://pubmed.ncbi.nlm.nih.gov/19561988.
[24] Feola A, Marella GL, Carfora A, Della Pietra B, Zangani P, Pietro CC. Snakebite envenoming a challenging diagnosis for the forensic pathologist: a systematic review. Toxins (Basel) 2020;12(11):1–19.
[25] Goswami PK, Samant M, Srivastava RS. Snake venom, anti-snake venom & potential of snake venom. Int J Pharm Pharm Sci 2014;6(5):4–7.
[26] Gutiérrez JM, Calvete JJ, Habib AG, Harrison RA, Williams DJ, Warrell DA. Snakebite envenoming. Nat Rev Dis Primers 2017 Sep 14;3:17063.
[27] Slagboom J, Kool J, Harrison RA, Casewell NR. Haemotoxic snake venoms: their functional activity, impact on snakebite victims and pharmaceutical promise. Br J Haematol 2017;177(6):947–59.
[28] Chakrabarty D, Sarkar A. Cytotoxic effects of snake venoms 14. Snake Venoms 2017;327.
[29] Kini RM, Rao VS, Joseph JS. Procoagulant proteins from snake venoms. Haemostasis 2001;31(3–6):218–24.
[30] Berling I, Isbister GK. Hematologic effects and complications of snake envenoming. Transfus Med Rev [Internet] 2015;29(2):82–9. Available from: https://doi.org/10.1016/j.tmrv.2014.09.005.
[31] Maduwage K, Buckley NA, de Silva HJ, Lalloo DG, Isbister G. Snake antivenom for snake venom induced consumption coagulopathy. Cochrane Database Syst Rev 2014;2014(12).
[32] Rathnayaka RMMKN, Ranathunga PEAN, Kularatne SAM. Systemic bleeding including pulmonary haemorrhage following hump-nosed pit viper (Hypnale hypnale) envenoming: a case report from Sri Lanka. Toxicon [Internet] 2019;170(September):21–8. Available from: https://doi.org/10.1016/j.toxicon.2019.09.009.
[33] Hati R, Mitra P, Sarker S, Bhattacharyya KK. Snake venom hemorrhagins. Crit Rev Toxicol 1999;29(1):1–19.
[34] Powell R. Snakes. In: Encyclopedia of toxicology. Elsevier; 2005. p. 57–60.

[35] Menezes MC, Furtado MF, Travaglia-Cardoso SR, Camargo ACM, Serrano SMT. Sex-based individual variation of snake venom proteome among eighteen Bothrops jararaca siblings. Toxicon 2006 Mar 1;47(3):304–12.

[36] Aguilar I, Guerrero B, Maria Salazar A, Girón ME, Pérez JC, Sánchez EE, et al. Individual venom variability in the south American rattlesnake Crotalus durissus cumanensis. Toxicon 2007;50(2):214–24.

[37] Alape-Girón A, Sanz L, Escolano J, Flores-Díaz M, Madrigal M, Sasa M, et al. Snake venomics of the lancehead pitviper bothrops asper. Geographic, individual, and ontogenetic variations. J Proteome Res 2008;7(8):3556–71.

[38] Casewell NR, Cook DAN, Wagstaff SC, Nasidi A, Durfa N, Wüster W, et al. Pre-clinical assays predict Pan-African Echis viper efficacy for a species-specific antivenom. PLoS Negl Trop Dis [Internet] 2010;4(10). Available from: www.plosntds.org.

[39] Zelanis A, Tashima AK, Pinto AFM, Leme AFP, Stuginski DR, Furtado MF, et al. Bothrops jararaca venom proteome rearrangement upon neonate to adult transition. Proteomics 2011;11(21):4218–28.

[40] Zelanis A, Andrade-Silva D, Rocha MM, Furtado MF, Serrano SMT, Junqueira-de-Azevedo ILM, et al. A transcriptomic view of the proteome variability of newborn and adult Bothrops jararaca snake venoms. PLoS Negl Trop Dis 2012;6(3):15–8.

[41] Zelanis A, Tashima AK, Rocha MMT, Furtado MF, Camargo ACM, Ho PL, et al. Analysis of the ontogenetic variation in the venom proteome/peptidome of Bothrops jararaca reveals different strategies to deal with prey. J Proteome Res [Internet] 2010;9(5):2278–91. Available from: https://doi.org/10.1021/pr901027r.

[42] Tasoulis T, Isbister GK. A review and database of snake venom proteomes. Toxins (Basel) 2017;9(9):290.

[43] Bjarnason JB, Fox JW. Snake venom metalloendopeptidases: reprolysins. Methods Enzymol 1995;248(C):345–68.

[44] Takeda S, Takeya H, Iwanaga S. Snake venom metalloproteinases: structure, function and relevance to the mammalian ADAM/ADAMTS family proteins. Biochim Biophys Acta-Proteins Proteomics [Internet] 2012;1824(1):164–76. Available from: https://doi.org/10.1016/j.bbapap.2011.04.009.

[45] Fox JW, Serrano SMT. Insights into and speculations about snake venom metalloproteinase (SVMP) synthesis, folding and disulfide bond formation and their contribution to venom complexity. FEBS J 2008;275(12):3016–30.

[46] Gutiérrez JM, Escalante T, Rucavado A, Herrera C, Fox JW. A comprehensive view of the structural and functional alterations of extracellular matrix by snake venom metalloproteinases (SVMPs): novel perspectives on the pathophysiology of envenoming. Toxins (Basel) 2016;8(10).

[47] Escalante T, Rucavado A, Fox JW, Gutiérrez JM. Key events in microvascular damage induced by snake venom hemorrhagic metalloproteinases. J Proteomics [Internet] 2011 Aug 24;74(9):1781–94. Available from: https://doi.org/10.1016/j.jprot.2011.03.026.

[48] Gutiérrez JM, Escalante T, Rucavado A, Herrera C. Hemorrhage caused by snake venom metalloproteinases: a journey of discovery and understanding. Toxins (Basel) 2016 Mar 26;8(4).

[49] Serrano SMT, Jia L-G, Wang D, Shannon JD, Fox JW. Function of the cysteine-rich domain of the haemorrhagic metalloproteinase atrolysin a: targeting adhesion proteins collagen I and von Willebrand factor. Biochem J 2005;391:69–76.

[50] Serrano SMT, Kim J, Wang D, Dragulev B, Shannon JD, Mann HH, et al. The cysteine-rich domain of snake venom metalloproteinases is a ligand for von Willebrand factor a domains: role in substrate targeting. J Biol Chem [Internet] 2006;281(52):39746–56. Available from: https://doi.org/10.1074/jbc.M604855200.

[51] Serrano SMT, Wang D, Shannon JD, Pinto AFM, Polanowska-Grabowska RK, Fox JW. Interaction of the cysteine-rich domain of snake venom metalloproteinases with the A1 domain of von Willebrand factor promotes site-specific proteolysis of von Willebrand factor and inhibition of von Willebrand factor-mediated platelet aggregation. FEBS J 2007;274(14):3611–21.

[52] Pinto AFM, Ma L, Dragulev B, Guimaraes JA, Fox JW. Use of SILAC for exploring sheddase and matrix degradation of fibroblasts in culture by the PIII SVMP atrolysin A: identification of two novel substrates with functional relevance. Arch Biochem Biophys 2006 Sep 1;465(1):11–5.

[53] Menezes MC, Paes Leme AF, Melo RL, Silva CA, Della Casa M, Bruni FM, et al. Activation of leukocyte rolling by the cysteine-rich domain and the hyper-variable region of HF3, a snake venom hemorrhagic metalloproteinase. FEBS Lett 2008 Nov 26;582(28):3915–21.

[54] Oliveira AK, Paes Leme AF, Asega AF, Camargo ACM, Fox JW, Serrano SMT. New insights into the structural elements involved in the skin haemorrhage induced by snake venom metalloproteinases. Thromb Haemost 2010 Sep;104(3):485–97.

[55] Menezes MC, De Oliveira AK, Melo RL, Lopes-Ferreira M, Rioli V, Balan A, et al. Disintegrin-like/cysteine-rich domains of the reprolysin HF3: site-directed mutagenesis reveals essential role of specific residues. Biochimie [Internet] 2011;93(2):345–51. Available from: https://doi.org/10.1016/j.biochi.2010.10.007.

[56] Bjarnason JB, Fox JW. Hemorrhagic metalloproteinases from snake venoms. In: Pharmacology and therapeutics, vol 62. Pergamon; 1994. p. 325–72.

[57] Ohsaka A. Hemorrhagic, necrotizing and edema-forming effects of snake venoms. In: Lee C-Y, editor. Snake venoms [Internet]. Berlin, Heidelberg: Springer Berlin Heidelberg; 1979. p. 480–546. Available from: https://doi.org/10.1007/978-3-642-66913-2_14.

[58] Fox JW, Serrano SMT. Structural considerations of the snake venom metalloproteinases, key members of the M12 reprolysin family of metalloproteinases. Toxicon 2005;45(8):969–85.

[59] Franceschi A, Rucavado A, Mora N, Gutiérrez JM. Purification and characterization of BaH4, a hemorrhagic metalloproteinase from the venom of the snake Bothrops asper. Toxicon 2000;38(1):63–77.

[60] Gutiérrez JM, Rucavado A. Snake venom metalloproteinases: their role in the pathogenesis of local tissue damage. Biochimie 2000;82(9–10):841–50.

[61] Gutiérrez JM, Rucavado A, Escalante T, Díaz C. Hemorrhage induced by snake venom metalloproteinases: biochemical and biophysical mechanisms involved in microvessel damage. Toxicon 2005;45(8):997–1011.

[62] Gutiérrez JM, Núñez J, Escalante T, Rucavado A. Blood flow is required for rapid endothelial cell damage induced by a snake venom hemorrhagic metalloproteinase. Microvasc Res 2006;71(1):55–63.

[63] Escalante T, Shannon J, Moura-da-Silva AM, María Gutiérrez J, Fox JW. Novel insights into capillary vessel basement membrane damage by snake venom hemorrhagic metalloproteinases: a biochemical and immunohistochemical study. Arch Biochem Biophys 2006;455(2):144–53.

[64] Paes Leme AF, Sherman NE, Smalley DM, Sizukusa LO, Oliveira AK, Menezes MC, et al. Hemorrhagic activity of HF3, a snake venom metalloproteinase: insights from the proteomic analysis of mouse skin and blood plasma. J Proteome Res 2012;11(1):279–91.

[65] Asega AF, Menezes MC, Trevisan-Silva D, Cajado-Carvalho D, Bertholim L, Oliveira AK, et al. Cleavage of proteoglycans, plasma proteins and the platelet-derived growth factor receptor in the hemorrhagic process induced by snake venom metalloproteinases. Sci Rep [Internet] 2020;10(1):1–17. Available from: https://doi.org/10.1038/s41598-020-69396-y.

[66] De Almeida MT, Freitas-De-Sousa LA, Colombini M, Gimenes SNC, Kitano ES, Faquim-Mauro EL, et al. Inflammatory reaction induced by two metalloproteinases isolated from *Bothrops atrox* venom and by fragments generated from the hydrolysis of basement membrane components. Toxins (Basel) 2020;12(2).

[67] Silva CCF, Menezes MC, Palomino M, Oliveira AK, Iwai LK, Faria M, et al. Peptides derived from plasma proteins released by bothropasin, a metalloprotease present in the Bothrops jararaca venom. Toxicon 2017 Oct 1;137:65–72.

[68] Ferreira AK, Cristofaro B, Menezes MC, de Oliveira AK, Tashima AK, de Melo RL, et al. Alphastatin-C a new inhibitor of endothelial cell activation is a pro-arteriogenic agent in vivo and retards B16-F10 melanoma growth in a preclinical model. Oncotarget 2020 Dec 22;11(51):4770–87.

[69] Paes Leme AF, Escalante T, Pereira JGC, Oliveira AK, Sanchez EF, Gutiérrez JM, et al. High resolution analysis of snake venom metalloproteinase (SVMP) peptide bond cleavage specificity using proteome based peptide libraries and mass spectrometry. J Proteomics [Internet] 2011;74(4):401–10. Available from: https://doi.org/10.1016/j.jprot.2010.12.002.

[70] Zelanis A, Oliveira AK, Prudova A, Huesgen PF, Tashima AK, Kizhakkedathu J, et al. Deep profiling of the cleavage specificity and human substrates of snake venom metalloprotease HF3 by proteomic identification

of cleavage site specificity (PICS) using proteome derived peptide libraries and terminal amine isotopic labeling of substrates. J Proteome Res 2019;18(9):3419–28.

[71] Bertholim L, Zelanis A, Oliveira AK, Serrano SMT. Proteome-derived peptide library for the elucidation of the cleavage specificity of HF3, a snake venom metalloproteinase. Amino Acids 2016;48(5):1331–5.

[72] Escalante T, Núñez J, Moura da Silva AM, Rucavado A, RDG T, Gutiérrez JM. Pulmonary hemorrhage induced by jararhagin, a metalloproteinase from Bothrops jararaca snake venom. Toxicol Appl Pharmacol 2003 Nov 15;193(1):17–28.

[73] Loría GD, Rucavado A, Kamiguti AS, Theakston RDG, Fox JW, Alape A, et al. Characterization of "basparin A," a prothrombin-activating metalloproteinase, from the venom of the snake *Bothrops asper* that inhibits platelet aggregation and induces defibrination and thrombosis. Arch Biochem Biophys 2003 Oct 1;418(1):13–24.

[74] Baldo C, Jamora C, Yamanouye N, Zorn TM, Moura-da-Silva AM. Mechanisms of vascular damage by hemorrhagic snake venom metalloproteinases: tissue distribution and in situ hydrolysis. PLoS Negl Trop Dis 2010 Jun;4(6).

[75] Pinto AFM, Terra RMS, Guimaraes JA, Fox JW. Mapping von Willebrand factor a domain binding sites on a snake venom metalloproteinase cysteine-rich domain. Arch Biochem Biophys 2007;457(1):41–6.

[76] Herrera C, Escalante T, Voisin MB, Rucavado A, Morazán D, Macêdo JKA, et al. Tissue localization and extracellular matrix degradation by PI, PII and PIII snake venom metalloproteinases: clues on the mechanisms of venom-induced hemorrhage. PLoS Negl Trop Dis 2015 Apr 24;9(4):1–20.

[77] Herrera C, Escalante T, Rucavado A, Gutiérrez JM. Hemorrhagic and procoagulant P-III snake venom metalloproteinases differ in their binding to the microvasculature of mouse cremaster muscle. Toxicon 2020 Apr 30;178:1–3.

[78] McKay DG, Moroz C, De Vries A, Csavossy I, Cruse V. The action of hemorrhagin and phospholipase derived from Vipera palestinae venom on the microcirculation. Lab Invest 1970 May;22(5):387–99.

[79] Kamiguti AS, Cardoso JLC, Theakston RDG, Sano-Martins IS, Hutton RA, Rugman FP, et al. Coagulopathy and haemorrhage in human victims of Bothrops jararaca envenoming in Brazil. Toxicon 1991;29(8):961–72.

[80] Anderson SG, Ownby CL. Systemic hemorrhage induced by proteinase H from Crotalus adamanteus (eastern diamondback rattlesnake) venom. Toxicon 1997 Aug 1;35(8):1301–13.

[81] Rahmy T, Tu AT, El Banhawey M, El-Asmar MF, Hassan FM. Cytopathologic effect of Cerastes cerastes (Egyptian sand viper) venom and isolated hemorrhagic toxin on liver and kidney: an electron microscopic study. J Nat Toxins 1992;1:45–58.

[82] Gutiérrez JM, Romero M, Núñez J, Chaves F, Borkow G, Ovadia M. Skeletal muscle necrosis and regeneration after injection of BaH1, a hemorrhagic metalloproteinase isolated from the venom of the snake Bothrops asper (terciopelo). Exp Mol Pathol 1995 Feb 1;28–41. Academic Press.

[83] Rucavado A, Soto M, Escalante T, Loría GD, Arni R, Gutiérrez JM. Thrombocytopenia and platelet hypoaggregation induced by Bothrops asper snake venom. Toxins involved and their contribution to metalloproteinase-induced pulmonary hemorrhage. Thromb Haemost 2005;94(1):123–31.

[84] Lomonte B, Rangel J. Snake venom Lys49 myotoxins: from phospholipases A 2 to non-enzymatic membrane disruptors. Toxicon 2012 Sep 15;60(4):520–30.

[85] Rucavado A, Escalante T, Franceschi A, Chaves F, León G, Cury Y, et al. Inhibition of local hemorrhage and dermonecrosis induced by Bothrops asper snake venom: effectiveness of early in situ administration of the peptidomimetic metalloproteinase inhibitor batimastat and the chelating agent CaNa2EDTA. Am J Trop Med Hyg 2000;63(5–6):313–9.

[86] Gutiérrez JM, Escalante T, Hernández R, Gastaldello S, Saravia-Otten P, Rucavado A. Why is skeletal muscle regeneration impaired after myonecrosis induced by viperid snake venoms? Toxins (Basel) 2018;10(5):1–21.

[87] Williams HF, Mellows BA, Mitchell R, Sfyri P, Layfield HJ, Salamah M, et al. Mechanisms underpinning the permanent muscle damage induced by snake venom metalloprotease. PLoS Negl Trop Dis [Internet] 2019;13(1):1–20. Available from: https://doi.org/10.1371/journal.pntd.0007041.

[88] Gutiérrez JM, Rucavado A, Escalante T, Herrera C, Fernández J, Lomonte B, et al. Unresolved issues in the understanding of the pathogenesis of local tissue damage induced by snake venoms. In: Toxicon, vol 148. Elsevier Ltd; 2018. p. 123–31.
[89] Moura-da-Silva A, Butera D, Tanjoni I. Importance of snake venom metalloproteinases in cell biology: effects on platelets,inflammatory and endothelial cells. Curr Pharm Des 2007;13(28):2893–905.
[90] Macêdo JKA, Fox JW. Biological activities and assays of the snake venom metalloproteinases (SVMPs). In: Venom genomics and proteomics. Dordrecht, The Netherlands: Springer; 2014. p. 1–24.
[91] Zhao X, Guan JL. Focal adhesion kinase and its signaling pathways in cell migration and angiogenesis. In: Advanced drug delivery reviews, vol 63. Elsevier; 2011. p. 610–5.
[92] Michael KE, Dumbauld DW, Burns KL, Hanks SK, García AJ. Focal adhesion kinase modulates cell adhesion strengthening via integrin activation. Mol Biol Cell [Internet] 2009;20(9):2508–19. Available from: https://doi.org/10.1091/mbc.e08-01-0076.
[93] Baldo C, Tanjoni I, León IR, Batista IFC, Della-Casa MS, Clissa PB, et al. BnP1, a novel P-I metalloproteinase from Bothrops neuwiedi venom: biological effects benchmarking relatively to jararhagin, a P-III SVMP. Toxicon 2008;51(1):54–65.
[94] Baldo C, Lopes DS, Faquim-Mauro EL, Jacysyn JF, Niland S, Eble JA, et al. Jararhagin disruption of endothelial cell anchorage is enhanced in collagen enriched matrices. Toxicon 2015 Dec 15;108:240–8.
[95] Tanjoni I, Weinlich R, Della-Casa MS, Clissa PB, Saldanha-Gama RF, De Freitas MS, et al. Jararhagin, a snake venom metalloproteinase, induces a specialized form of apoptosis (anoikis) selective to endothelial cells. Apoptosis 2005;10(4):851–61.
[96] Kakanj M, Ghazi-Khansari M, Mirakabadi AZ, Daraei B, Vatanpour H. Cytotoxic effect of Iranian Vipera lebetina snake venom on HUVEC cells. Iran J Pharm Res 2015;14(September 2014):109–14.
[97] Zigrino P, Kamiguti AS, Eble J, Drescher C, Nischt R, Fox JW, et al. The reprolysin Jararhagin, a snake venom metalloproteinase, functions as a fibrillar collagen agonist involved in fibroblast cell adhesion and signaling. J Biol Chem 2002 Oct 25;277(43):40528–35.
[98] Gallagher P, Bao Y, Serrano SMT, Laing GD, Theakston RDG, Gutiérrez JM, et al. Role of the snake venom toxin jararhagin in proinflammatory pathogenesis: in vitro and in vivo gene expression analysis of the effects of the toxin. Arch Biochem Biophys 2005 Sep 1;441(1):1–15.
[99] Menezes MC, Kitano ES, Bauer VC, Oliveira AK. Early response of C2C12 myotubes to a sub-cytotoxic dose of hemorrhagic metalloproteinase HF3 from Bothrops jararaca venom. J Proteomics [Internet] 2019;198(October 2018):163–76. Available from: https://doi.org/10.1016/j.jprot.2018.12.006.
[100] LeBlanc NJ, Menet R, Picard K, Parent G, Tremblay M-È, ElAli A. Canonical Wnt pathway maintains blood-brain barrier integrity upon ischemic stroke and its activation ameliorates tissue plasminogen activator therapy. Mol Neurobiol 2019;56(9):6521–38.
[101] Seo T, Sakon T, Nakazawa S, Nishioka A, Watanabe K, Matsumoto K, et al. Haemorrhagic snake venom metalloproteases and human ADAMs cleave LRP5/6, which disrupts cell–cell adhesions in vitro and induces haemorrhage in vivo. FEBS J 2017;284(11):1657–71.
[102] Zhong X, Rescorla FJ. Cell surface adhesion molecules and adhesion-initiated signaling: Understanding of anoikis resistance mechanisms and therapeutic opportunities. In: Cellular signalling, vol 24. Pergamon; 2012. p. 393–401.
[103] Díaz C, Valverde L, Brenes O, Rucavado A, Gutiérrez JM. Characterization of events associated with apoptosis/anoikis induced by snake venom metalloproteinase BaP1 on human endothelial cells. J Cell Biochem 2005;94(3):520–8.
[104] Araki S, Masuda S, Maeda H, Ying MJ, Hayashi H. Involvement of specific integrins in apoptosis induced by vascular apoptosis-inducing protein 1. Toxicon 2002 May 1;40(5):535–42.
[105] Akiyama M, Takeda S, Kokame K, Takagi J, Miyata T. n.d. Crystal structures of the noncatalytic domains of ADAMTS13 reveal multiple discontinuous exosites for von Willebrand factor [Internet]. Available from: www.pnas.org/cgi/content/full/.
[106] Seals DF, Courtneidge SA. The ADAMs family of metalloproteases: multidomain proteins with multiple functions. Genes Dev 2003;17(1):7–30.

[107] Shimokawa KI, Shannon JD, Jia LG, Fox JW. Sequence and biological activity of catrocollastatin-C: a disintegrin- like/cysteine-rich two-domain protein from Crotalus atrox venom. Arch Biochem Biophys 1997;343(1):35–43.

[108] Souza DHF, Iemma MRC, Ferreira LL, Faria JP, Oliva MLV, Zingali RB, et al. The disintegrin-like domain of the snake venom metalloprotease alternagin inhibits $\alpha 2\beta 1$ integrin-mediated cell adhesion. Arch Biochem Biophys 2000;384(2):341–50.

[109] Zeltz C, Gullberg D. The integrin—collagen connection—a glue for tissue repair? J Cell Sci 2016;129(4):653–64.

[110] Zolkiewska A. Disintegrin-like/cysteine-rich region of ADAM 12 is an active cell adhesion domain. Exp Cell Res 1999;252(2):423–31.

[111] Smith KM, Gaultier A, Cousin H, Alfandari D, White JM, DeSimone DW. The cysteine-rich domain regulates ADAM protease function in vivo. J Cell Biol 2002;159(5):893–902.

[112] Alfandari D, Cousin H, Gaultier A, Smith K, White JM, Darribère T, et al. Xenopus ADAM 13 is a metalloprotease required for cranial neural crest-cell migration. Curr Biol 2001;11(12):918–30.

[113] Gaultier A, Cousin H, Darribère T, Alfandari D. ADAM13 disintegrin and cysteine-rich domains bind to the second heparin-binding domain of fibronectin. J Biol Chem 2002 Jun 28;277(26):23336–44.

[114] Janes PW, Saha N, Barton WA, Kolev MV, Wimmer-Kleikamp SH, Nievergall E, et al. Adam meets Eph: an ADAM substrate recognition module acts as a molecular switch for ephrin cleavage in trans. Cell 2005;123(2):291–304.

[115] de Groot R, Lane DA, Crawley JTB. The role of the ADAMTS13 cysteine-rich domain in VWF binding and proteolysis. Blood J Am Soc Hematol 2015;125(12):1968–75.

[116] Ainsworth S, Slagboom J, Alomran N, Pla D, Alhamdi Y, King SI, et al. The paraspecific neutralisation of snake venom induced coagulopathy by antivenoms. Commun Biol 2018;1(1).

[117] Gómez-Betancur I, Gogineni V, Salazar-Ospina A, León F. Perspective on the therapeutics of anti-snake venom [Internet]. In: Molecules, vol 24; 2019. Available from: http://www.sciencedirect.

[118] Bermúdez-Méndez E, Fuglsang-Madsen A, Føns S, Lomonte B, María Gutiérrez J, Laustsen AH. Toxins innovative immunization strategies for antivenom development. Available from: www.mdpi.com/journal/toxins; 2018.

[119] Calvete JJ, Sanz L, Angulo Y, Lomonte B, Gutiérrez JM. Venoms, venomics, antivenomics. FEBS Lett 2009;583:1736–43. No longer published by Elsevier.

[120] O'Leary MA, Isbister GK. Commercial monovalent antivenoms in Australia are polyvalent. Toxicon 2009 Aug 1;54(2):192–5.

[121] Ferraz CR, Arrahman A, Xie C, Casewell NR, Lewis RJ, Kool J, et al. Multifunctional toxins in snake venoms and therapeutic implications: from pain to hemorrhage and necrosis. Front Ecol Evol 2019;7(Jun):1–19.

[122] Gutiérrez JM, Rucavado A, Chaves F, Díaz C, Escalante T. Experimental pathology of local tissue damage induced by Bothrops asper snake venom. Toxicon 2009;54(7):958–75.

[123] Alangode A, Rajan K, Nair BG. Snake antivenom: challenges and alternate approaches. Biochem Pharmacol 2020;181, 114135. Elsevier Inc.

[124] Gutiérrez JM, León G, Lomonte B. Pharmacokinetic-pharmacodynamic relationships of immunoglobulin therapy for envenomation. Clin Pharmacokinet 2003;42(8):721–41.

[125] da Silva NMV, Arruda EZ, Murakami YLB, Moraes RAM, El-Kik CZ, Tomaz MA, et al. Evaluation of three Brazilian antivenom ability to antagonize myonecrosis and hemorrhage induced by Bothrops snake venoms in a mouse model. Toxicon 2007 Aug 1;50(2):196–205.

[126] Xie C, Albulescu L-O, Bittenbinder MA, Somsen GW, Vonk FJ, Casewell NR, et al. Neutralizing effects of small molecule inhibitors and metal chelators on coagulopathic viperinae snake venom toxins. Biomedicine 2020;8:297. Available from: www.mdpi.com/journal/biomedicines.

[127] Mors WB, Célia do Nascimento M, Ruppelt Pereira BM, Alvares Pereira N. Plant natural products active against snake bite—the molecular approach. Phytochemistry 2000 Nov 30;55(6):627–42.

[128] Matos F de A. Plantas medicinais: guia de seleção e emprego de plantas usadas em fitoterapia no Nordeste do Brasil. 2000.

[129] Xiao J. Dietary flavonoid aglycones and their glycosides: which show better biological significance? Crit Rev Food Sci Nutr [Internet] 2017;57(9):1874–905. Available from: https://doi.org/10.1080/10408398.2015.1032400.

[130] Sachetto ATA, Rosa JG, Santoro ML, et al. Billiald P, editor, PLoS Negl Trop Dis [Internet] 2018 Oct 11;12(10), e0006774. Available from: https://doi.org/10.1371/journal.pntd.0006774.

[131] Gómez-Betancur I, Pereañez JA, Patiño AC, Benjumea D. Inhibitory effect of pinostrobin from Renealmia alpinia, on the enzymatic and biological activities of a PLA2. Int J Biol Macromol 2016 Aug 1;89:35–42.

[132] Soares A, Ticli F, Marcussi S, Lourenco M, Januario A, Sampaio S, et al. Medicinal plants with inhibitory properties against snake venoms. Curr Med Chem 2005;12(22):2625–41.

[133] Toyama DO, Marangoni S, Diz-Filho EBS, Oliveira SCB, Toyama MH. Effect of umbelliferone (7-hydroxycoumarin, 7-HOC) on the enzymatic, edematogenic and necrotic activities of secretory phospholipase A2 (sPLA2) isolated from Crotalus durissus collilineatus venom. Toxicon 2009 Mar 15;53(4):417–26.

[134] Knudsen C, Ledsgaard L, Dehli RI, Ahmadi S, Sørensen CV, Laustsen AH. Engineering and design considerations for next-generation snakebite antivenoms [Internet]. Toxicon 2019;167:67–75. Elsevier Ltd. [cited 2021 Jan 31]. Available from: https://doi.org/10.1016/j.toxicon.2019.06.005.

[135] Bramhall SR, Hallissey MT, Whiting J, Scholefield J, Tierney G, Stuart RC, et al. Marimastat as maintenance therapy for patients with advanced gastric cancer: a randomised trial. Br J Cancer 2002;86(12):1864–70.

[136] Rasmussen HS, PP MC. Matrix metalloproteinase inhibition as a novel anticancer strategy: a review with special focus on Batimastat and Marimastat. In: Pharmacology and therapeutics, vol 75. Pergamon; 1997. p. 69–75.

[137] Underwood CK, Min D, Lyons JG, Hambley TW. The interaction of metal ions and Marimastat with matrix metalloproteinase 9. J Inorg Biochem 2003 Jun 1;95(2–3):165–70.

[138] Arias AS, Rucavado A, Gutiérrez JM. Peptidomimetic hydroxamate metalloproteinase inhibitors abrogate local and systemic toxicity induced by Echis ocellatus (saw-scaled) snake venom. Toxicon 2017 Jun 15;132:40–9.

[139] Lewin M, Samuel S, Merkel J, Bickler P. Varespladib (LY315920) appears to be a potent, broad-spectrum, inhibitor of snake venom phospholipase A2 and a possible pre-referral treatment for envenomation. Toxins (Basel) [Internet] 2016;8(9). Available from: www.mdpi.com/journal/toxins.

[140] Fox JW, Bjarnason JB. Atrolysins: metalloproteinases from Crotalus atrox venom. Methods Enzymol 1995;248(C):368–87.

[141] Gay CC, Maruñak SL, Teibler P, Ruiz R. Acosta de Pérez OC, Leiva LC. Systemic alterations induced by a Bothrops alternatus hemorrhagic metalloproteinase (baltergin) in mice. Toxicon 2009 Jan 1;53(1):53–9.

[142] Freitas-De-Sousa LA, Amazonas DR, Sousa LF, Sant'Anna SS, Nishiyama MY, Serrano SMT, et al. Comparison of venoms from wild and long-term captive Bothrops atrox snakes and characterization of Batroxrhagin, the predominant class PIII metalloproteinase from the venom of this species. Biochimie 2015 Nov 27;118:60–70.

[143] Oliveira AK, Paes Leme AF, Assakura MT, Menezes MC, Zelanis A, Tashima AK, et al. Simplified procedures for the isolation of HF3, bothropasin, disintegrin-like/cysteine-rich protein and a novel P-I metalloproteinase from Bothrops jararaca venom. Toxicon 2009 Jun 1;53(7–8):797–801.

[144] Mandelbaum FR, Reichel AP, Assakura MT. Isolation and characterization of a proteolytic enzyme from the venom of the snake Bothrops jararaca (Jararaca). Toxicon 1982 Jan 1;20(6):955–72.

[145] Muniz JRC, Ambrosio ALB, Selistre-de-Araujo HS, Cominetti MR, Moura-da-Silva AM, Oliva G, et al. The three-dimensional structure of bothropasin, the main hemorrhagic factor from Bothrops jararaca venom: Insights for a new classification of snake venom metalloprotease subgroups. Toxicon 2008 Dec 1;52(7):807–16.

[146] Tachoua W, Boukhalfa-Abib H, Laraba-Djebari F. Hemorrhagic metalloproteinase, Cc HSM-III, isolated from Cerastes cerastes venom: Purification and biochemical characterization. J Biochem Mol Toxicol 2017;31(7):1–10.

[147] Omori-Satoh T, Sadahiro S. Resolution of the major hemorrhagic component of Trimeresurus flavoviridis venom into two parts. BBA Protein Struct 1979 Oct 24;580(2):392–404.

[148] Chen RQ, Jin Y, Wu JB, Zhou XD, Li DS, Lu QM, et al. A novel high molecular weight metalloproteinase cleaves fragment F1 of activated human prothrombin. Toxicon 2004 Sep 1;44(3):281–7.

[149] Assakura MT, Reichl AP, Mandelbaum FR. Comparison of immunological, biochemical and biophysical properties of three hemorrhagic factors isolated from the venom of Bothrops jararaca (jararaca). Toxicon 1986 Jan 1;24(9):943–6.

[150] Mandelbaum FR, Assakura MT, Reichl AP. Characterization of two hemorrhagic factors isolated from the venom of Bothrops neuwiedi (jararaca pintada). Toxicon 1984 Jan 1;22(2):193–206.

[151] Silva CA, Zuliani JP, Assakura MT, Mentele R, Camargo ACM, Teixeira CFP, et al. Activation of α Mβ 2-mediated phagocytosis by HF3, a P-III class metalloproteinase isolated from the venom of Bothrops jararaca. Biochem Biophys Res Commun 2004 Sep 24;322(3):950–6.

[152] Oyama E, Takahashi H. Purification and characterization of two high molecular mass snake venom metalloproteinases (P-III SVMPs), named SV-PAD-2 and HR-Ele-1, from the venom of Protobothrops elegans (Sakishima-habu). Toxicon [Internet] 2015;103(June):30–8. Available from: https://doi.org/10.1016/j.toxicon.2015.06.010.

[153] Moura-da-Silva AM, Baldo C. Jararhagin, a hemorrhagic snake venom metalloproteinase from Bothrops jararaca. Toxicon 2012 Sep 1;60(3):280–9.

[154] Paine MJI, Desmond HP, Theakston RDG, Crampton JM. Purification, cloning, and molecular characterization of a high molecular weight hemorrhagic metalloprotease, jararhagin, from Bothrops jararaca venom. Insights into the disintegrin gene family. J Biol Chem 1992 Nov 15;267(32):22869–76.

[155] Sanchez EF, Gabriel LM, Gontijo S, Gremski LH, Veiga SS, Evangelista KS, et al. Structural and functional characterization of a P-III metalloproteinase, leucurolysin-B, from Bothrops leucurus venom. Arch Biochem Biophys 2007;468(2):193–204.

[156] Hamza L, Gargioli C, Castelli S, Rufini S, Laraba-Djebari F. Purification and characterization of a fibrinogenolytic and hemorrhagic metalloproteinase isolated from Vipera lebetina venom. Biochimie 2010;92(7):797–805.

[157] Kikushima E, Nakamura S, Oshima Y, Shibuya T, Miao JY, Hayashi H, et al. Hemorrhagic activity of the vascular apoptosis inducing proteins VAP1 and VAP2 from Crotalus atrox. Toxicon 2008 Sep 15;52(4):589–93.

[158] Howes J-M, Kamiguti AS, Theakston RDG, Wilkinson MC, Laing GD. Effects of three novel metalloproteinases from the venom of the West African saw-scaled viper, Echis ocellatus on blood coagulation and platelets. Biochim Biophys Acta (BBA) Gen Subj 2005;1724(1–2):194–202.

[159] Nikai T, Taniguchi K, Komori Y, Masuda K, Fox JW, Sugihara H. Primary structure and functional characterization of bilitoxin-1, a novel dimeric P-II snake venom metalloproteinase from Agkistrodon bilineatus venom. Arch Biochem Biophys 2000;378(1):6–15.

[160] Kurtović T, Brgles M, Leonardi A, Balija ML, Križaj I, Allmaier G, et al. Ammodytagin, a heterodimeric metalloproteinase from Vipera ammodytes ammodytes venom with strong hemorrhagic activity. Toxicon 2011 Nov 1;58(6–7):570–82.

[161] Stroka A, Donato JL, Bon C, Hyslop S, de Araújo AL. Purification and characterization of a hemorrhagic metalloproteinase from Bothrops lanceolatus (Fer-de-lance) snake venom. Toxicon 2005 Mar 15;45(4):411–20.

[162] Camacho E, Villalobos E, Sanz L, Pérez A, Escalante T, Lomonte B, et al. Understanding structural and functional aspects of PII snake venom metalloproteinases: characterization of BlatH1, a hemorrhagic dimeric enzyme from the venom of Bothriechis lateralis. Biochimie [Internet] 2014;101(1):145–55. Available from: https://doi.org/10.1016/j.biochi.2014.01.008.

[163] Boukhalfa-Abib H, Meksem A, Laraba-Djebari F. Purification and biochemical characterization of a novel hemorrhagic metalloproteinase from horned viper (Cerastes cerastes) venom. Compar Biochem Physiol C Toxicol Pharmacol 2009 Aug 1;150(2):285–90.

[164] Gong W, Zhu X, Liu S, Teng M, Niu L. Crystal structures of acutolysin A, a three-disulfide hemorrhagic zinc metalloproteinase from the snake venom of Agkistrodon acutus. J Mol Biol 1998 Oct 30;283(3):657–68.
[165] Xu X, Wang C, Liu J, Lu Z. Purification and characterization of hemorrhagic components from Agkistrodon acutus (hundred pace snake) venom. Toxicon 1981 Jan 1;19(5):633–44.
[166] Wei CB, Chen J, Li JH. Acutolysin C, a weak hemorrhagic toxin from the venom of Agkistrodon acutus with leucoagglutination activity. J Venom Anim Toxins Incl Trop Dis 2011;17(1):34–41.
[167] Shannon JD, Baramova EN, Bjarnason JB, Fox JW. Amino acid sequence of a Crotalus atrox venom metalloproteinase which cleaves type IV collagen and gelatin. J Biol Chem [Internet] 1989;264(20):11575–83. Available from: https://doi.org/10.1016/S0021-9258(18)80102-8.
[168] Freitas-de-Sousa LA, Colombini M, Lopes-Ferreira M, Serrano SMT, Moura-da-Silva AM. Insights into the mechanisms involved in strong hemorrhage and dermonecrosis induced by atroxlysin-Ia, a PI-class snake venom metalloproteinase. Toxins 2017;9(8):239.
[169] Cintra ACO, de Toni LGB, Sartim MA, Franco JJ, Caetano RC, Murakami MT, et al. Batroxase, a new metalloproteinase from B. atrox snake venom with strong fibrinolytic activity. Toxicon 2012 Jul 1;60(1):70–82.
[170] Gutiérrez J, Romero M, Díaz C, Borkow G, Ovadia M. Isolation and characterization of a metalloproteinase with weak hemorrhagic activity from the venom of the snake Bothrops asper (terciopelo). Toxicon 1995 Jan 1;33(1):19–29.
[171] Patiño AC, Pereañez JA, Núñez V, Benjumea DM, Fernandez M, Rucavado A, et al. Isolation and biological characterization of Batx-I, a weak hemorrhagic and fibrinogenolytic PI metalloproteinase from Colombian Bothrops atrox venom. Toxicon 2010 Nov 1;56(6):936–43.
[172] Torres-Huaco FD, Ponce-Soto LA, Martins-de-Souza D, Marangoni S. Purification and characterization of a new weak hemorrhagic metalloproteinase BmHF-1 from Bothrops marajoensis snake venom. Protein J 2010;29(6):407–16.
[173] Bernardes CP, Menaldo DL, Camacho E, Rosa JC, Escalante T, Rucavado A, et al. Proteomic analysis of Bothrops pirajai snake venom and characterization of BpirMP, a new P-I metalloproteinase. J Proteomics 2013 Mar 7;80:250–67.
[174] Bernardes CP, Menaldo DL, Mamede CCN, Zoccal KF, Cintra ACO, Faccioli LH, et al. Evaluation of the local inflammatory events induced by BpirMP, a metalloproteinase from Bothrops pirajai venom. Mol Immunol 2015 Dec 1;68(2):456–64.
[175] Torres-Huaco FD, Maruñak S, Teibler P, Bustillo S, Acosta de Pérez O, Leiva LC, et al. Local and systemic effects of BtaMP-1, a new weakly hemorrhagic Snake Venom Metalloproteinase purified from Bothriopsis taeniata Snake Venom. Int J Biol Macromol 2019 Dec 1;141:1044–54.
[176] Gomes MSR, Mendes MM, de Oliveira F, de Andrade RM, Bernardes CP, Hamaguchi A, et al. BthMP: a new weakly hemorrhagic metalloproteinase from Bothrops moojeni snake venom. Toxicon 2009 Jan 1;53(1):24–32.
[177] Santos MA, Pando LA, Rodrigues VDM, Castro MDS, Rocha Gomes MSR. Purificação e avaliação da interferência de BthMP (uma metaloprotease da peçonha de Bothrops moojeni) na coagulação sanguínea. Jornal Interdisciplinar de Biociências 2016;1(2):7.
[178] Miyata T, Takeya H, Ozeki Y, Arakawa M, Tokunaga F, Iwanaga S, et al. Primary structure of hemorrhagic protein, HR2a, isolated from the venom of Trimeresurus flavoviridis. J Biochem 1989;105(5):847–53.
[179] Takahashi T, Ohsaka A. Purification and some properties of two hemorrhagic principles (HR2a and HR2b) in the venom of Trimeresurus flavoviridis; complete separation of the principles from proteolytic activity. BBA Protein Struct 1970 Apr 28;207(1):65–75.
[180] Mori N, Nikai T, Sugihara H, Tu AT. Biochemical characterization of hemorrhagic toxins with fibrinogenase activity isolated from Crotalus ruber ruber venom. Arch Biochem Biophys 1987 Feb 15;253(1):108–21.
[181] Maruyama M, Sugiki M, Yoshida E, Mihara H, Nakajima N. Purification and characterization of two fibrinolytic enzymes from Bothrops jararaca (jararaca) venom. Toxicon 1992 Aug 1;30(8):853–64.

[182] Sanchez EF, Flores-Ortiz RJ, Alvarenga VG, Eble JA. Direct fibrinolytic snake venom metalloproteinases affecting hemostasis: structural, biochemical features and therapeutic potential. Toxins 2017;9(12):392.
[183] Rucavado A, Flores-Sánchez E, Franceschi A, Magalhaes A, Gutiérrez JM. Characterization of the local tissue damage induced by LHF-II, a metalloproteinase with weak hemorrhagic activity isolated from Lachesis muta muta snake venom. Toxicon 1999 Sep 1;37(9):1297–312.
[184] Rodrigues VM, Soares AM, Guerra-Sá R, Rodrigues V, Fontes MRM, Giglio JR. Structural and functional characterization of neuwiedase, a nonhemorrhagic fibrin(ogen)olytic metalloprotease from Bothrops neuwiedi snake venom. Arch Biochem Biophys 2000 Sep 15;381(2):213–24.
[185] Girón ME, Estrella A, Sánchez EE, Galán J, Tao WA, Guerrero B, et al. Purification and characterization of a metalloproteinase, Porthidin-1, from the venom of Lansberg's hog-nosed pitvipers (Porthidium lansbergii hutmanni). Toxicon Off J Int Soc Toxinol [Internet] 2011 Mar 15;57(4):608–18. 2011/01/19. Available from: https://pubmed.ncbi.nlm.nih.gov/21255600.

Index

Note: Page numbers followed by *f* indicate figures and *t* indicate tables.

G

H

I

K

L

M

Printed in the United States
by Baker & Taylor Publisher Services